Mathematics Study Resources

Volume 26

Series Editors

Kolja Knauer, Departament de Matemàtiques i Informàtica, Universitat de Barcelona, Barcelona, Spain

Elijah Liflyand, Departament of Mathematics, Bar-Ilan University, Ramat-Gan, Israel

This series comprises direct translations of successful foreign language titles, especially from the German language.

Powered by advances in automated translation, these books draw on global teaching excellence to provide students and lecturers with diverse materials for teaching and study.

Andreas Meister

Numerical Methods for Linear Systems of Equations

An Introduction to Modern Methods with MATLAB® Implementations by C. Vömel

 Springer

Andreas Meister
University of Kassel
Kassel, Germany

ISSN 2731-3824　　　　　　ISSN 2731-3832　(electronic)
Mathematics Study Resources
ISBN 978-3-658-50259-1　　　ISBN 978-3-658-50260-7　(eBook)
https://doi.org/10.1007/978-3-658-50260-7

This book is a translation of the original German edition "Numerik linearer Gleichungssysteme," 6th edition, by Prof. Dr. Andreas Meister, published by Springer Fachmedien Wiesbaden GmbH in 2025. The translation was done with the help of an artificial intelligence machine translation tool. A subsequent human revision was done primarily in terms of content, so that the book will read stylistically differently from a conventional translation. Springer Nature works continuously to further the development of tools for the production of books and on the related technologies to support the authors.

Translation from the German language edition: "Numerik linearer Gleichungssysteme" by Andreas Meister, © Der/die Herausgeber bzw. der/die Autor(en), exklusiv lizenziert an Springer Fachmedien Wiesbaden GmbH, ein Teil von Springer Nature 2025. Published by Springer Fachmedien Wiesbaden. All Rights Reserved.

Responsible Editor: Andreas Rüdinger

This Springer imprint is published by the registered company Springer Fachmedien Wiesbaden GmbH, part of Springer Nature.
The registered company address is: Abraham-Lincoln-Str. 46, 65189 Wiesbaden, Germany

If disposing of this product, please recycle the paper.

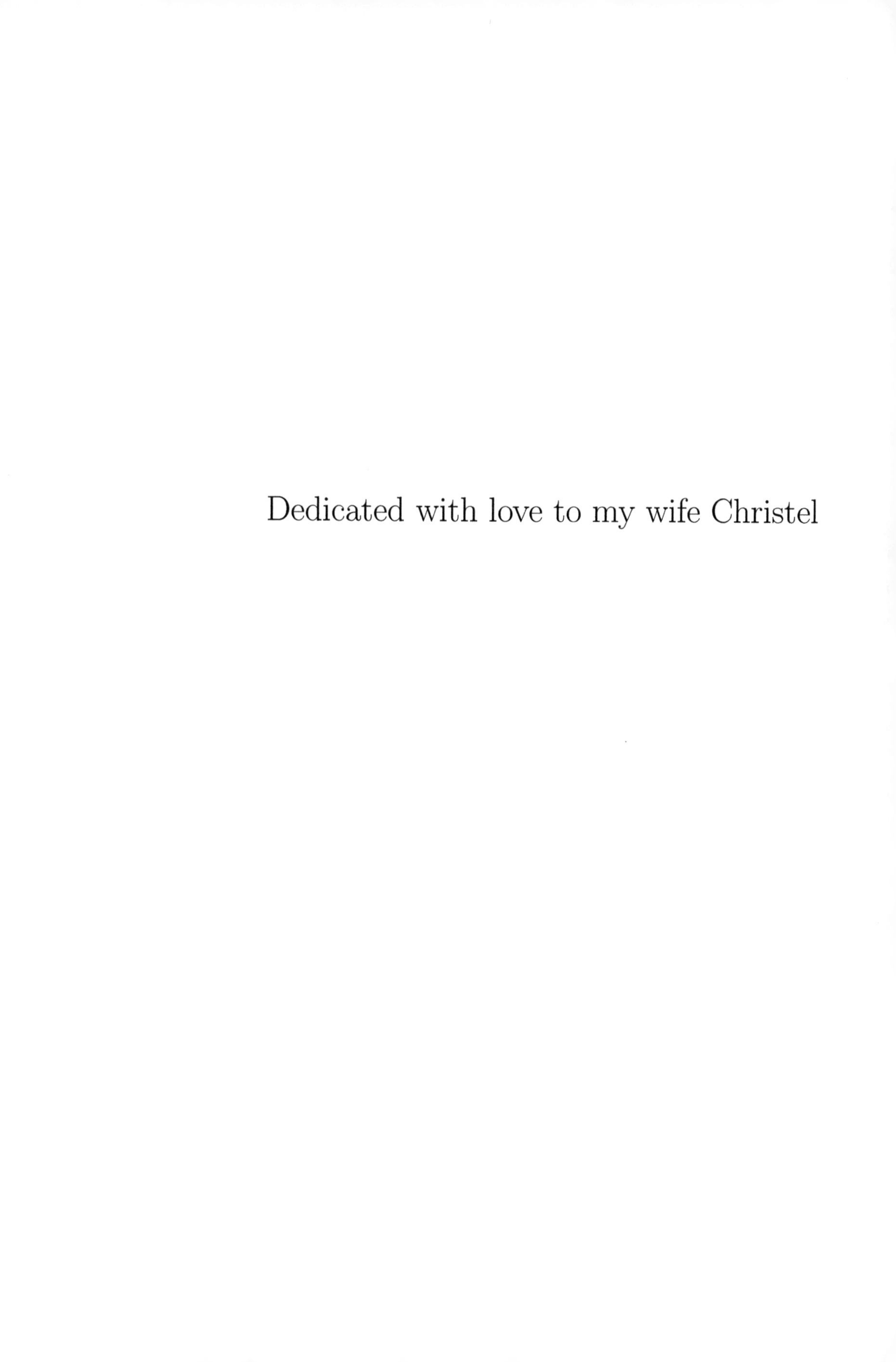

Dedicated with love to my wife Christel

Preface

In the field of numerical methods for linear systems of equations, we deal with the efficient solution of large linear systems, thus addressing an important subfield of Numerical Linear Algebra, which has gained increasing significance in many decades.

The dramatic increase in the performance of personal computers, laptops, and mainframe systems over the past decades has led to the widespread development of numerical methods for the simulation of practically relevant problems in medicine, physics, engineering, and many other areas. In addition to the finite element method, which inherently leads to a linear system, the frequently used finite difference and finite volume methods also require, in combination with an implicit time-stepping scheme, an algorithm for solving linear systems. It is therefore not surprising that research activities in the field of linear system solvers have experienced a significant upswing, leading to the development of a multitude of efficient methods.

This book is based on the content of a four-hour lecture given by the author at different universities, which, in the context of the aforementioned developments, dealt with the derivation and analysis of both classical and modern methods for solving linear systems.

Due to the large number of different direct and iterative methods in this field, it is certainly not possible to present all existing algorithms within a four-hour lecture. The aim of this manuscript is therefore to provide the interested reader with an overview of broad areas of this field, to discuss the methods important for practical applications, to integrate related algorithms through remarks and references, and to facilitate the study of further methods.

The required background knowledge is deliberately limited to the usual content of calculus and linear algebra courses in the first two semesters of a university study program, since the methods described are of great interest far beyond the boundaries of mathematics studies. To support self-study, all necessary fundamentals are also provided in a separate chapter.

After describing the occurrence of linear systems in the first chapter using some model examples, we provide in the second chapter the fundamentals of linear algebra needed for the subsequent methods. The third chapter is devoted to direct methods, which are often involved in modern linear system solvers or are partially used in incomplete versions as preconditioners. The focus is on the description of iterative methods, which are presented in the following fourth chapter. Particular emphasis is always placed on motivation as well as a clear, consistent, and mathematically sound derivation of the iterative system solvers. In addition to splitting methods, such as the Jacobi and Gauss-Seidel methods, Richardson iteration, and relaxation methods, we first describe the two-grid method and then the multigrid method as well as its full version. The derivation of the CG method is carried out by combining the previously described methods of steepest descent and conjugate directions. Furthermore, we consider a wide range of modern Krylov subspace methods, from the GMRES method to the BiCG method and the QMRCGSTAB method, which are suitable for solving systems with a nonsymmetric and indefinite matrix. The final fifth chapter is devoted to a detailed description and investigation of possible preconditioning techniques, since the algorithms derived generally only yield a stable and efficient overall method for practically relevant problems when combined with a suitable preconditioner.

The methods presented in this book find application in a wide variety of numerical procedures, which have been developed in various programming languages. Therefore, a presentation of the algorithms was deliberately chosen that allows implementation in any computer program. Additionally, MATLAB [1] implementations of common Krylov subspace methods are listed in the supplementary appendix. This provides interested users with the opportunity to directly utilize the discussed methods, offering deeper insight into the specific properties of each method in the context of individual problems. Many of the presented methods are already also available in software packages such as LAPACK [3], LINSOL [77], and Templates [8].

I would especially like to thank Prof. Dr. Thomas Sonar, whose encouraging support and years of professional guidance have contributed significantly to the creation of this book. Furthermore, I would like to express my sincere thanks to PD Dr. Christof Vömel for the development of the MATLAB-Code. Finally, my thanks go to Dr. Stefan Kopecz, Dr. Thomas Izgin and M.Sc. Janina Bender for their support during the creation of this first English edition.

The supplementary materials provided are available online on SpringerLink.

Kassel, September 2025   ANDREAS MEISTER

Contents

1 Examples of the Occurrence of Linear Systems of Equations

In this chapter, we deal with the emergence of linear systems of equations based on physically and technically relevant application problems. The selected examples illustrate, on the one hand, the occurrence of sparse (Examples 1.1 and 1.4) as well as dense (Examples 1.2, 1.3, and 1.5) matrices and, on the other hand, serve as model problems (Examples 1.1 and 1.4) for the analysis of the methods described in detail in the following sections.

Example 1.1 The Poisson Equation

The Poisson equation is an elliptic partial differential equation of second order, which has become a standard problem in the study of partial differential equations and their numerical treatment. It arises, among other things, in the numerical simulation of incompressible viscous flow fields, that is, in the discretization of the well-known incompressible Navier-Stokes equations [19, 24].

Using the Laplace operator

$$\Delta = \frac{\partial^2}{\partial x^2} + \frac{\partial^2}{\partial y^2}$$

the Poisson equation can be written as a boundary value problem on the domain $\Omega \subset \mathbb{R}^2$ in the form

$$
\begin{aligned}
-\Delta u(x,y) &= f(x,y) \text{ for } (x,y) \in \Omega \subset \mathbb{R}^2, \\
u(x,y) &= \varphi(x,y) \text{ for } (x,y) \in \partial\Omega,
\end{aligned}
\tag{1.0.1}
$$

where $\partial\Omega$ denotes the boundary of Ω.

Given a function $f \in C(\Omega;\mathbb{R})$ and given boundary values $\varphi \in C(\partial\Omega;\mathbb{R})$, we seek a function $u \in C^2(\Omega;\mathbb{R}) \cap C(\overline{\Omega};\mathbb{R})$ that satisfies the boundary value problem (1.0.1).

To discretize the Poisson equation on the unit square $\Omega = (0,1) \times (0,1)$ using a central finite difference method, $\overline{\Omega} = \Omega \cup \partial\Omega$ is equipped with a grid Ω_h of a constant step size $h = 1/(N+1)$ with $N \in \mathbb{N}$.

Supplementary Information The online version contains supplementary material available at https://doi.org/10.1007/978-3-658-50260-7_1.

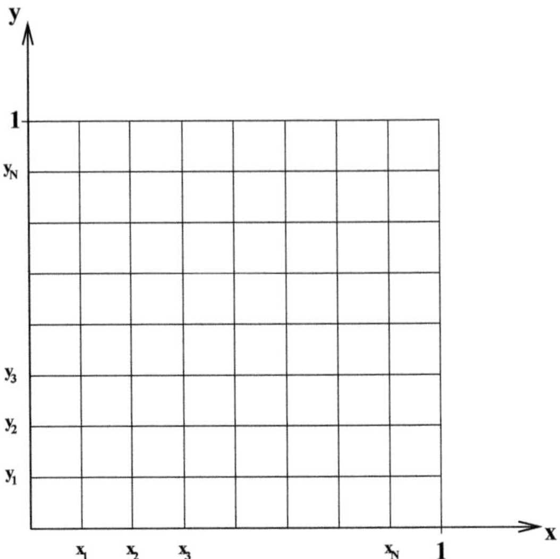

Figure 1.1   Equidistant discretization of the unit square

We write

$$(x_i,y_j) = (ih,jh) \text{ for } i,j = 0,\dots,N+1$$

as well as

$$u_{ij} = u(x_i,y_j) \text{ and } f_{ij} = f(x_i,y_j) \text{ for } i,j = 0,\dots,N+1.$$

Furthermore, we approximate

$$\frac{\partial^2 u}{\partial x^2}(x_i,y_j) \approx \frac{1}{h}\left(\underbrace{\frac{u_{i+1,j} - u_{i,j}}{h}}_{\approx \frac{\partial u}{\partial x}(x_{i+1/2},y_j)} - \frac{u_{i,j} - u_{i-1,j}}{h} \right)$$

$$= \frac{1}{h^2}(u_{i+1,j} - 2u_{ij} + u_{i-1,j}). \tag{1.0.2}$$

With an analogous approach for $\dfrac{\partial^2 u}{\partial y^2}$, we obtain for the Laplace operator

$$-\Delta u(x_i,y_j) \approx \frac{1}{h^2}(4u_{ij} - u_{i-1,j} - u_{i+1,j} - u_{i,j-1} - u_{i,j+1})$$

and thus the discrete form of equation (1.0.1)

$$4u_{i,j} - u_{i-1,j} - u_{i+1,j} - u_{i,j-1} - u_{i,j+1} = h^2 f_{ij} \qquad \text{for} \qquad 1 \le i,j \le N,$$

$$u_{0,j} = \varphi_{0,j}, \; u_{N+1,j} = \varphi_{N+1,j} \qquad \text{for} \qquad j = 0,\dots,N+1,$$

$$u_{i,0} = \varphi_{i,0}, \; u_{i,N+1} = \varphi_{i,N+1} \qquad \text{for} \qquad i = 0,\dots,N+1.$$

A row-wise renumbering, corresponding to a lexicographical ordering of the interior grid
points, yields

$$u_1 = u_{1,1}, \ u_2 = u_{2,1}, \ u_3 = u_{3,1}, \ldots, \ u_N = u_{N,1}, \ u_{N+1} = u_{1,2}, \ldots, u_{N^2} = u_{N,N}.$$

The renumbering is also carried out for f, and thus we obtain a system of equations $\boldsymbol{A}\boldsymbol{u} = \boldsymbol{g}$ for determining the solution vector $\boldsymbol{u} := (u_1, \ldots, u_{N^2})^T$. Here, the matrix reads

$$\boldsymbol{A} = \begin{pmatrix} \boldsymbol{B} & -\boldsymbol{I} & & \\ -\boldsymbol{I} & \ddots & \ddots & \\ & \ddots & \ddots & -\boldsymbol{I} \\ & & -\boldsymbol{I} & \boldsymbol{B} \end{pmatrix} \in \mathbb{R}^{N^2 \times N^2} \tag{1.0.3}$$

with

$$\boldsymbol{B} = \begin{pmatrix} 4 & -1 & & \\ -1 & \ddots & \ddots & \\ & \ddots & \ddots & -1 \\ & & -1 & 4 \end{pmatrix} \in \mathbb{R}^{N \times N} \quad \text{and} \quad \boldsymbol{I} = \begin{pmatrix} 1 & & \\ & \ddots & \\ & & 1 \end{pmatrix} \in \mathbb{R}^{N \times N}.$$

The right-hand side, for the special case $\varphi \equiv 0$, has the form

$$\boldsymbol{g} = h^2 \begin{pmatrix} f_1 \\ \vdots \\ f_{N^2} \end{pmatrix} \in \mathbb{R}^{N^2}. \tag{1.0.4}$$

In the case $\varphi \not\equiv 0$, the corresponding boundary values must be taken into account in the right-hand side vector. For the first component, we obtain in this case

$$g_1 = h^2 f_1 + \varphi_{0,1} + \varphi_{1,0}.$$

Remark:
The matrix $\boldsymbol{A}$ is symmetric, positive definite, and sparse. Its structure also depends on the chosen arrangement of the interior grid points.

Example 1.2 Linear Integral Equation of the Second Kind

Linear integral equations of the first and second kind arise in the determination of density functions by means of which the solution of a partial differential equation (for example, the Helmholtz equation) can be represented in the form of a single- or double-layer potential [45, 37].

We consider the integral equation of the second kind

$$u(x) = v(x) + \int_0^1 k(x,y)u(y)dy \text{ for } x \in [0,1]. \tag{1.0.5}$$

Given an integral kernel $k : [0,1] \times [0,1] \to \mathbb{R}$ with $k(x,0) = 0$ and a given function $v : [0,1] \to \mathbb{R}$, we will derive a linear system of equations for the computation of an approximate solution of the sought function $u : [0,1] \to \mathbb{R}$.

To discretize the integral equation, we use a Nyström method. For this, we divide the interval $[0,1]$ into N subintervals of length $h = 1/N$ with $N \in \mathbb{N}$. Let

$$x_j = \frac{j}{N} = jh \text{ for } j = 0,\ldots,N,$$

then, using numerical integration, we obtain

$$\int_0^1 k(x,y)u(y)dy \approx h \sum_{j=1}^{N} k(x,x_j)u(x_j) = h \sum_{j=0}^{N} k(x,x_j)u(x_j)$$

from (1.0.5) and hence the approximate representation

$$u(x) \approx v(x) + \frac{1}{N} \sum_{j=0}^{N} k(x,x_j)u(x_j) \text{ for } x \in [0,1]. \tag{1.0.6}$$

Let us consider equation (1.0.6) at the nodes x_i, $i = 0,\ldots,N$ and define $u_i = u(x_i)$, $v_i = v(x_i)$, then, with $\boldsymbol{v} = (v_0,\ldots,v_N)^T$, we obtain the linear system $\boldsymbol{Au} = \boldsymbol{v}$ for the determination of the vector $\boldsymbol{u} = (u_0,\ldots,u_N)^T$. Here, the matrix can be written in the form

$$\boldsymbol{A} = (a_{ij})_{i,j=0,\ldots,N} \in \mathbb{R}^{(N+1)\times(N+1)}$$

with

$$a_{ij} = \delta_{ij} - \frac{1}{N}k(x_i,x_j) \tag{1.0.7}$$

using the Kronecker symbol

$$\delta_{ij} = \left\{ \begin{array}{ll} 0, & i \neq j \\ 1, & i = j. \end{array} \right.$$

Remark:
From the representation of the matrix coefficients (1.0.7), it is immediately apparent that the sparsity structure as well as all other properties of the matrix (symmetry, definiteness) depend on the underlying integral kernel. In general, the matrix $\boldsymbol{A}$ is dense.

Example 1.3 Function Definition from Measured Values

Suppose it is known that a physical quantity z can be written as a polynomial of degree n in time t:

$$z(t) = \sum_{i=0}^{n} \alpha_i t^i \text{ for } t \in \mathbb{R}_0^+.$$

From measurements, the physical quantity is known at the points in time $0 \leq t_0 < \ldots < t_N$, $N \geq n$. The corresponding measured values are $z_0,\ldots,z_N$. The task is to determine the coefficients $\alpha_0,\ldots,\alpha_n$ such that the sum of the squared errors is minimized (method of least squares). That is,

$$f(\boldsymbol{\alpha}) = \sum_{j=0}^{N} \left[\sum_{i=0}^{n} \alpha_i t_j^i - z_j \right]^2 = \sum_{j=0}^{N} [z(t_j) - z_j]^2$$

is to be minimized over all $\boldsymbol{\alpha} = (\alpha_0,\ldots,\alpha_n)^T \in \mathbb{R}^{n+1}$. The solution vector $\boldsymbol{\alpha}$ thus satisfies

$$\frac{\partial f(\boldsymbol{\alpha})}{\partial \alpha_k} = 0 \text{ for } k = 0,\ldots,n.$$

Due to the form of the function f, it follows that

$$\sum_{j=0}^{N} 2[z(t_j) - z_j]\, t_j^k \;=\; 0 \qquad\qquad \text{for } k = 0,\ldots,n$$

$$\Leftrightarrow \quad \sum_{j=0}^{N} z(t_j)\, t_j^k \;=\; \sum_{j=0}^{N} z_j\, t_j^k \quad \text{for } k = 0,\ldots,n$$

$$\Leftrightarrow \quad \sum_{j=0}^{N} \left(\sum_{i=0}^{n} \alpha_i t_j^i \right) t_j^k \;=\; \sum_{j=0}^{N} z_j\, t_j^k \quad \text{for } k = 0,\ldots,n$$

$$\Leftrightarrow \quad \sum_{i=0}^{n} \left(\sum_{j=0}^{N} t_j^{i+k} \right) \alpha_i \;=\; \sum_{j=0}^{N} z_j\, t_j^k \quad \text{for } k = 0,\ldots,n.$$

Thus, the linear system of equations $\boldsymbol{A}\boldsymbol{\alpha} = \boldsymbol{b}$ remains to be solved, where the matrix $\boldsymbol{A}$ and the right-hand side $\boldsymbol{b}$ are given by

$$\boldsymbol{A} = (a_{ki})_{k,i=0,\ldots,n} \in \mathbb{R}^{(n+1)\times(n+1)} \text{ with } a_{ki} = \sum_{j=0}^{N} t_j^{k+i} \text{ for } k,i = 0,\ldots,n$$

and

$$\boldsymbol{b} = (b_0,\ldots,b_n)^T \in \mathbb{R}^{n+1} \text{ with } b_k = \sum_{j=0}^{N} z_j t_j^k \text{ for } k = 0,\ldots,n.$$

Remark:
The matrix $\boldsymbol{A}$ is always dense and symmetric. The present system of equations is usually referred to as the normal equation associated with the linear least squares problem $\min_{\boldsymbol{\alpha} \in \mathbb{R}^{n+1}} f(\boldsymbol{\alpha})$.

Example 1.4 The Convection-Diffusion Equation

The convection-diffusion equation is often used as a model problem in the development of novel methods within fluid mechanics, since it exhibits a structural similarity to the Euler and Navier-Stokes equations. Following the works [69] and [52], we consider the stationary convection-diffusion equation on the unit square $\Omega = (0,1) \times (0,1)$ in the form

$$\begin{aligned} \boldsymbol{\beta} \cdot \nabla u(x,y) - \varepsilon \Delta u(x,y) &= 0 & \text{for } &(x,y) \in \Omega \subset \mathbb{R}^2, \\ u(x,y) &= x^2 + y^2 & \text{for } &(x,y) \in \partial\Omega \end{aligned} \tag{1.0.8}$$

with $\boldsymbol{\beta} = a \cdot (\cos\alpha, \sin\alpha)^T$, $\alpha = 45°$ and $a,\varepsilon \in \mathbb{R}_0^+$.

We use the equidistant discretization of the unit square introduced in Example 1.1. In order to avoid a decoupling of the discrete system in the form of a checkerboard pattern in the limiting case of a vanishing diffusion parameter ε, the gradient ∇u within the convective term is discretized using a one-sided difference according to

$$\frac{\partial u}{\partial x}(x_{i+1},y_j) \approx \frac{u_{i+1,j} - u_{i,j}}{h} \quad \text{and} \quad \frac{\partial u}{\partial y}(x_i,y_{j+1}) \approx \frac{u_{i,j+1} - u_{i,j}}{h}.$$

This strategy is referred to as the upwind method in fluid mechanics. The use of the approximation (1.0.2) for the Laplace operator, in combination with a lexicographical ordering, yields a system of equations of the form $\boldsymbol{Au} = \boldsymbol{g}$ for the computation of the vector $\boldsymbol{u} := (u_1,\ldots,u_{N^2})^T$. The sparsity pattern of the matrix here coincides with the structure of the matrix in the example of the Poisson equation. The explicit representation of the matrix is

$$\boldsymbol{A} = \begin{pmatrix} \boldsymbol{B} & -\varepsilon \boldsymbol{I} & & \\ \boldsymbol{D} & \ddots & \ddots & \\ & \ddots & \ddots & -\varepsilon \boldsymbol{I} \\ & & \boldsymbol{D} & \boldsymbol{B} \end{pmatrix} \in \mathbb{R}^{N^2 \times N^2} \tag{1.0.9}$$

with the matrix $\boldsymbol{B} \in \mathbb{R}^{N \times N}$ given by

$$\boldsymbol{B} = \begin{pmatrix} 4\varepsilon + h \cdot a \cdot (\cos\alpha + \sin\alpha) & -\varepsilon & & \\ -\varepsilon - h \cdot a \cdot \cos\alpha & \ddots & \ddots & \\ & \ddots & \ddots & -\varepsilon \\ & & -\varepsilon - h \cdot a \cdot \cos\alpha & 4\varepsilon + h \cdot a \cdot (\cos\alpha + \sin\alpha) \end{pmatrix}$$

and the matrix $\boldsymbol{D} \in \mathbb{R}^{N \times N}$ according to

$$\boldsymbol{D} = \begin{pmatrix} -\varepsilon - h \cdot a \cdot \sin\alpha & & \\ & \ddots & \\ & & -\varepsilon - h \cdot a \cdot \sin\alpha \end{pmatrix},$$

where $\boldsymbol{I} \in \mathbb{R}^{N \times N}$ again represents the identity matrix. The right-hand side $\boldsymbol{g}$ in this case depends exclusively on the boundary conditions given via $(1.0.8)_2$, the step size h , and the input parameters a , ε , as well as the angle α .

Remark:
The matrix $\boldsymbol{A}$ is always sparse and, in the case $a \neq 0$, also nonsymmetric.

Example 1.5 Polynomial Interpolation

Many real-world problems yield, within mathematical modeling, a system of partial or ordinary differential equations. Since the use of a line method in the discretization of partial differential equations leads to a system of ordinary differential equations, the approximate solution of such systems is of central importance. One possibility for the numerical approximation of systems of ordinary differential equations

$$\boldsymbol{y}'(t) = \boldsymbol{f}(\boldsymbol{y}(t),t)$$

lies in the successive integration of the differential equation over a time interval $[t^n,t^{n+1} = t^n + \Delta t]$

$$\boldsymbol{y}(t^{n+1}) - \boldsymbol{y}(t^n) = \int_{t^n}^{t^{n+1}} \boldsymbol{y}'(t) \, dt = \int_{t^n}^{t^{n+1}} \boldsymbol{f}(\boldsymbol{y}(t),t) \, dt$$

with $n = 0,1,2,\ldots$ in combination with an approximation of the integral

$$\int_{t^n}^{t^{n+1}} \boldsymbol{f}(\boldsymbol{y}(t),t) \, dt$$

on the basis of numerical quadrature formulas. The latter can, in turn, be based on the approximate representation of the integrand by an interpolation polynomial. Polynomial interpolation thus provides a fundamental tool for many applications. In addition to this area of application, there are numerous additional uses. For example, interpolation can be used to approximately determine data for times or locations for which no measurements are available, given time- or location-dependent measured values. Interpolation techniques can also be used in the construction of ship hulls or railway tracks.

Let us denote by $\mathcal{P}_n$, $n \in \mathbb{N}_0$, the space of all polynomials of degree at most n. Every polynomial $p \in \mathcal{P}_n$ can be written in the form

$$p(x) = a_0 + a_1 x + a_2 x^2 + \cdots + a_n x^n \tag{1.0.10}$$

with real coefficients $a_0,\ldots,a_n$. With this, we can formulate the so-called interpolation problem as follows. We seek a polynomial $p \in \mathcal{P}_n$ that, for given $n + 1$ points $(x_0,f_0),\ldots,(x_n,f_n) \in \mathbb{R}^2$ at pairwise distinct nodes $x_0,\ldots,x_n \in \mathbb{R}$, satisfies the interpolation conditions

$$p(x_k) = f_k \ \text{ for } \ k = 0,1,\ldots,n.$$

A polynomial that solves the interpolation problem is called an interpolation polynomial or interpolating polynomial.

If we consider the sought polynomial in the above representation (1.0.10), then the $n+1$ degrees of freedom $a_0,\ldots,a_n \in \mathbb{R}$ are determined by the $n+1$ conditions

$$\sum_{i=0}^{n} a_i x_k^i = p(x_k) = f_k, \ \ k = 0,\ldots,n.$$

The resulting $n+1$ equations can be clearly written in the form of a linear system $\boldsymbol{A}\boldsymbol{x} = \boldsymbol{b}$ with

$$\boldsymbol{A} = \begin{pmatrix} 1 & x_0 & x_0^2 & \cdots & x_0^n \\ 1 & x_1 & x_1^2 & \cdots & x_1^n \\ \vdots & \vdots & \vdots & & \vdots \\ 1 & x_n & x_n^2 & \cdots & x_n^n \end{pmatrix}, \ \boldsymbol{x} = \begin{pmatrix} a_0 \\ a_1 \\ \vdots \\ a_n \end{pmatrix} \ \text{and} \ \boldsymbol{b} = \begin{pmatrix} f_0 \\ f_1 \\ \vdots \\ f_n \end{pmatrix}.$$

Remark:
The matrix $\boldsymbol{A} \in \mathbb{R}^{(n+1)\times(n+1)}$ is always nonsymmetric and dense.

2   Fundamentals of Linear Algebra

The derivation and analysis of the methods for solving linear systems of equations considered in the following are based on the fundamentals presented in this chapter. Here, we always consider mappings between real or complex linear spaces, which are often also referred to as vector spaces over $\mathbb{R}$ or $\mathbb{C}$, respectively. Such linear spaces, for example $\mathbb{R}^n$ or $\mathbb{C}^n$, are non-empty sets whose elements can be combined by addition and multiplied by scalars $\lambda \in \mathbb{R}$ or $\mathbb{C}$, satisfying the vector space axioms. These axioms merely guarantee that addition, subtraction, and multiplication can be performed as usual. In addition to various vector and matrix norms, the scalar product as well as the important concepts of eigenvalue, eigenvector, and spectral radius are introduced. Of central importance with regard to the investigation of the convergence of some iterative methods is also the Banach fixed point theorem presented in this section.

2.1   Vector Norms and Inner Product

Necessary concepts such as orthogonality, length, distance, and convergence are based on inner products or norms. Therefore, we will introduce these mappings in this section and present some essential statements.

Definition 2.1 Let X be a complex or real linear space. A mapping

$$\|\cdot\| : X \longrightarrow \mathbb{R}$$

with the properties

(N1)	$\|x\| \geq 0$	(Positivity)
(N2)	$\|x\| = 0 \Leftrightarrow x = 0$	(Definiteness)
(N3)	$\|\alpha \cdot x\| = |\alpha| \cdot \|x\| \quad \forall\, x \in X, \forall\, \alpha \in \mathbb{C} \ (\text{or } \mathbb{R})$	(Homogeneity)
(N4)	$\|x + y\| \leq \|x\| + \|y\| \quad \forall\, x,y \in X$	(Triangle inequality)

is called a norm on X. A linear space X with a norm is called a normed space. If $X = \mathbb{C}^n$ or $X = \mathbb{R}^n$, the norm is also called a vector norm.

Supplementary Information   The online version contains supplementary material available at https://doi.org/10.1007/978-3-658-50260-7_2.

The positivity of a norm can also be derived from axioms (N3) and (N4) and therefore, for formal reasons, does not need to be explicitly included in the definition. Nevertheless, we use the above formulation to have this property of the norm explicitly stated.

On $\mathbb{C}^n$ or $\mathbb{R}^n$, norms are, for example, given by

(a) $\|\boldsymbol{x}\|_1 := \sum\limits_{i=1}^{n} |x_i|$ $\qquad\qquad$ (absolute sum norm),

(b) $\|\boldsymbol{x}\|_2 := \left(\sum\limits_{i=1}^{n} |x_i|^2 \right)^{\frac{1}{2}}$ $\qquad\qquad$ (Euclidean norm),

(c) $\|\boldsymbol{x}\|_\infty := \max\limits_{i=1,\dots,n} |x_i|$ $\qquad\qquad$ (maximum norm).

We now proceed to introduce the required concept of convergence.

Definition 2.2 A sequence $\{\boldsymbol{x}_n\}_{n\in\mathbb{N}}$ of elements from a normed space X is called convergent with limit element $\boldsymbol{x} \in X$, if for every $\varepsilon > 0$ there exists a number $N = N(\varepsilon) \in \mathbb{N}$ such that

$$\|\boldsymbol{x}_n - \boldsymbol{x}\| < \varepsilon \quad \forall\, n \geq N$$

holds. A sequence that does not converge is called divergent.

Definition 2.3 Let X be a normed space. A sequence $\{\boldsymbol{x}_n\}_{n\in\mathbb{N}}$ in X is called a Cauchy sequence, if for every $\varepsilon > 0$ there exists a number $N = N(\varepsilon) \in \mathbb{N}$ such that

$$\|\boldsymbol{x}_n - \boldsymbol{x}_m\| < \varepsilon \quad \forall n,m \geq N$$

holds.

The following theorem provides a general relationship between convergent sequences and Cauchy sequences.

Theorem 2.4 *Let X be a normed space. Every convergent sequence $\{\boldsymbol{x}_n\}_{n\in\mathbb{N}}$ in X is a Cauchy sequence.*

Proof:
Let $\varepsilon > 0$ be given and $\boldsymbol{x}$ the limit element of the sequence $\{\boldsymbol{x}_n\}_{n\in\mathbb{N}}$

$$\Rightarrow \exists N = N\left(\frac{\varepsilon}{2}\right) \in \mathbb{N} \text{ with } \|\boldsymbol{x}_n - \boldsymbol{x}\| < \tfrac{\varepsilon}{2} \;\; \forall n \geq N$$

$$\Rightarrow \|\boldsymbol{x}_n - \boldsymbol{x}_m\| \leq \|\boldsymbol{x}_n - \boldsymbol{x}\| + \|\boldsymbol{x} - \boldsymbol{x}_m\| < \varepsilon \;\; \forall n,m \geq N\,. \qquad\qquad \square$$

Definition 2.5 Let X be a complex or real linear space. A mapping

$$(.,.) : X \times X \longrightarrow \mathbb{C}$$

with the properties

$$
\begin{array}{lll}
\text{(H1)} & (\boldsymbol{x},\boldsymbol{x}) \in \mathbb{R}_0^+ \quad \forall\, \boldsymbol{x} \in X & \text{(Positivity)} \\[4pt]
\text{(H2)} & (\boldsymbol{x},\boldsymbol{x}) = 0 \Leftrightarrow \boldsymbol{x} = \boldsymbol{0} & \text{(Definiteness)} \\[4pt]
\text{(H3)} & (\boldsymbol{x},\boldsymbol{y}) = \overline{(\boldsymbol{y},\boldsymbol{x})} \quad \forall\, \boldsymbol{x},\boldsymbol{y} \in X & \text{(Symmetry)} \\[4pt]
\text{(H4)} & (\alpha\boldsymbol{x} + \beta\boldsymbol{y},\boldsymbol{z}) = \alpha(\boldsymbol{x},\boldsymbol{z}) + \beta(\boldsymbol{y},\boldsymbol{z}) \quad \forall \boldsymbol{x},\boldsymbol{y},\boldsymbol{z} \in X \ \ \alpha,\beta \in \mathbb{C} & \text{(Linearity)}
\end{array}
$$

is called an inner product or scalar product on X. A linear space X equipped with an inner product is called a pre-Hilbert space.

Remark:

It also holds that

$$
\text{(H4')} \quad (\boldsymbol{x},\alpha\boldsymbol{y} + \beta\boldsymbol{z}) \overset{(H3)}{=} \overline{(\alpha\boldsymbol{y} + \beta\boldsymbol{z},\boldsymbol{x})} \overset{(H4)}{=} \overline{\alpha(\boldsymbol{y},\boldsymbol{x}) + \beta(\boldsymbol{z},\boldsymbol{x})}
$$

$$
\overset{(H3)}{=} \overline{\alpha}(\boldsymbol{x},\boldsymbol{y}) + \overline{\beta}(\boldsymbol{x},\boldsymbol{z}).
$$

This property is called antilinearity.

Theorem 2.6 *On every pre-Hilbert space X, the following defines a norm:*

$$
\|\boldsymbol{x}\| := \sqrt{(\boldsymbol{x},\boldsymbol{x})}, \quad \boldsymbol{x} \in X
$$

Proof:

The first three norm properties follow by direct calculation. Property (N4) follows by application of the Cauchy-Schwarz inequality

$$
|(\boldsymbol{x},\boldsymbol{y})| \le \|\boldsymbol{x}\|\,\|\boldsymbol{y}\|.
$$

$\square$

Usually, on $\mathbb{C}^n$ the Euclidean inner product is used, defined by

$$
(\boldsymbol{x},\boldsymbol{y})_2 := \sum_{i=1}^{n} x_i \overline{y_i} = \boldsymbol{y}^*\boldsymbol{x}
$$

Here, the following relation holds:

$$
\|\boldsymbol{x}\|_2 = \sqrt{(\boldsymbol{x},\boldsymbol{x})_2}.
$$

Theorem 2.7 *Let X be a normed space and $\{\boldsymbol{x}_n\}_{n\in\mathbb{N}}$ a convergent sequence in X, then the limit element is uniquely determined.*

Proof:

Let $\boldsymbol{x}$ and $\boldsymbol{y}$ be two limit elements, i.e., $\boldsymbol{x}_n \longrightarrow \boldsymbol{x}$ for $n \longrightarrow \infty$ and $\boldsymbol{x}_n \longrightarrow \boldsymbol{y}$ for $n \longrightarrow \infty$, then it follows that

$$
\begin{aligned}
0 \le \|\boldsymbol{x} - \boldsymbol{y}\| &= \|\boldsymbol{x} - \boldsymbol{x}_n + \boldsymbol{x}_n - \boldsymbol{y}\| \\
&\le \underbrace{\|\boldsymbol{x} - \boldsymbol{x}_n\|}_{\xrightarrow{n\to\infty} 0} + \underbrace{\|\boldsymbol{x}_n - \boldsymbol{y}\|}_{\xrightarrow{n\to\infty} 0} \xrightarrow{n\to\infty} 0
\end{aligned}
$$

$$\Rightarrow \quad 0 \leq \|\boldsymbol{x} - \boldsymbol{y}\| \leq 0$$

$$\Rightarrow \quad \|\boldsymbol{x} - \boldsymbol{y}\| = 0 \stackrel{(N2)}{\Rightarrow} \boldsymbol{x} - \boldsymbol{y} = 0 \Rightarrow \boldsymbol{x} = \boldsymbol{y}.$$

$\square$

Definition 2.8 Two norms $\|.\|_a$ and $\|.\|_b$ on a linear space X are called *equivalent*, if a sequence converges with respect to $\|.\|_a$ if and only if it converges with respect to $\|.\|_b$.

Theorem and Definition 2.9 *Two norms $\|.\|_a$ and $\|.\|_b$ on a linear space X are equivalent if and only if there exist real numbers $\alpha, \beta > 0$ such that*

$$\alpha\|\boldsymbol{x}\|_b \leq \|\boldsymbol{x}\|_a \leq \beta\|\boldsymbol{x}\|_b \quad \forall\, \boldsymbol{x} \in X \tag{2.1.1}$$

holds. The largest such number α and the smallest such number β are called equivalence constants.

Proof:

" $\Leftarrow$ "

Let $\{\boldsymbol{x}_n\}_{n \in \mathbb{N}}$ be an arbitrary sequence with $\boldsymbol{x}_n \xrightarrow{\|.\|_b} \boldsymbol{x}$ for $n \to \infty$, then by (2.1.1) it follows that

$$\|\boldsymbol{x}_n - \boldsymbol{x}\|_a \leq \beta\|\boldsymbol{x}_n - \boldsymbol{x}\|_b \to 0 \text{ for } n \to \infty$$

and thus the required convergence of the sequence in the norm $\|.\|_a$. The second inequality in (2.1.1) yields in an analogous way the convergence of a sequence in the norm $\|.\|_b$ from the convergence of the sequence in the norm $\|.\|_a$.

" $\Rightarrow$ "

Assume: There does not exist a $\beta \in \mathbb{R}^+$ with $\|\boldsymbol{x}\|_a \leq \beta$ for all $\boldsymbol{x} \in X$ with $\|\boldsymbol{x}\|_b = 1$. Then there exists a sequence $\{\boldsymbol{x}_n\}_{n \in \mathbb{N}}$ with

$$\|\boldsymbol{x}_n\|_a \geq n^2 \quad \text{and} \quad \|\boldsymbol{x}_n\|_b = 1.$$

Let

$$\boldsymbol{y}_n := \frac{\boldsymbol{x}_n}{n},$$

then we obtain

$$\|\boldsymbol{y}_n\|_a = \frac{1}{n}\|\boldsymbol{x}_n\|_a \geq n \quad \text{and} \quad \|\boldsymbol{y}_n\|_b = \frac{1}{n}. \tag{2.1.2}$$

Thus, it follows that $\boldsymbol{y}_n \xrightarrow{\|.\|_b} \boldsymbol{0}$ as $n \to \infty$, which means that, due to the assumed equivalence of the norms $\|.\|_a$ and $\|.\|_b$, the sequence also converges in the norm $\|.\|_a$. This directly yields the desired contradiction to property (2.1.2).

Consequently, there exists a $\beta \in \mathbb{R}^+$ such that $\|\boldsymbol{x}\|_a \leq \beta$ for all $\boldsymbol{x} \in X$ with $\|\boldsymbol{x}\|_b = 1$. Now let $\boldsymbol{y} \in X \setminus \{\boldsymbol{0}\}$, then we obtain by

$$\|\boldsymbol{y}\|_a = \left\|\|\boldsymbol{y}\|_b \frac{\boldsymbol{y}}{\|\boldsymbol{y}\|_b}\right\|_a = \|\boldsymbol{y}\|_b \left\|\frac{\boldsymbol{y}}{\|\boldsymbol{y}\|_b}\right\|_a \leq \beta\|\boldsymbol{y}\|_b \tag{2.1.3}$$

the extension of the property to all vectors from $X \setminus \{\mathbf{0}\}$. Since the zero element always satisfies equation (2.1.1), it follows that

$$\|\mathbf{y}\|_a \leq \beta \|\mathbf{y}\|_b \quad \text{for all} \quad \mathbf{y} \in X.$$

The second estimate follows analogously. $\hspace{10cm}$ $\square$

Based on Theorem 2.9, the definition of equivalent norms is often given on the basis of the inequalities (2.1.1). As a direct consequence of these inequalities, we obtain the following statement.

Corollary 2.10 *The limit elements of a sequence coincide with respect to two equivalent norms.*

Theorem 2.11 *On a finite-dimensional real or complex linear space, all norms are equivalent.*

Proof:
To prove the above statement, it suffices to show that a particular norm is equivalent to any norm. Let $X := \operatorname{span}\{\mathbf{u}_1, \ldots, \mathbf{u}_m\}$ be an m-dimensional linear space, then every $\mathbf{x} \in X$ can be written in the form

$$\mathbf{x} = \sum_{i=1}^{m} \alpha_i \mathbf{u}_i, \quad \alpha_i \in \mathbb{C}.$$

Define on X the norm

$$\|\mathbf{x}\|_{\max} := \max_{i=1,\ldots,m} |\alpha_i|,$$

then for any other norm $\|.\|$ on X we have the estimate

$$\|\mathbf{x}\| \;=\; \left\| \sum_{i=1}^{m} \alpha_i \mathbf{u}_i \right\| \overset{(N4)}{\leq} \sum_{i=1}^{m} \|\alpha_i \mathbf{u}_i\| \overset{(N3)}{=} \sum_{i=1}^{m} |\alpha_i| \, \|\mathbf{u}_i\|$$

$$\leq \; \max_{i=1,\ldots,m} |\alpha_i| \underbrace{\sum_{i=1}^{m} \|\mathbf{u}_i\|}_{\beta :=} = \beta \|\mathbf{x}\|_{\max}. \tag{2.1.4}$$

We will prove the second inequality by means of the following proof by contradiction.

Assumption: There does not exist any $\alpha \in \mathbb{R}^+$ such that $\|\mathbf{x}\|_{\max} \leq \alpha \|\mathbf{x}\|$ for all $\mathbf{x} \in X$.

As we have already seen in equation (2.1.3), this property is equivalent to the statement:

There does not exist any $\alpha \in \mathbb{R}^+$ such that $\|\mathbf{x}\|_{\max} \leq \alpha$ for all $\mathbf{x} \in X$ with $\|\mathbf{x}\| = 1$.

Thus, there exists a sequence $\{\mathbf{x}_n\}_{n \in \mathbb{N}}$ with

$$\|\mathbf{x}_n\|_{\max} \geq n \text{ and } \|\mathbf{x}_n\| = 1.$$

Let

$$\boldsymbol{y}_n := \frac{\boldsymbol{x}_n}{\|\boldsymbol{x}_n\|_{\max}}, \quad \text{then} \quad \|\boldsymbol{y}_n\|_{\max} = 1 \quad \text{for all} \quad n \in \mathbb{N}. \tag{2.1.5}$$

We write

$$\boldsymbol{y}_n = \sum_{i=1}^{m} \alpha_{i,n} \boldsymbol{u}_i$$

and obtain with (2.1.5) $\max\limits_{i=1,\ldots,m} |\alpha_{i,n}| = 1$ for all $n \in \mathbb{N}$. Thus, $\{\alpha_{1,n}\}_{n \in \mathbb{N}}, \ldots, \{\alpha_{m,n}\}_{n \in \mathbb{N}}$ are m bounded sequences of complex numbers, so that for $i = 1, \ldots, m$, by the Bolzano-Weierstrass theorem, there exists a convergent subsequence

$$\alpha_{i,n(j)} \xrightarrow{j \to \infty} \widetilde{\alpha}_i \text{ with } n(j+1) > n(j).$$

Define

$$\boldsymbol{y} := \sum_{i=1}^{m} \widetilde{\alpha}_i \boldsymbol{u}_i,$$

then it follows for $\boldsymbol{y}_{n(j)} := \sum\limits_{i=1}^{m} \alpha_{i,n(j)} \boldsymbol{u}_i$

$$\|\boldsymbol{y}_{n(j)} - \boldsymbol{y}\|_{\max} = \max_{i=1,\ldots,m} |\alpha_{i,n(j)} - \widetilde{\alpha}_i| \xrightarrow{j \to \infty} 0. \tag{2.1.6}$$

Using the inequality (2.1.4) we obtain

$$\boldsymbol{y}_{n(j)} \xrightarrow{\|\cdot\|} \boldsymbol{y} \text{ for } j \to \infty,$$

so that with (2.1.5) we have

$$\|\boldsymbol{y}_{n(j)}\| = \left\| \frac{\boldsymbol{x}_{n(j)}}{\|\boldsymbol{x}_{n(j)}\|_{\max}} \right\| = \frac{1}{\|\boldsymbol{x}_{n(j)}\|_{\max}} \leq \frac{1}{n(j)} \xrightarrow{j \to \infty} 0$$

and consequently $\boldsymbol{y} = \boldsymbol{0}$. In this way, using $\|\boldsymbol{y}_{n(j)}\|_{\max} \xrightarrow{j \to \infty} 0$ we obtain a contradiction to $\|\boldsymbol{y}_n\|_{\max} = 1$ for all $n \in \mathbb{N}$. $\qquad\square$

For the vector norms listed, we obtain the equivalence constants for $\boldsymbol{x} \in \mathbb{R}^n$ as follows:

$$\|\boldsymbol{x}\|_2 \leq \|\boldsymbol{x}\|_1 \leq \sqrt{n}\|\boldsymbol{x}\|_2, \tag{2.1.7}$$

$$\|\boldsymbol{x}\|_\infty \leq \|\boldsymbol{x}\|_2 \leq \sqrt{n}\|\boldsymbol{x}\|_\infty, \tag{2.1.8}$$

and

$$\frac{1}{n}\|\boldsymbol{x}\|_1 \leq \|\boldsymbol{x}\|_\infty \leq \|\boldsymbol{x}\|_1. \tag{2.1.9}$$

Definition 2.12 A subset V of a normed space X is called complete if every Cauchy sequence in V converges to a limit element in V. A complete normed space is called a Banach space.

For the linear systems of equations considered, we always deal with finite-dimensional normed spaces. For these spaces, the following useful property holds.

Theorem 2.13 *Every finite-dimensional normed space is a Banach space.*

Proof:
Let X be a finite-dimensional normed space. According to Theorem 2.11, all norms on X are equivalent, so the proof can be given for any norm.

Let $X := \text{span}\{u_1, \dots, u_k\}$, then every $x \in X$ can be written as

$$x = \sum_{i=1}^{k} \alpha_i u_i \text{ with } \alpha_i \in \mathbb{C}.$$

We now consider again the norm

$$\|x\|_{\max} := \max_{i=1,\dots,k} |\alpha_i|.$$

Let $\{x_n\}_{n \in \mathbb{N}}$ be a Cauchy sequence in X, then we write

$$x_n = \sum_{i=1}^{k} \alpha_{i,n} u_i \text{ with } \alpha_{i,n} \in \mathbb{C}$$

and obtain

$$\max_{i=1,\dots,k} |\alpha_{i,n} - \alpha_{i,m}| = \|x_n - x_m\|_{\max} \to 0 \text{ for } n,m \to \infty.$$

Consequently, the k sequences of coefficients are each Cauchy sequences in $\mathbb{C}$, so due to the completeness of the space of complex numbers

$$\alpha_{i,n} \xrightarrow{n \to \infty} \tilde{\alpha}_i \in \mathbb{C} \text{ for } i = 1, \dots, k$$

holds. With $\tilde{x} = \sum_{i=1}^{k} \tilde{\alpha}_i u_i$ we have the desired vector, because of

$$\|x_n - \tilde{x}\|_{\max} = \max_{i=1,\dots,k} |\alpha_{i,n} - \tilde{\alpha}_i| \xrightarrow{n \to \infty} 0.$$

$\square$

2.2 Linear Operators, Matrices, and Matrix Norms

The properties of the matrix of a linear system of equations are essential for the selection of a suitable iterative method and generally also determine the convergence behavior of the chosen method. In addition to introducing special classes of matrices and defining different matrix norms, in this section we will explain the spectral radius and discuss its relationship to matrix norms, which will prove to be fundamental for convergence statements of splitting methods.

Definition 2.14 Let X,Y be normed spaces with norms $\|.\|_X$ and $\|.\|_Y$, respectively. An operator $A : X \to Y$ is called

(a) continuous at the point $x \in X$ if for all sequences $\{x_n\}_{n \in \mathbb{N}}$ from X with $x_n \xrightarrow{\|\cdot\|_X} x$ as $n \to \infty$

$$Ax_n \xrightarrow{\|\cdot\|_Y} Ax \quad \text{for } n \to \infty$$

follows.

(b) continuous, if A is continuous at all points $x \in X$.

(c) linear, if

$$A(\alpha x + \beta y) = \alpha A x + \beta A y \quad \forall x, y \in X \quad \forall \alpha, \beta \in \mathbb{C}$$

holds.

(d) bounded, if A is linear and there exists a $C \geq 0$ such that

$$\|Ax\|_Y \leq C \|x\|_X \quad \forall x \in X$$

holds. Any number C with this property is called a bound of A.

In the following, we will deal with so-called induced matrix norms, which are each defined on the basis of a vector norm by the following theorem.

Theorem and Definition 2.15 *A linear operator* $A : X \to Y$ *is bounded if and only if*

$$\|A\| := \sup_{\|x\|_X = 1} \|Ax\|_Y < \infty$$

holds. $\|A\|$ *is the smallest bound of* A *and is called the norm of the operator.*

Proof:

" $\Rightarrow$ "

Suppose A is bounded, then in particular there exists a $C \geq 0$ such that

$$\|Ax\|_Y \leq C \text{ for all } x \in X \text{ with } \|x\|_X = 1$$

and we obtain

$$\|A\| = \sup_{\|x\|_X = 1} \|Ax\|_Y \leq C < \infty.$$

In particular, $\|A\|$ thus represents the smallest bound of A.

" $\Leftarrow$ "

Suppose A is linear and $\|A\| < \infty$

Consider any $x \in X \setminus \{0\}$, then it follows that

$$
\begin{aligned}
\|Ax\|_Y &= \left\| \|x\|_X A\left(\frac{x}{\|x\|_X}\right) \right\|_Y \\[2mm]
&= \|x\|_X \left\| A\left(\frac{x}{\|x\|_X}\right) \right\|_Y \\[2mm]
&\leq \|x\|_X \sup_{\|z\|_X = 1} \|Az\|_Y \\[2mm]
&= \|x\|_X \|A\|.
\end{aligned}
$$

Thus, $\boldsymbol{A}$ is bounded with bound $\|\boldsymbol{A}\|$. $\qquad\qquad\square$

The property

$$\|\boldsymbol{A}\boldsymbol{x}\| \le \|\boldsymbol{A}\|\|\boldsymbol{x}\|$$

is called the compatibility condition.

Continuity and boundedness are equivalent concepts in the context of linear operators. This fact is demonstrated by the following theorem.

Theorem 2.16 *For a linear operator* $\boldsymbol{A} : X \to Y$, *the following three properties are equivalent:*

   (a) $\boldsymbol{A}$ *is continuous at the point* $\boldsymbol{x} = \boldsymbol{0}$.

   (b) $\boldsymbol{A}$ *is continuous.*

   (c) $\boldsymbol{A}$ *is bounded.*

Proof:
" $(a) \Rightarrow (b)$ "

Let $\boldsymbol{x} \in X$ be given and $\{\boldsymbol{x}_n\}_{n\in\mathbb{N}}$ an arbitrary sequence from X with $\boldsymbol{x}_n \to \boldsymbol{x}$, $n \to \infty$, then by continuity at the point $\boldsymbol{x} = \boldsymbol{0}$ and the linearity of the operator, we have the equation

$$\boldsymbol{A}\boldsymbol{x}_n = \underbrace{\boldsymbol{A}(\boldsymbol{x}_n - \boldsymbol{x})}_{\xrightarrow{\|\cdot\|_Y}0,\ n\to\infty} + \boldsymbol{A}\boldsymbol{x} \xrightarrow{\|\cdot\|_Y} \boldsymbol{A}\boldsymbol{x} \text{ for } n \to \infty.$$

" $(b) \Rightarrow (c)$ "

Assumption: $\boldsymbol{A}$ is continuous and not bounded.

Then there exists a sequence $\{\boldsymbol{x}_n\}_{n\in\mathbb{N}}$ from X with $\|\boldsymbol{x}_n\|_X = 1$ and $\|\boldsymbol{A}\boldsymbol{x}_n\|_Y \ge n$. We define

$$\boldsymbol{y}_n := \frac{\boldsymbol{x}_n}{\|\boldsymbol{A}\boldsymbol{x}_n\|_Y}$$

and thus obtain

$$\|\boldsymbol{y}_n\|_X = \frac{\|\boldsymbol{x}_n\|_X}{\|\boldsymbol{A}\boldsymbol{x}_n\|_Y} = \frac{1}{\|\boldsymbol{A}\boldsymbol{x}_n\|_Y} \le \frac{1}{n}.$$

Consequently, the sequence $\boldsymbol{y}_n$ converges in the norm on X to the zero element, and it follows from the continuity of the operator that

$$\boldsymbol{A}\boldsymbol{y}_n \xrightarrow{\|\cdot\|_Y} \boldsymbol{A}(\boldsymbol{0}) = \boldsymbol{0} \text{ for } n \to \infty,$$

which yields

$$\|\boldsymbol{A}\boldsymbol{y}_n\|_Y = \frac{\|\boldsymbol{A}\boldsymbol{x}_n\|_Y}{\|\boldsymbol{A}\boldsymbol{x}_n\|_Y} = 1 \text{ for all } n \in \mathbb{N}$$

and thus we obviously have a contradiction.

" $(c) \Rightarrow (a)$ "

Let $\boldsymbol{A}$ be bounded and $\{\boldsymbol{x}_n\}_{n\in\mathbb{N}}$ a sequence from X with $\boldsymbol{x}_n \xrightarrow{\|\cdot\|_X} \boldsymbol{0}$ for $n \to \infty$. Then it follows

$$\|\boldsymbol{A}\boldsymbol{x}_n\|_Y \le \|\boldsymbol{A}\|\|\boldsymbol{x}_n\|_X \to 0 \text{ for } n \to \infty$$

and thus with

$$\boldsymbol{A}\boldsymbol{x}_n \xrightarrow{\|\cdot\|_Y} \boldsymbol{0} = \boldsymbol{A}\,(\boldsymbol{0}) \text{ for } n \to \infty$$

the continuity of the operator at the point $\boldsymbol{x} = \boldsymbol{0}$ is proven. $\square$

When considering compositions of linear mappings, the following theorem for estimating the norm of the composition proves to be helpful.

Theorem 2.17 *Let $\boldsymbol{A} : X \to Y$ and $\boldsymbol{B} : Y \to Z$ be bounded linear operators, then $\boldsymbol{BA} : X \to Z$ is bounded with*

$$\|\boldsymbol{BA}\| \le \|\boldsymbol{B}\|\,\|\boldsymbol{A}\|. \tag{2.2.1}$$

Proof:
With

$$\|\boldsymbol{BA}\boldsymbol{x}\|_Z \le \|\boldsymbol{B}\|\,\|\boldsymbol{A}\boldsymbol{x}\|_Y \le \|\boldsymbol{B}\|\,\|\boldsymbol{A}\|\,\|\boldsymbol{x}\|_X$$

the proof follows directly from

$$\|\boldsymbol{BA}\| = \sup_{\|\boldsymbol{x}\|_X=1} \|\boldsymbol{BA}\boldsymbol{x}\|_Z \le \sup_{\|\boldsymbol{x}\|_X=1} \|\boldsymbol{B}\|\|\boldsymbol{A}\|\|\boldsymbol{x}\|_X = \|\boldsymbol{B}\|\|\boldsymbol{A}\|.$$

$\square$

The property (2.2.1) is also called the submultiplicativity of the underlying norm. In the following, unless explicitly mentioned otherwise, we will always restrict ourselves to norms that are defined on the basis of Definition 2.15. In the context of the matrices considered, however, other norms are also conceivable that do not satisfy the inequality (2.2.1).

Every linear mapping between finite-dimensional vector spaces can be represented by a matrix

$$\boldsymbol{A} = \begin{pmatrix} a_{11} & \cdots & a_{1n} \\ \vdots & \ddots & \vdots \\ a_{m1} & \cdots & a_{mn} \end{pmatrix} \in \mathbb{C}^{m\times n}.$$

We now want to preliminarily divide the set of all matrices into different classes.

Definition 2.18 For a given matrix

$$\boldsymbol{A} = \begin{pmatrix} a_{11} & \cdots & a_{1n} \\ \vdots & \ddots & \vdots \\ a_{n1} & \cdots & a_{nn} \end{pmatrix} \in \mathbb{R}^{n\times n}$$

the matrix

$$\boldsymbol{A}^T = \begin{pmatrix} a_{11} & \cdots & a_{n1} \\ \vdots & \ddots & \vdots \\ a_{1n} & \cdots & a_{nn} \end{pmatrix} \in \mathbb{R}^{n\times n}$$

is called the transpose of $\boldsymbol{A}$.

Definition 2.19 A matrix $\boldsymbol{A} \in \mathbb{R}^{n\times n}$ is called

(a) symmetric, if $\boldsymbol{A}^T = \boldsymbol{A}$ holds,

(b) orthogonal, if $\boldsymbol{A}^T \boldsymbol{A} = \boldsymbol{I}$ holds.

Definition 2.20 Given the matrix

$$
\boldsymbol{A} = \begin{pmatrix} a_{11} & \cdots & a_{1n} \\ \vdots & \ddots & \vdots \\ a_{n1} & \cdots & a_{nn} \end{pmatrix} \in \mathbb{C}^{n \times n}
$$

the matrix

$$
\boldsymbol{A}^* = \begin{pmatrix} \overline{a_{11}} & \cdots & \overline{a_{n1}} \\ \vdots & \ddots & \vdots \\ \overline{a_{1n}} & \cdots & \overline{a_{nn}} \end{pmatrix} \in \mathbb{C}^{n \times n}
$$

is called the adjoint matrix of $\boldsymbol{A}$.

Definition 2.21 A matrix $\boldsymbol{A} \in \mathbb{C}^{n \times n}$ is called

(a) Hermitian, if $\boldsymbol{A}^* = \boldsymbol{A}$ holds,

(b) unitary, if $\boldsymbol{A}^* \boldsymbol{A} = \boldsymbol{I}$ holds,

(c) normal, if $\boldsymbol{A}^* \boldsymbol{A} = \boldsymbol{A} \boldsymbol{A}^*$ holds,

(d) similar to the matrix $\boldsymbol{B} \in \mathbb{C}^{n \times n}$, if there exists a non-singular matrix $\boldsymbol{C} \in \mathbb{C}^{n \times n}$ such that $\boldsymbol{B} = \boldsymbol{C}^{-1} \boldsymbol{A} \boldsymbol{C}$,

(e) lower left triangular matrix, if $a_{ij} = 0 \quad \forall j > i$,

(f) upper right triangular matrix, if $a_{ij} = 0 \quad \forall j < i$,

(g) diagonal matrix, if $a_{ij} = 0 \quad \forall j \neq i$.

Definition 2.22 Let $X = \mathbb{R}^n$ or $\mathbb{C}^n$. A matrix $\boldsymbol{A} : X \to X$ is called

(a) positive semidefinite, if $(\boldsymbol{A}\boldsymbol{x},\boldsymbol{x})_2 \geq 0$ for all $\boldsymbol{x} \in X$,

(b) positive definite, if $(\boldsymbol{A}\boldsymbol{x},\boldsymbol{x})_2 > 0$ for all $\boldsymbol{x} \in X \backslash \{\boldsymbol{0}\}$,

(c) negative semidefinite, if $-\boldsymbol{A}$ is positive semidefinite,

(d) negative definite, if $-\boldsymbol{A}$ is positive definite.

Definition 2.23 Two systems of equations are called equivalent, if their solution sets are identical.

Lemma 2.24 *Let $\boldsymbol{P} \in \mathbb{C}^{n \times n}$ be invertible and $\boldsymbol{A} \in \mathbb{C}^{n \times n}$, then the systems of equations $\boldsymbol{A}\boldsymbol{x} = \boldsymbol{y}$ and $\boldsymbol{P}\boldsymbol{A}\boldsymbol{x} = \boldsymbol{P}\boldsymbol{y}$ are equivalent.*

Proof:

Due to the invertibility of the matrix $\boldsymbol{P}$ it follows that

$$\boldsymbol{P}\boldsymbol{x} = \boldsymbol{0} \quad \Leftrightarrow \quad \boldsymbol{x} = \boldsymbol{0}.$$

Thus, we obtain the claim directly from

$$\boldsymbol{P}(\boldsymbol{A}\boldsymbol{x} - \boldsymbol{y}) = 0 \quad \Leftrightarrow \quad \boldsymbol{A}\boldsymbol{x} - \boldsymbol{y} = \boldsymbol{0}.$$

$\square$

Lemma and Definition 2.25 *The linear system of equations* $\boldsymbol{A}\boldsymbol{x} = \boldsymbol{b}$ *with* $\boldsymbol{A} \in \mathbb{C}^{m \times n}$ $(\mathbb{R}^{m \times n})$ *is solvable if and only if* $\mathrm{rang}(\boldsymbol{A}) = \mathrm{rang}(\boldsymbol{A},\boldsymbol{b})$ *where* $\mathrm{rang}(\boldsymbol{A})$ *denotes the dimension of the image of* $\boldsymbol{A}$ *, that is,*

$$\mathrm{rang}(\boldsymbol{A}) = \dim \mathrm{image}(\boldsymbol{A})$$

with

$$\mathrm{image}(\boldsymbol{A}) = \{\boldsymbol{y} \in \mathbb{C}^m \,(\mathbb{R}^m) \,|\, \exists \, \boldsymbol{x} \in \mathbb{C}^n \,(\mathbb{R}^n) \text{ such that } \boldsymbol{y} = \boldsymbol{A}\boldsymbol{x}\}.$$

Proof:

" $\Rightarrow$ "

Let $\boldsymbol{A}\boldsymbol{x} = \boldsymbol{b}$, then $\boldsymbol{b} \in \mathrm{image}(\boldsymbol{A})$. Thus we obtain $\mathrm{rang}(\boldsymbol{A}) = \mathrm{rang}(\boldsymbol{A},\boldsymbol{b})$.

" $\Leftarrow$ "

Let $\mathrm{rang}(\boldsymbol{A}) = \mathrm{rang}(\boldsymbol{A},\boldsymbol{b})$, then $\boldsymbol{b}$ can be written in the form $\boldsymbol{b} = \sum_{i=1}^{n} x_i \boldsymbol{a}_i$ with $x_i \in \mathbb{C}$, where $\boldsymbol{a}_i$ for $i = 1,\ldots,n$ denotes the i-th column of the matrix $\boldsymbol{A}$. From this it follows directly

$$\boldsymbol{A}\boldsymbol{x} = \boldsymbol{b} \text{ with } \boldsymbol{x} = (x_1,\ldots,x_n)^T.$$

$\square$

Lemma 2.26 *Let* $\boldsymbol{L},\widetilde{\boldsymbol{L}} \in \mathbb{C}^{n \times n}$ *be lower left and* $\boldsymbol{R},\widetilde{\boldsymbol{R}} \in \mathbb{C}^{n \times n}$ *upper right triangular matrices, then*

$$\boldsymbol{L}\widetilde{\boldsymbol{L}} \text{ and } \boldsymbol{R}\widetilde{\boldsymbol{R}}$$

are also lower left and upper right triangular matrices, respectively.

Proof:

Let $\overline{\boldsymbol{L}} = \left(\overline{l}_{ij}\right)_{i,j=1,\ldots,n} = \boldsymbol{L}\widetilde{\boldsymbol{L}}$, then for $j > i$ it follows that

$$\overline{l}_{ij} = \sum_{m=1}^{n} l_{im}\widetilde{l}_{mj} = \sum_{m=1}^{j-1} l_{im} \underbrace{\widetilde{l}_{mj}}_{=0} + \sum_{m=j}^{n} \underbrace{l_{im}}_{=0} \widetilde{l}_{mj} = 0.$$

Analogously, the claim holds for upper right triangular matrices. $\square$

Lemma 2.27 *Let* $\boldsymbol{Q},\widetilde{\boldsymbol{Q}} \in \mathbb{R}^{n \times n}\,(\mathbb{C}^{n \times n})$ *be orthogonal (unitary) matrices, then*

$$\boldsymbol{Q}\widetilde{\boldsymbol{Q}}$$

is also orthogonal (unitary).

Proof:

In the proof, we restrict ourselves to orthogonal matrices. The proof of the claim for unitary matrices proceeds analogously. Due to the orthogonality of the matrices $\widetilde{Q}$ and Q, it follows that

$$\left(Q\widetilde{Q}\right)^T Q\widetilde{Q} = \widetilde{Q}^T Q^T Q\widetilde{Q} = \widetilde{Q}^T I\widetilde{Q} = \widetilde{Q}^T \widetilde{Q} = I.$$

$\square$

From the following theorem, we can deduce a central property of unitary as well as orthogonal matrices.

Theorem 2.28 *Let* $Q \in \mathbb{C}^{n\times n}$ $(\mathbb{R}^{n\times n})$ *be a unitary (orthogonal) matrix, then*

$$\|Qx\|_2 = \|x\|_2 \text{ for all } x \in \mathbb{C}^{n\times n} \left(\mathbb{R}^{n\times n}\right).$$

Proof:

The proof follows directly, taking into account that $Q^*Q = I$, from

$$\|Qx\|_2^2 = (Qx,Qx)_2 = (x,Q^*Qx)_2 = (x,x)_2 = \|x\|_2^2.$$

$\square$

Orthogonal as well as unitary matrices therefore always represent length-preserving mappings with respect to the Euclidean norm.

As already mentioned earlier, a matrix norm can be defined by means of a given vector norm. If $A \in \mathbb{C}^{n\times n}$ and $\|.\|_a : \mathbb{C}^n \to \mathbb{R}$ is a norm, then

$$\|A\|_a := \sup_{\|x\|_a=1} \|Ax\|_a \tag{2.2.2}$$

is called the matrix norm induced by the vector norm. According to the compatibility condition mentioned on page 17, the inequality

$$\|Ax\|_a \leq \|A\|_a\|x\|_a$$

holds for all $x \in \mathbb{C}^n$. At this point, the natural question arises as to whether there is a simple computational rule for an induced matrix norm. As an example, we will now derive an explicit representation of the corresponding matrix norm for the absolute column sum norm. As a preliminary consideration, for any $x \in \mathbb{C}^n$, using the norm properties (N3) and (N4), we write the estimate

$$\|Ax\|_1 = \left\| x_1 \begin{pmatrix} a_{11} \\ \vdots \\ a_{n1} \end{pmatrix} + \ldots + x_n \begin{pmatrix} a_{1n} \\ \vdots \\ a_{nn} \end{pmatrix} \right\|_1$$

$$\leq |x_1| \left\| \begin{pmatrix} a_{11} \\ \vdots \\ a_{n1} \end{pmatrix} \right\|_1 + \ldots + |x_n| \left\| \begin{pmatrix} a_{1n} \\ \vdots \\ a_{nn} \end{pmatrix} \right\|_1.$$

Using

$$\max_{j=1,\ldots,n} \left\| \begin{pmatrix} a_{1j} \\ \vdots \\ a_{nj} \end{pmatrix} \right\|_1 = \max_{j=1,\ldots,n} \sum_{i=1}^{n} |a_{ij}|$$

we thus obtain

$$\|A\|_1 = \sup_{\|x\|_1=1} \|Ax\|_1 \leq \sup_{\|x\|_1=1} \max_{j=1,\ldots,n} \sum_{i=1}^{n} |a_{ij}| \underbrace{(|x_1| + \ldots + |x_n|)}_{=\|x\|_1} = \max_{j=1,\ldots,n} \sum_{i=1}^{n} |a_{ij}|.$$

Consequently, we have derived an upper bound for the 1-norm of the matrix A given by $\max_{j=1,\ldots,n} \sum_{i=1}^{n} |a_{ij}|$. Let $m \in \{1,\ldots,n\}$ with

$$\sum_{i=1}^{n} |a_{im}| = \max_{j=1,\ldots,n} \sum_{i=1}^{n} |a_{ij}|$$

given, then with the m-th unit vector e_m we obtain the estimate

$$\|A\|_1 = \sup_{\|x\|_1=1} \|Ax\|_1 \geq \|Ae_m\|_1 = \left\| \begin{pmatrix} a_{1m} \\ \vdots \\ a_{nm} \end{pmatrix} \right\|_1 = \sum_{i=1}^{n} |a_{im}| = \max_{j=1,\ldots,n} \sum_{i=1}^{n} |a_{ij}|,$$

which shows that $\max_{j=1,\ldots,n} \sum_{i=1}^{n} |a_{ij}|$ is also a lower bound, and thus, in summary, the so-called absolute column sum norm

$$\|A\|_1 = \max_{j=1,\ldots,n} \sum_{i=1}^{n} |a_{ij}|$$

holds. Accordingly, the absolute row sum norm is given by

$$\|A\|_\infty = \max_{i=1,\ldots,n} \sum_{j=1}^{n} |a_{ij}|.$$

For the matrix norm induced by the Euclidean vector norm, such a simple computable representation for an arbitrary matrix unfortunately cannot be given. However, we will deal with the relationships of this matrix norm to the spectral radius of the corresponding matrix and also provide here an estimate using

$$\|A\|_F = \sqrt{\sum_{i,k=1}^{n} |a_{ik}|^2}$$

Here, $\|.\|_F : \mathbb{C}^{n \times n} \to \mathbb{R}$ is called the Frobenius norm. By applying the Cauchy-Schwarz inequality, we obtain

$$|(Ax)_i|^2 = \left| \sum_{j=1}^{n} a_{ij} x_j \right|^2 \leq \sum_{j=1}^{n} |a_{ij}|^2 \sum_{j=1}^{n} |x_j|^2,$$

so that the announced estimate of the Euclidean matrix norm by the Frobenius norm is given by

$$\|A\|_2^2 = \sup_{\|x\|_2=1} \|Ax\|_2^2 = \sup_{\|x\|_2=1} \sum_{i=1}^{n} |(Ax)_i|^2$$

$$\leq \sup_{\|x\|_2=1} \left(\sum_{i,j=1}^{n} |a_{ij}|^2 \sum_{j=1}^{n} |x_j|^2 \right) = \sum_{i,j=1}^{n} |a_{ij}|^2 = \|A\|_F^2$$

and can be written as

$$\|A\|_2 \leq \|A\|_F.$$

If we consider the identity matrix $I \in \mathbb{C}^{n \times n}$, we obtain $\|I\|_F = \sqrt{n}$. From (2.2.2) it follows that

$$\|I\|_a = \sup_{\|x\|_a=1} \|Ix\|_a = \sup_{\|x\|_a=1} \|x\|_a = 1,$$

which immediately shows that the Frobenius norm is not an induced matrix norm. Also, the norm given by $\|A\| := \max_{i,k=1,\ldots,n} |a_{ik}|$ does not possess a corresponding vector norm. Simple examples show that this norm is not submultiplicative. At this point, it should be emphasized again that in the following considerations, unless explicitly stated otherwise, we will always restrict ourselves to induced matrix norms, although some statements (for example, Theorem 2.29) are generally valid. The only exception is the Frobenius norm considered in Section 5.7 in the derivation of the incomplete Frobenius inverses.

Theorem 2.29 *A matrix* $A \in \mathbb{C}^{n \times n}$ *is bounded in every norm.*

Proof:
Let $x \in X$ be arbitrary, then it follows that

$$\|Ax\|_\infty = \max_{i=1,\ldots,n} |(Ax)_i| = \max_{i=1,\ldots,n} \left| \sum_{j=1}^{n} a_{ij} x_j \right|$$

$$\leq \max_{i=1,\ldots,n} \sum_{j=1}^{n} |a_{ij}||x_j| \leq S\|x\|_\infty$$

with $S = \max_{i=1,\ldots,n} \sum_{j=1}^{n} |a_{ij}|$. This statement holds for any norm due to the equivalence of norms (Theorem 2.11).

$$\square$$

Consequently, by Theorem 2.29, we also obtain the continuity of a linear mapping represented by a matrix $A \in \mathbb{C}^{n \times n}$.

Definition 2.30 A complex number $\lambda \in \mathbb{C}$ is called an eigenvalue of the matrix $A \in \mathbb{C}^{n \times n}$ if there exists a vector $x \in \mathbb{C}^n \setminus \{0\}$ such that

$$Ax = \lambda x.$$

The vector x is called an eigenvector corresponding to the eigenvalue λ.

The set
$$\sigma(A) = \{\lambda \mid \lambda \text{ is an eigenvalue of } A\}$$
is called the spectrum of A.

The number
$$\rho(A) = \max\{|\lambda| \mid \lambda \in \sigma(A)\}$$
is called the spectral radius of A.

The eigenvalues $\lambda \in \sigma(A)$ are, due to
$$(A - \lambda I)\, x = 0 \text{ with } x \in \mathbb{C}^n \setminus \{0\}$$
the roots of the characteristic polynomial
$$p(\lambda) = \det(A - \lambda I).$$

Since every polynomial over $\mathbb{C}$ splits into linear factors by the Fundamental Theorem of Algebra, i.e., $p(\lambda) = \prod_{i=1}^{n}(\lambda - \lambda_i)$, we directly obtain $\sigma(A) \neq \emptyset$ for all $A \in \mathbb{C}^{n \times n}$.

The following theorem of Schur provides the essential tool for proving Theorem 2.37, which, in combination with Theorem 2.36, yields a direct connection between a norm and the spectral radius of a matrix. These statements will later prove to be crucial in the convergence analysis of splitting methods.

Theorem 2.31 *For every matrix $A \in \mathbb{C}^{n \times n}$ $(\mathbb{R}^{n \times n}$ with $\sigma(A) \subset \mathbb{R})$ there exists a unitary (orthogonal) matrix $U \in \mathbb{C}^{n \times n}$ $(\mathbb{R}^{n \times n})$ such that*

$$U^* A U$$

is an upper triangular matrix.

Proof:
We restrict the proof of the statement initially to the complex case.

The proof is carried out by complete induction.

For $n = 1$, $U = I$ satisfies the statement.

Assume the statement holds for $j = 1, \ldots, n$, then choose a $\lambda \in \sigma(A)$ with corresponding eigenvector $\widetilde{v}_1 \in \mathbb{C}^{n+1} \setminus \{0\}$. By extending $v_1 = \widetilde{v}_1/\|\widetilde{v}_1\|_2$ with $v_2, \ldots, v_{n+1}$ to form an orthonormal basis of $\mathbb{C}^{n+1}$, we obtain with

$$\mathbb{C}^{(n+1) \times (n+1)} \ni V = (v_1 \ldots v_{n+1})$$

the equation
$$V^* A V e_1 = V^* A v_1 = V^* \lambda v_1 = \lambda e_1,$$
where $e_1 = (1, 0, \ldots, 0)^T \in \mathbb{C}^{n+1}$. From this it follows that

$$
\boldsymbol{V}^*\boldsymbol{A}\boldsymbol{V} = \begin{pmatrix} \lambda & \tilde{\boldsymbol{a}}^T \\ 0 & \\ \vdots & \widetilde{\boldsymbol{A}} \\ 0 & \end{pmatrix} \quad \text{with } \widetilde{\boldsymbol{A}} \in \mathbb{C}^{n\times n} \text{ and } \tilde{\boldsymbol{a}} \in \mathbb{C}^n.
$$

By the induction hypothesis, there exists a unitary matrix $\widetilde{\boldsymbol{W}} \in \mathbb{C}^{n\times n}$ such that $\widetilde{\boldsymbol{W}}^* \widetilde{\boldsymbol{A}} \widetilde{\boldsymbol{W}}$ is an upper triangular matrix. With $\widetilde{\boldsymbol{W}}$, also

$$
\boldsymbol{W} = \begin{pmatrix} 1 & 0 \cdots 0 \\ 0 & \\ \vdots & \widetilde{\boldsymbol{W}} \\ 0 & \end{pmatrix}
$$

is unitary. Hence, according to Lemma 2.27 $\boldsymbol{U} := \boldsymbol{V}\boldsymbol{W} \in \mathbb{C}^{(n+1)\times(n+1)}$ represents a unitary matrix, for which a simple calculation shows that $\boldsymbol{U}^*\boldsymbol{A}\boldsymbol{U}$ is an upper right triangular matrix.

We now consider the case of a non-singular matrix $\boldsymbol{A} \in \mathbb{R}^{(n+1)\times(n+1)}$. Due to $\sigma(\boldsymbol{A}) \subset \mathbb{R}$, every eigenvector $\tilde{\boldsymbol{v}}_1 \in \mathbb{C}^{n+1} \setminus \{\boldsymbol{0}\}$ corresponds to an eigenvalue $\lambda \in \mathbb{R}$. Since $\tilde{\boldsymbol{v}}_1$ can be written in the form $\tilde{\boldsymbol{v}}_1 = \tilde{\boldsymbol{x}}_1 + i\tilde{\boldsymbol{y}}_1$ with $\tilde{\boldsymbol{x}}_1, \tilde{\boldsymbol{y}}_1 \in \mathbb{R}^{n+1}$, we obtain

$$
\boldsymbol{A}\tilde{\boldsymbol{x}}_1 + i\boldsymbol{A}\tilde{\boldsymbol{y}}_1 = \boldsymbol{A}\tilde{\boldsymbol{v}}_1 = \lambda\tilde{\boldsymbol{v}}_1 = \lambda\tilde{\boldsymbol{x}}_1 + i\lambda\tilde{\boldsymbol{y}}_1
$$

such that the property

$$
\boldsymbol{A}\tilde{\boldsymbol{x}}_1 = \lambda\tilde{\boldsymbol{x}}_1 \text{ and } \boldsymbol{A}\tilde{\boldsymbol{y}}_1 = \lambda\tilde{\boldsymbol{y}}_1,
$$

holds, proving that both $\tilde{\boldsymbol{x}}_1$ and $\tilde{\boldsymbol{y}}_1$ represent an eigenvector from $\mathbb{R}^{n+1} \setminus \{\boldsymbol{0}\}$. Taking this property into account, the proof in the case of a real matrix follows analogously to the above procedure. $\qquad\square$

Lemma 2.32 *For every Hermitian matrix* $\boldsymbol{A} \in \mathbb{C}^{n\times n}$*, it holds that* $\sigma(\boldsymbol{A}) \subset \mathbb{R}$*.*

Proof:
Let $\boldsymbol{x} \in \mathbb{C}^n \setminus \{\boldsymbol{0}\}$ be an eigenvector corresponding to the eigenvalue $\lambda \in \sigma(\boldsymbol{A})$. Then, using

$$
\lambda \underbrace{\|\boldsymbol{x}\|_2}_{\in \mathbb{R}^+} = \lambda(\boldsymbol{x},\boldsymbol{x})_2 = (\lambda\boldsymbol{x},\boldsymbol{x})_2 = (\boldsymbol{A}\boldsymbol{x},\boldsymbol{x})_2 = (\boldsymbol{x},\boldsymbol{A}\boldsymbol{x})_2 = (\boldsymbol{x},\lambda\boldsymbol{x})_2 = \overline{\lambda}(\boldsymbol{x},\boldsymbol{x})_2 = \overline{\lambda} \underbrace{\|\boldsymbol{x}\|_2}_{\in \mathbb{R}^+}
$$

we directly obtain $\lambda = \overline{\lambda}$ and thus $\lambda \in \mathbb{R}$. $\qquad\square$

As a direct consequence of the previous theorem, we obtain the following statement for Hermitian and, due to Lemma 2.32, also for symmetric matrices.

Corollary 2.33 *Let* $\boldsymbol{A} \in \mathbb{C}^{n\times n}$ *(*$\mathbb{R}^{n\times n}$*) be Hermitian (symmetric), then there exists a unitary (orthogonal) matrix* $\boldsymbol{U} \in \mathbb{C}^{n\times n}$ *(*$\mathbb{R}^{n\times n}$*) such that*

$$
\boldsymbol{U}^*\boldsymbol{A}\boldsymbol{U} = \operatorname{diag}\{\lambda_1,\ldots,\lambda_n\} \in \mathbb{R}^{n\times n}
$$

holds. Here, for $i = 1,\ldots,n$*, each* $\lambda_i \in \mathbb{R}$ *is an eigenvalue of the matrix* $\boldsymbol{A}$ *with the* i*-th column of* $\boldsymbol{U}$ *as the corresponding eigenvector.*

Analogous to the above considerations, we can formulate an important statement regarding the eigenvalues of every positive definite matrix.

Theorem 2.34 *Let $A \in \mathbb{C}^{n \times n}$ be a positive definite matrix, then $\sigma(A) \in \mathbb{R}^+$.*

Proof:
Let $\lambda \in \sigma(A)$ and $x \neq 0$ the corresponding eigenvector, then due to the positive definiteness of the matrix A, the inequality

$$0 < (Ax,x)_2 = \lambda(x,x)_2 = \lambda \underbrace{\|x\|_2^2}_{>0}$$

holds, and consequently $\lambda > 0$. $\qquad\square$

As already mentioned above, a simple formula for the matrix norm induced by the Euclidean vector norm can unfortunately not be given in general. However, the following relationship to the spectral radius exists.

Theorem 2.35 *Let $A \in \mathbb{C}^{n \times n}$, then*

$$\|A\|_2 = \sqrt{\rho(A^* A)}. \tag{2.2.3}$$

Proof:
Since $A^* A$ is Hermitian, according to Corollary 2.33 there exists a unitary matrix $U \in \mathbb{C}^{n \times n}$ such that $U^* A^* A U = \mathrm{diag}\{\lambda_1, \ldots, \lambda_n\} \in \mathbb{R}^{n \times n}$ holds. Every $x \in \mathbb{C}^n$ can therefore be represented using the columns $u_1, \ldots, u_n$ of the unitary matrix U in the form $x = \sum_{i=1}^{n} \alpha_i u_i$ with $\alpha_i \in \mathbb{C}$, and it holds that

$$A^* A x = \sum_{i=1}^{n} \lambda_i \alpha_i u_i.$$

With this we obtain

$$\begin{aligned}
\|Ax\|_2^2 &= (Ax, Ax)_2 = (x, A^* A x)_2 \\
&= \left(\sum_{i=1}^{n} \alpha_i u_i, \sum_{i=1}^{n} \lambda_i \alpha_i u_i \right)_2 = \sum_{i=1}^{n} (\alpha_i u_i, \lambda_i \alpha_i u_i)_2 \\
&= \sum_{i=1}^{n} \lambda_i |\alpha_i|^2 \\
&\leq \rho(A^* A) \sum_{i=1}^{n} |\alpha_i|^2 = \rho(A^* A) \|x\|_2^2.
\end{aligned} \tag{2.2.4}$$

Consequently, we can write

$$\frac{\|Ax\|_2^2}{\|x\|_2^2} \leq \rho(A^* A).$$

Equality is achieved by considering the eigenvector u_j corresponding to the eigenvalue λ_j of largest modulus. With (2.2.4) it follows that

$$0 \leq \|\boldsymbol{A}\boldsymbol{u}_k\|_2^2 = \lambda_k, \; k = 1,\dots,n$$

and hence we obtain

$$\frac{\|\boldsymbol{A}\boldsymbol{u}_j\|_2^2}{\|\boldsymbol{u}_j\|_2^2} = \frac{\lambda_j \|\boldsymbol{u}_j\|_2^2}{\|\boldsymbol{u}_j\|_2^2} = \lambda_j = \rho\left(\boldsymbol{A}^*\boldsymbol{A}\right).$$

$\square$

Due to the property (2.2.3), $\|\boldsymbol{A}\|_2$ is also called the spectral norm of the matrix $\boldsymbol{A}$. The following two theorems are of crucial importance for the convergence statements of iterative methods that are based on a splitting of the matrix of the system of equations.

Theorem 2.36 *For every matrix $\boldsymbol{A} \in \mathbb{C}^{n \times n}$ the following holds*

  (a) $\rho(\boldsymbol{A}) \leq \|\boldsymbol{A}\|$ for every induced matrix norm,

  (b) $\|\boldsymbol{A}\|_2 = \rho(\boldsymbol{A})$ if $\boldsymbol{A}$ is Hermitian.

Proof:
Let $\lambda \in \mathbb{C}$ be the eigenvalue of $\boldsymbol{A}$ with largest modulus, with corresponding eigenvector $\boldsymbol{u} \in \mathbb{C}^n \setminus \{\boldsymbol{0}\}$ for which, without loss of generality, $\|\boldsymbol{u}\| = 1$ holds. Then it follows that

$$\|\boldsymbol{A}\| = \sup_{\|\boldsymbol{x}\|=1} \|\boldsymbol{A}\boldsymbol{x}\| \geq \|\boldsymbol{A}\boldsymbol{u}\| = |\lambda|\|\boldsymbol{u}\| = |\lambda|,$$

which directly yields $\rho(\boldsymbol{A}) \leq \|\boldsymbol{A}\|$.

With Corollary 2.33 we obtain for Hermitian matrices the property $\rho(\boldsymbol{A}^2) = \rho(\boldsymbol{A})^2$, so that for Hermitian matrices, using Theorem 2.35, the following chain of equalities

$$\|\boldsymbol{A}\|_2 = \sqrt{\rho(\boldsymbol{A}^*\boldsymbol{A})} = \sqrt{\rho(\boldsymbol{A}^2)} = \sqrt{\rho(\boldsymbol{A})^2} = \rho(\boldsymbol{A})$$

hold as claimed. $\square$

Theorem 2.37 *For every matrix $\boldsymbol{A} \in \mathbb{C}^{n \times n}$ and every $\varepsilon > 0$ there exists a norm on $\mathbb{C}^{n \times n}$ such that*

$$\|\boldsymbol{A}\| \leq \rho(\boldsymbol{A}) + \varepsilon$$

holds.

Proof:
For $n = 1$ the statement is as trivial as for any zero matrix $\boldsymbol{A} \in \mathbb{C}^{n \times n}$. Therefore, let $n \geq 2$ and $\boldsymbol{A}$ not equal to the zero matrix. Schur's theorem guarantees the existence of a unitary matrix $\boldsymbol{U} \in \mathbb{C}^{n \times n}$ such that

$$\boldsymbol{R} = \boldsymbol{U}^*\boldsymbol{A}\boldsymbol{U} = \begin{pmatrix} r_{11} & \cdots & r_{1n} \\ & \ddots & \vdots \\ & & r_{nn} \end{pmatrix},$$

where $\lambda_i = r_{ii}$, $i = 1,\dots,n$ are the eigenvalues of $\boldsymbol{A}$. Let $\alpha := \max\limits_{i,k=1,\dots,n} |r_{ik}| > 0$, then for a given $\varepsilon > 0$ we define

$$\delta := \min\left(1, \frac{\varepsilon}{(n-1)\alpha}\right) > 0. \tag{2.2.5}$$

With

$$\boldsymbol{D} = \begin{pmatrix} 1 & & & \\ & \delta & & \\ & & \ddots & \\ & & & \delta^{n-1} \end{pmatrix}$$

we obtain

$$\boldsymbol{C} := \boldsymbol{D}^{-1}\boldsymbol{R}\boldsymbol{D} = \begin{pmatrix} r_{11} & \delta r_{12} & \cdots & \cdots & \delta^{n-1}r_{1n} \\ & \ddots & \ddots & & \vdots \\ & & \ddots & \ddots & \vdots \\ & & & \ddots & \delta r_{n-1,n} \\ & & & & r_{n,n} \end{pmatrix}.$$

Using the Definition (2.2.5), it follows that

$$\begin{aligned} \|\boldsymbol{C}\|_\infty &\leq \max_{i=1,\ldots,n} |r_{ii}| + (n-1)\delta\alpha \\ &= \rho(\boldsymbol{A}) + (n-1)\delta\alpha \\ &\leq \rho(\boldsymbol{A}) + \varepsilon. \end{aligned} \tag{2.2.6}$$

Since $\boldsymbol{D}$ and $\boldsymbol{U}$ are non-singular matrices, the following defines a norm on $\mathbb{C}^n$:

$$\|\boldsymbol{x}\| := \|\boldsymbol{D}^{-1}\boldsymbol{U}^{-1}\boldsymbol{x}\|_\infty.$$

Let $\boldsymbol{y} := \boldsymbol{D}^{-1}\boldsymbol{U}^{-1}\boldsymbol{x}$, then using (2.2.6) we obtain

$$\begin{aligned} \|\boldsymbol{A}\| &= \sup_{\|\boldsymbol{x}\|=1} \|\boldsymbol{A}\boldsymbol{x}\| = \sup_{\|\boldsymbol{D}^{-1}\boldsymbol{U}^{-1}\boldsymbol{x}\|_\infty=1} \|\boldsymbol{D}^{-1}\boldsymbol{U}^{-1}\boldsymbol{A}\boldsymbol{x}\|_\infty \\ &= \sup_{\|\boldsymbol{y}\|_\infty=1} \|\boldsymbol{D}^{-1}\boldsymbol{U}^{-1}\boldsymbol{A}\boldsymbol{U}\boldsymbol{D}\boldsymbol{y}\|_\infty \\ &= \sup_{\|\boldsymbol{y}\|_\infty=1} \|\boldsymbol{C}\boldsymbol{y}\|_\infty \\ &= \|\boldsymbol{C}\|_\infty \leq \rho(\boldsymbol{A}) + \varepsilon. \end{aligned}$$

$\square$

With the two theorems above, it thus follows that for all matrices $\boldsymbol{A}$ and all $\varepsilon > 0$, there always exists a norm such that

$$\rho(\boldsymbol{A}) \leq \|\boldsymbol{A}\| \leq \rho(\boldsymbol{A}) + \varepsilon$$

holds. In this way, convergence statements that are based on the norm of a matrix and are independent of the specific choice of the norm can be directly transferred to the spectral radius of the matrix.

Let us now consider normal matrices, which are sometimes also defined via their diagonalizability by means of a unitary matrix [25]. The possibility of such a definition of normal matrices is demonstrated by the following theorem, which also provides us in the following with a special representation of the condition number of such matrices.

Theorem 2.38 *A matrix $A \in \mathbb{C}^{n \times n}$ is normal if and only if there exists a unitary matrix $U \in \mathbb{C}^{n \times n}$ such that $D = U^* A U$ is a diagonal matrix.*

Proof:

" $\Rightarrow$ "

By Theorem 2.31, there exists a unitary matrix U such that $R = (r_{ij})_{i,j=1,\ldots,n} = U^* A U$ is an upper triangular matrix. With

$$R^* R = U^* A^* U U^* A U = U^* A^* A U = U^* A A^* U = U^* A U U^* A^* U = R R^*$$

R is also normal. For the diagonal elements of the matrix $B = (b_{ij})_{i,j=1,\ldots,n} = R^* R = R R^*$ we have

$$\sum_{j=1}^{i} |r_{ji}|^2 = b_{ii} = \sum_{j=i}^{n} |r_{ij}|^2.$$

A successive evaluation of this equation for $i = 1,\ldots,n$ yields $r_{ij} = 0$ for all $i \neq j$. Thus, $D = R$ is the desired diagonal matrix.

" $\Leftarrow$ "

Let $U \in \mathbb{C}^{n \times n}$ be unitary with $D = \text{diag}\{d_{11}, \ldots, d_{nn}\} = U^* A U$, then it follows that

$$A^* A = U D^* D U^* = U D D^* U^* = A A^*.$$

$\square$

2.3   Condition Number and Singular Values

In addition to the spectral radius, the condition number of a matrix is also an interesting quantity that can be used to estimate the influence of errors as well as the convergence rates of iterative methods. Therefore, in this section, we will introduce the concept of the condition number and study its significance with respect to the solution of linear systems of equations. It will turn out that the condition number of a normal matrix $A \in \mathbb{R}^{n \times n}$ with respect to the spectral norm can be expressed in terms of its eigenvalues. In the general case of a non-singular matrix, its singular value distribution must be considered.

Definition 2.39 Let $A \in \mathbb{C}^{n \times n}$ be invertible, then

$$\text{cond}_a(A) := \|A\|_a \|A^{-1}\|_a$$

is called the condition number of the matrix A with respect to the induced matrix norm $\|\cdot\|_a$.

It is easy to see that the condition number of a non-singular matrix is bounded below independently of the underlying induced matrix norm. The mathematical proof of this is provided by the following lemma.

Lemma 2.40 *Let* $A \in \mathbb{C}^{n \times n}$ *be invertible, then*

$$\operatorname{cond}(A) \geq \operatorname{cond}(I) = 1$$

holds for $\operatorname{cond}(A) = \|A\| \, \|A^{-1}\|$ *with an induced matrix norm* $\|.\|$.

Proof:
The proof follows directly from the following inequality

$$\operatorname{cond}(I) = \|I\| \|I^{-1}\| = 1 = \|I\| = \|AA^{-1}\| \overset{\text{Theorem 2.17}}{\leq} \|A\| \|A^{-1}\| = \operatorname{cond}(A).$$

$\square$

We will now illustrate the relevance of the condition number with respect to the iterative solution of linear systems of equations using two theorems. In practical applications, it is common to use the norm of the residual vector $r_m = b - Ax_m$ as a measure of the quality of the current approximate solution within an iterative method, since the error vector $e_m = A^{-1}b - x_m$ is not available due to the unknown exact solution. One reason for this is the property that the residual, like the error, vanishes identically if and only if the iterate coincides with the exact solution. It will be shown that for a small condition number of the matrix of the linear system, a convergence estimate based on the residual is meaningful. However, if the condition number is large, then despite a reduction in the residual, the error norm can increase significantly.

Theorem 2.41 *Let an iterative method for solving* $Ax = b$ *with a non-singular matrix* A *be given. Let* $e_k = A^{-1}b - x_k$ *denote the error vector and* $r_k = b - Ax_k$ *the residual vector of the* k *-th iteration step, then*

$$\frac{1}{\operatorname{cond}(A)} \frac{\|r_k\|}{\|r_0\|} \leq \frac{\|e_k\|}{\|e_0\|} \leq \operatorname{cond}(A) \frac{\|r_k\|}{\|r_0\|} \leq \operatorname{cond}(A)^2 \frac{\|e_k\|}{\|e_0\|}. \tag{2.3.1}$$

Proof:
With

$$\|r_k\| = \|b - A x_k\| = \|A e_k\| \leq \|A\| \, \|e_k\|$$

and

$$\|e_k\| = \|A^{-1}b - x_k\| \leq \|A^{-1}\| \, \|b - Ax_k\| = \|A^{-1}\| \, \|r_k\|$$

the first part of the assertion follows from

$$\frac{1}{\operatorname{cond}(A)} \frac{\|r_k\|}{\|r_0\|} = \frac{1}{\|A\| \|A^{-1}\|} \frac{\|r_k\|}{\|r_0\|} \leq \frac{1}{\|A\|} \frac{\|r_k\|}{\|A^{-1}r_0\|} = \frac{1}{\|A\|} \frac{\|A e_k\|}{\|e_0\|} \leq \frac{\|e_k\|}{\|e_0\|}.$$

The remaining inequalities follow analogously.

$\square$

From the above theorem, we can see that in the case of an orthogonal matrix A, we always have

$$\frac{\|e_k\|_2}{\|e_0\|_2} = \frac{\|r_k\|_2}{\|r_0\|_2},$$

which allows for a direct convergence analysis based on the residual.

Measurement data from physical experiments naturally exhibit inaccuracies. The effect of such or other erroneous input data on the solution can also be estimated using the condition number of the matrix. Here, too, a small condition number proves advantageous, since small relative errors in the data b will also result in small relative errors in the solutions x.

Theorem 2.42 *Let A be non-singular, x the solution of the system of equations $Ax = b$ and $x + \Delta x$ the solution of $A(x + \Delta x) = b + \Delta b$, then*

$$\frac{\|\Delta x\|}{\|x\|} \leq \operatorname{cond}(A)\frac{\|\Delta b\|}{\|b\|}.$$

Proof:
Due to the linearity of A, it follows that

$$\Delta b = (b + \Delta b) - b = A(x + \Delta x) - Ax = A(\Delta x).$$

With this we obtain

$$\|\Delta x\| = \|A^{-1}(\Delta b)\| \leq \|A^{-1}\|\,\|\Delta b\|,$$

so that with

$$\|b\| = \|Ax\| \leq \|A\|\,\|x\|$$

the claimed inequality with

$$\frac{\|\Delta x\|}{\|x\|} \leq \frac{\|A^{-1}\|\|\Delta b\|}{\|A\|^{-1}\|b\|} = \operatorname{cond}(A)\frac{\|\Delta b\|}{\|b\|}$$

follows. $\qquad\qquad\square$

If a system of equations is given in which the matrix has a very large condition number, then, according to Theorems 2.41 and 2.42, it is advisable to first reformulate it into an equivalent system $\tilde{A}x = \tilde{b}$ with $\operatorname{cond}(\tilde{A}) \ll \operatorname{cond}(A)$ and then apply an iterative method to the transformed system. Such an equivalent reformulation can, for example, be carried out by multiplication with a non-singular matrix P according to Lemma 2.24. This raises the question of the choice of the matrix P. For computational efficiency, the matrix should be easy to compute, and for effectiveness, it should represent a good approximation of the inverse of the matrix A, so that $\operatorname{cond}(PA) \ll \operatorname{cond}(A)$ holds. Such equivalent transformations are called preconditionings and are discussed in detail in Chapter 5. It should already be mentioned here that, although the stabilization of numerical methods is one reason for using preconditioning, the main objective of such techniques is to accelerate iterative methods.

The following section provides a representation of the condition number of a non-singular matrix A with respect to the spectral norm in terms of its singular values or its eigenvalues. The condition numbers with respect to the absolute row or absolute column sum norm can then be estimated via $\operatorname{cond}_2(A)$.

Theorem and Definition 2.43 *For every invertible matrix $A \in \mathbb{R}^{n\times n}$ there exist orthogonal matrices $U, V \in \mathbb{R}^{n\times n}$ such that*

$$U^T AV = \operatorname{diag}\{\sigma_1, \ldots, \sigma_n\} \tag{2.3.2}$$

with $0 < \sigma_1 \le \ldots \le \sigma_n$. The listed real numbers σ_i $(i = 1, \ldots, n)$ are called the singular values of the matrix $\boldsymbol{A}$ and each satisfies the equation

$$\det\left(\boldsymbol{A}^T\boldsymbol{A} - \sigma_i^2\boldsymbol{I}\right) = 0. \tag{2.3.3}$$

The i-th column of $\boldsymbol{U}$ or $\boldsymbol{V}$ is called the i-th left or right singular vector of the matrix $\boldsymbol{A}$.

Proof:
Since $\boldsymbol{A}$ is invertible, $\boldsymbol{A}^T\boldsymbol{A}$ is a symmetric, positive definite matrix. Thus, according to Corollary 2.33, there exists an orthogonal matrix $\boldsymbol{V} \in \mathbb{R}^{n\times n}$ such that

$$\boldsymbol{V}^T\boldsymbol{A}^T\boldsymbol{A}\boldsymbol{V} = \mathrm{diag}\{\lambda_1, \ldots, \lambda_n\}$$

with $0 < \lambda_1 \le \ldots \le \lambda_n$. If we define $\sigma_i = \sqrt{\lambda_i}$, we obtain $0 < \sigma_1 \le \ldots \le \sigma_n$ with $\det\left(\boldsymbol{A}^T\boldsymbol{A} - \sigma_i^2\boldsymbol{I}\right) = 0$. To prove the remaining statement (2.3.2), we define $\boldsymbol{U} = \boldsymbol{A}\boldsymbol{V}\boldsymbol{D}^{-1}$ with $\boldsymbol{D} = \mathrm{diag}\{\sigma_1, \ldots, \sigma_n\}$. Since

$$\boldsymbol{U}^T\boldsymbol{U} = \boldsymbol{D}^{-T}\boldsymbol{V}^T\boldsymbol{A}^T\boldsymbol{A}\boldsymbol{V}\boldsymbol{D}^{-1} = \boldsymbol{D}^{-1}\mathrm{diag}\{\lambda_1, \ldots, \lambda_n\}\boldsymbol{D}^{-1} = \boldsymbol{I}$$

$\boldsymbol{U}$ represents an orthogonal matrix, and we obtain the claimed representation by

$$\boldsymbol{U}^T\boldsymbol{A}\boldsymbol{V} = \boldsymbol{D}^{-T}\boldsymbol{V}^T\boldsymbol{A}^T\boldsymbol{A}\boldsymbol{V} = \boldsymbol{D}^{-T}\mathrm{diag}\{\lambda_1, \ldots, \lambda_n\} = \boldsymbol{D}.$$

$\square$

Theorem 2.44 *Let $\boldsymbol{A} \in \mathbb{R}^{n\times n}$, then with the notation from Theorem 2.43 the following statements hold:*

$$\sup_{\boldsymbol{x}\neq 0} \frac{\|\boldsymbol{A}\boldsymbol{x}\|_2}{\|\boldsymbol{x}\|_2} = \sigma_n$$

and

$$\inf_{\boldsymbol{x}\neq 0} \frac{\|\boldsymbol{A}\boldsymbol{x}\|_2}{\|\boldsymbol{x}\|_2} = \sigma_1.$$

Proof:
From the definition of the Euclidean norm and the invertibility of $\boldsymbol{V}$ it follows that

$$\sup_{\boldsymbol{x}\neq 0} \frac{\|\boldsymbol{A}\boldsymbol{x}\|_2^2}{\|\boldsymbol{x}\|_2^2} = \sup_{\boldsymbol{V}\boldsymbol{x}\neq 0} \frac{\boldsymbol{x}^T\boldsymbol{V}^T\boldsymbol{A}^T\boldsymbol{A}\boldsymbol{V}\boldsymbol{x}}{\boldsymbol{x}^T\boldsymbol{V}^T\boldsymbol{V}\boldsymbol{x}} = \sup_{\boldsymbol{V}\boldsymbol{x}\neq 0} \frac{\boldsymbol{x}^T\boldsymbol{V}^T\boldsymbol{A}^T\boldsymbol{U}\boldsymbol{U}^T\boldsymbol{A}\boldsymbol{V}\boldsymbol{x}}{\boldsymbol{x}^T\boldsymbol{V}^T\boldsymbol{V}\boldsymbol{x}}$$

since $\boldsymbol{U}\boldsymbol{U}^T = \boldsymbol{I}$. With this, and since $\boldsymbol{V}^T\boldsymbol{V} = \boldsymbol{I}$, the first statement follows. Analogously, the validity of the second statement can be shown. $\square$

After these considerations, we can express the spectral norm of a non-singular matrix $\boldsymbol{A}$ and its inverse in terms of the singular values of the matrix as follows:

$$\|\boldsymbol{A}\|_2 = \sigma_n \quad \text{and} \quad \|\boldsymbol{A}^{-1}\|_2 = \frac{1}{\sigma_1}$$

Since normal matrices are, according to Theorem 2.38, unitarily diagonalizable, we obtain as an immediate consequence of the above equation the following relationship between the condition number and the singular values, respectively eigenvalues, of a non-singular matrix in the following form.

Corollary 2.45 *Let $A \in \mathbb{R}^{n \times n}$ be invertible, then*

$$\mathrm{cond}_2(A) = \frac{\sigma_n}{\sigma_1}, \tag{2.3.4}$$

*where σ_n is the largest and σ_1 the smallest singular value of the matrix.
If A is also normal, then*

$$\mathrm{cond}_2(A) = \frac{|\lambda_n|}{|\lambda_1|}, \tag{2.3.5}$$

where λ_n denotes the eigenvalue of largest modulus and λ_1 the eigenvalue of smallest modulus of the matrix.

Although a statement equivalent to Corollary 2.45 with respect to another matrix norm does not generally exist, error estimates and convergence statements can be transferred to the corresponding condition numbers. Analogous to the vector and matrix norms, we now define the concept of equivalent condition numbers.

Definition 2.46 Let $\|.\|_a$ and $\|.\|_b$ be two vector norms on $\mathbb{R}^n$. Then the condition numbers cond_a and cond_b are called equivalent, if there exist real numbers $\alpha, \beta > 0$ such that

$$\alpha \, \mathrm{cond}_b(A) \leq \mathrm{cond}_a(A) \leq \beta \, \mathrm{cond}_b(A) \tag{2.3.6}$$

holds for all invertible $A \in \mathbb{R}^{n \times n}$.

As a direct consequence of the equivalence of matrix and vector norms, we obtain for every invertible matrix $A \in \mathbb{R}^{n \times n}$ the inequalities

$$\frac{1}{n}\mathrm{cond}_2(A) \leq \mathrm{cond}_1(A) \leq n \, \mathrm{cond}_2(A), \tag{2.3.7}$$

$$\frac{1}{n}\mathrm{cond}_\infty(A) \leq \mathrm{cond}_2(A) \leq n \, \mathrm{cond}_\infty(A) \tag{2.3.8}$$

and

$$\frac{1}{n^2}\mathrm{cond}_1(A) \leq \mathrm{cond}_\infty(A) \leq n^2 \, \mathrm{cond}_1(A). \tag{2.3.9}$$

2.4 The Banach Fixed Point Theorem

Many iterative methods for solving a linear system of equations $Ax = b$ can be written in the form

$$x_{n+1} = F(x_n) \text{ for } n = 0,1,2,\ldots \tag{2.4.1}$$

where F is a mapping from the considered domain into itself. The solution sought by the procedure (2.4.1) must therefore be a fixed point of the mapping F. The iteration rule (2.4.1) is therefore also called fixed point iteration. For this class of methods, the Banach fixed point theorem provides statements on the existence and uniqueness of a fixed point as well as an a priori and an a posteriori error estimate. First, we introduce the necessary concepts of fixed point and contraction constant.

Definition 2.47 An element x of a set $D \subset X$ is called a fixed point of an operator $F : D \subset X \to X$ if

$$F(x) = x$$

holds.

Definition 2.48 Let X be a normed space. An operator

$$F : D \subset X \to X$$

is called contracting, if there exists a number $0 \leq q < 1$ such that

$$\|F(x) - F(y)\| \leq q \, \|x - y\| \quad \forall x,y \in D.$$

The number q is called the contraction constant of the operator F.

Theorem 2.49 *Contracting operators are continuous and have at most one fixed point.*

Proof:
We first consider the continuity of the operator. Let X be a normed space, $F : D \subset X \to X$ a contracting operator, and $\{x_n\}_{n \in \mathbb{N}}$ a sequence in D with $x_n \to x \in D$ for $n \to \infty$, then it follows that

$$0 \leq \|F(x_n) - F(x)\| \leq q \, \|x_n - x\| \to 0 \text{ for } n \to \infty,$$

which shows the continuity of the operator F.

Let $x,y \in D$ be fixed points of F, then we obtain

$$\|x - y\| = \|F(x) - F(y)\| \leq q\|x - y\|,$$

so that with

$$\underbrace{(1 - q)}_{>0} \underbrace{\|x - y\|}_{\geq 0} \leq 0$$

one can conclude $\|x - y\| = 0$ and thus $x = y$ follows. $\qquad\qquad\square$

Theorem 2.50 (Banach Fixed Point Theorem)
Let D be a complete subset of a normed space X and

$$F : D \to D$$

a contracting operator, then there exists exactly one fixed point $x \in D$ of F, and the sequence given by

$$x_{n+1} = F(x_n) \text{ for } n = 0,1,2\ldots$$

converges for every initial value $x_0 \in D$ to x. Furthermore, the a priori error estimate

$$\|x_n - x\| \leq \frac{q^n}{1 - q}\|x_1 - x_0\|$$

and the a posteriori error estimate

$$\|x_n - x\| \leq \frac{q}{1 - q}\|x_n - x_{n-1}\|,$$

hold, where q represents the contraction constant of the operator.

Proof:

Taking into account $x_0 \in D$ the successive application of F in the form $x_{n+1} = F(x_n)$, $n = 0,1,\ldots$ is well-defined since $F : D \to D$. It holds that

$$\|x_{n+1} - x_n\| \leq q \, \|x_n - x_{n-1}\| \leq \ldots \leq q^n \, \|x_1 - x_0\|. \tag{2.4.2}$$

Let $m > n$, then it follows that

$$
\begin{aligned}
\|x_n - x_m\| \quad &\leq \quad \|x_n - x_{n+1}\| + \|x_{n+1} - x_{n+2}\| + \ldots + \|x_{m-1} - x_m\| \\[2mm]
&\overset{(2.4.2)}{\leq} \quad (q^n + \ldots + q^{m-1})\|x_1 - x_0\| \\[2mm]
&\leq \quad q^n \sum_{i=0}^{\infty} q^i \, \|x_1 - x_0\| \\[2mm]
&= \quad \frac{q^n}{1-q}\|x_1 - x_0\|. \tag{2.4.3}
\end{aligned}
$$

Since $|q| < 1$ we obtain

$$\|x_n - x_m\| \to 0 \text{ for } n \to \infty,$$

so that $\{x_n\}_{n \in \mathbb{N}}$ is a Cauchy sequence. Due to the completeness of the subset D, there exists an $x \in D$ with $x_n \to x$ as $n \to \infty$. With

$$x = \lim_{n \to \infty} x_n = \lim_{n \to \infty} F(x_{n-1}) \overset{\text{Theorem 2.49}}{=} F\left(\lim_{n \to \infty} x_{n-1}\right) = F(x)$$

it is shown that x is a fixed point of F, which is uniquely determined according to Theorem 2.49.

From equation (2.4.3) we obtain the a priori error estimate

$$\|x_n - x\| = \lim_{m \to \infty} \|x_n - x_m\| \overset{(2.4.3)}{\leq} \frac{q^n}{1-q}\|x_1 - x_0\|.$$

From

$$
\begin{aligned}
\|x_n - x\| \quad &= \quad \|F(x_{n-1}) - F(x_n) + F(x_n) - F(x)\| \\[2mm]
&\leq \quad \|F(x_{n-1}) - F(x_n)\| + \|F(x_n) - F(x)\| \\[2mm]
&\leq \quad q\|x_{n-1} - x_n\| + q\|x_n - x\|
\end{aligned}
$$

one obtains, by simple rearrangement, the claimed a posteriori error estimate

$$\|x_n - x\| \leq \frac{q}{1-q}\|x_{n-1} - x_n\|.$$

$$\square$$

Example 2.51 We seek an $x \in \mathbb{R}$ with $x = 1 - \sin x$. We consider, with $D = [\varepsilon, 1]$, $0 < \varepsilon \leq 1 - \sin 1 < 1$, a subset of $\mathbb{R}$ that is complete with respect to the norm $\|x\| = |x|$. The function $f(x) = 1 - \sin x$ is arbitrarily often continuously differentiable and also satisfies $f : D \to D$ as well as $|f(x) - f(y)| \leq q|x - y|$ with a contraction constant $q = \max\limits_{x \in D}|f'(x)| = |\cos \varepsilon| < 1$. Thus, f has exactly one fixed point $x \in D$ and the iteration

$$x_{n+1} = f(x_n)$$

converges to x for all $x_0 \in D$. The following sketch illustrates geometrically the course of the iteration for $x_0 = 0.25$.

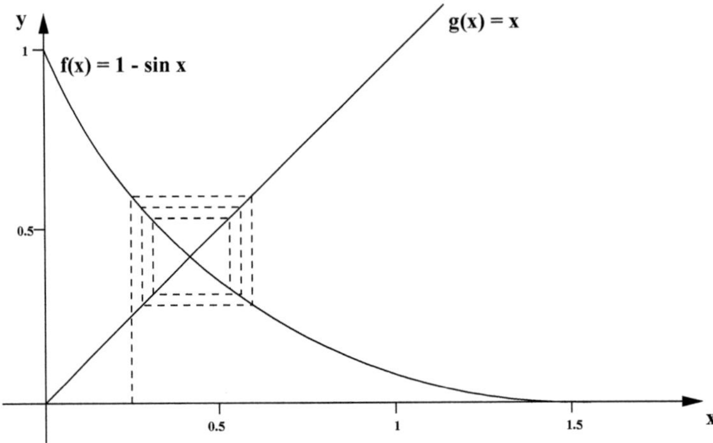

Figure 2.1   Iteration history for $f(x) = 1 - \sin x$ with $x_0 = 0.25$

2.5 Exercises

Problem 1:
Given a linear system of equations $Ax = b$ (1). Show that the operations

  (a) Multiplication of an equation in (1) by a complex number $(\neq 0)$,

  (b) Swapping two equations in (1)

  (c) Addition of the j-th equation from (1) to the i-th equation in (1),

always yield a system of equations equivalent to (1).
Hint: Express all of the above operations as a matrix multiplication and use Lemma 2.24.

Problem 2:
Let $A = (a_{ij})_{i,j=1,\ldots,n} \in \mathbb{R}^{n\times n}$ be a symmetric and positive definite matrix. Show:

  (a) $a_{ii} > 0, \quad i = 1,\ldots,n$,

  (b) $\displaystyle\max_{i,j=1,\ldots,n} |a_{ij}| = \max_{i=1,\ldots,n} |a_{ii}|$.

Problem 3:
Prove the following statement: If $A \in \mathbb{R}^{n\times n}$ is a symmetric, strictly diagonally dominant matrix with positive diagonal elements, then A is positive definite.

Problem 4:
Let $A \in \mathbb{C}^{n\times n}$. Show that:

$$A^j \to 0, \quad j \to \infty \qquad \Longleftrightarrow \qquad \rho(A) < 1.$$

Problem 5:
Let

$$A = \begin{pmatrix} 0.78 & 0.563 \\ 0.913 & 0.659 \end{pmatrix}.$$

Compute $\mathrm{cond}_\infty(A) = \|A\|_\infty \|A^{-1}\|_\infty$.

Problem 6:
Let $A \in \mathbb{C}^{n\times n}$ be non-singular and $U \in \mathbb{C}^{n\times n}$ unitary. Show that:

$$\mathrm{cond}_2(AU) = \mathrm{cond}_2(A) = \mathrm{cond}_2(UA).$$

Problem 7:

  (a) Let

$$A = \begin{pmatrix} \frac{1}{2} & 100 \\ 0 & \frac{1}{4} \end{pmatrix}$$

  be given. Determine $\rho(A)$, $\|A\|_\infty$, $\|A\|_1$, $\|A\|_F$.

  (b) Sketch the unit balls in $\mathbb{R}^2$ for the vector norms $\|.\|_\infty$, $\|.\|_1$, $\|.\|_2$ and determine the best equivalence constants for these three norms in $\mathbb{R}^n$.

Problem 8:
Let X be a Banach space and $\boldsymbol{A} : X \to X$ an operator whose power $\boldsymbol{A}^m$ for some $m \in \mathbb{N}$ is a contracting operator. Prove that there exists exactly one fixed point $\boldsymbol{x}$ of $\boldsymbol{A}$ and that the sequence generated by the iteration

$$\boldsymbol{x}_n = \boldsymbol{A}\boldsymbol{x}_{n-1}, \quad n = 1, 2, \ldots$$
$$\boldsymbol{x}_0 \in X \quad \text{arbitrary}$$

converges to $\boldsymbol{x}$.

Problem 9:
Let $\|.\|$ be an arbitrary induced matrix norm on $\mathbb{C}^{n \times n}$. Show: It holds that

$$\rho(\boldsymbol{A}) = \lim_{k \to \infty} \|\boldsymbol{A}^k\|^{1/k}$$

for all $\boldsymbol{A} \in \mathbb{C}^{n \times n}$.

Problem 10:
Show that the spectral radius does not define a norm on $\mathbb{R}^{n \times n}$. What about this property if the space $\mathbb{R}^{n \times n}_{\text{Symm}}$ of symmetric, real $n \times n$ matrices is considered?

Problem 11:
Given the function $f : \mathbb{R} \mapsto \mathbb{R}$ with

$$f(x) = 0.5 + \sin x - 2x.$$

(a) Show that f has exactly one root.

(b) Compute the root of f using the fixed point iteration method, verifying the assumptions of Banach's fixed point theorem in each case.

(c) For the computed approximation x_5, perform an a priori and an a posteriori error estimate.

Hint: A simple sampling of the function shows that the root is located in the interval $[0.4, 0.5]$.

Problem 12:
Given the function $f : \mathbb{R}^+ \mapsto \mathbb{R}$ with

$$f(x) = \frac{x}{4} - 2 - 3 \ln x.$$

(a) Show that f has exactly two roots.

(b) Compute both roots of f using the fixed point iteration method, verifying the assumptions of Banach's fixed point theorem in each case.

(c) For the computed approximation x_5, perform an a priori and an a posteriori error estimate in each case.

Note: A simple function sampling shows that the roots are located in the interval $[0.5, 0.6]$ as well as in the interval $[56, 57]$.

Problem 13:
Susanne and Frank are conducting a physics experiment. They are trying to determine the parameters x_1 and x_2 and know the functional relationship

$$\begin{pmatrix} 6 & 3 \\ 1 & 2 \end{pmatrix} x = \begin{pmatrix} a \\ 1 \end{pmatrix}.$$

The two perform an experiment and measure $a = 1$. As experienced experimenters, they know that errors always occur in experiments, but they do not have time to repeat the experiment.

Estimate for Susanne and Frank, as a function of the unknown measurement error ϵ, the upper bound for the relative error $\|\Delta x\|/\|x\|$ in the ∞-norm.

Problem 14:
Calculate the condition number of the 3×3 Hilbert-matrix

$$A = \begin{pmatrix} 1 & 1/2 & 1/3 \\ 1/2 & 1/3 & 1/4 \\ 1/3 & 1/4 & 1/5 \end{pmatrix}$$

in the ∞- and the 1-norm. If you wish, check your result with MATLAB. Use the commands $\mathrm{cond}(A,1)$ and $\mathrm{cond}(A,\mathrm{inf})$ for the condition numbers. An $n \times n$ Hilbert-matrix is generated using hilb(n).

Problem 15:
The solutions of the nonlinear system of equations are sought

$$\begin{aligned} x^2 + y^2 &= 4, \\ \tfrac{1}{16}x^2 + y^2 &= 1. \end{aligned}$$

(a) Draw a sketch that illustrates the location of the solutions. Determine a *good* integer starting value (x_0, y_0) for the first quadrant.

(b) The solution in the first quadrant is sought. Provide a suitable fixed point equation and verify the prerequisites of Banach's fixed point theorem for this equation.

3   Direct Methods

A direct method for solving a linear system of equations is an algorithm that, neglecting rounding errors, determines the exact solution in a finite number of arithmetic operations. The algorithms are often based on a multiplicative decomposition of the matrix A of the linear system into two matrices of the form $A = BC$, where the matrices B and C are either easily invertible or at least matrix-vector products with the inverse of these matrices can be easily computed. First, we will consider the LU decomposition of a matrix with and without pivoting and examine its existence and uniqueness. Afterwards, we will present three different approaches for computing a QR decomposition: the Gram-Schmidt scheme, the Givens method, and the Householder transformation. Nowadays, however, direct methods are rarely used for the immediate solution of large linear systems of equations. However, they are often used in an incomplete form as preconditioners within iterative methods and for solving subproblems.

3.1   Gaussian Elimination

The basic idea of the Gaussian elimination method lies formally in a successive transformation of the system $Ax = b$ into an equivalent system of the form

$$L\,R\,x = b$$

with an upper right triangular matrix R and a lower left triangular matrix L. Comparing the two systems above, it becomes clear that matrix A must be represented by the product $L\,R$. Thereafter, the system can be solved by simple forward and subsequent backward substitution. In the usual formulation of the method, the multiplication $\tilde{b} = L^{-1}b$ is carried out simultaneously with the computation of the matrix R. In the following, we denote by

Supplementary Information The online version contains supplementary material available at https://doi.org/10.1007/978-3-658-50260-7_3.

A. Meister, *Numerical Methods for Linear Systems of Equations*, Mathematics Study Resources 26, https://doi.org/10.1007/978-3-658-50260-7_3

$$
\boldsymbol{P}_{kj} \;=\;
\begin{pmatrix}
1 & & & & & & & & & \\
 & \ddots & & & & & & & & \\
 & & 1 & & & & & & & \\
 & & & 0 & \cdots & \cdots & \cdots & 1 & & \\
 & & & \vdots & 1 & & & \vdots & & \\
 & & & \vdots & & \ddots & & \vdots & & \\
 & & & \vdots & & & 1 & \vdots & & \\
 & & & 1 & \cdots & \cdots & \cdots & 0 & & \\
 & & & & & & & & 1 & \\
 & & & & & & & & & \ddots \\
 & & & & & & & & & & 1
\end{pmatrix}
\begin{array}{l}\\[6.5em] \leftarrow k\text{-th row} \\[4.5em] \leftarrow j\text{-th row}\end{array}
\qquad (3.1.1)
$$

always a permutation matrix, which is obtained from the identity matrix $\boldsymbol{I}$ by swapping the j-th and the k-th row $(j \geq k)$. For $k = j$ we have $\boldsymbol{P}_{kj} = \boldsymbol{I}$. In the following, we will represent the matrix $\boldsymbol{L}$ by a multiplicative combination of matrices of the form

$$
\boldsymbol{L}_k =
\begin{pmatrix}
1 & & & & & & \\
 & \ddots & & & & & \\
 & & 1 & & & & \\
 & & -\ell_{k+1,k} & \ddots & & & \\
 & & \vdots & & \ddots & & \\
 & & -\ell_{n,k} & & & 1 &
\end{pmatrix} . \qquad (3.1.2)
$$

Such matrices, which differ from the identity matrix in at most one column, are called Frobenius matrices.

Definition 3.1 The decomposition of a matrix $\boldsymbol{A} \in \mathbb{R}^{n \times n}$ into a product

$$
\boldsymbol{A} = \boldsymbol{L}\boldsymbol{R}
$$

of a lower left triangular matrix $\boldsymbol{L} \in \mathbb{R}^{n \times n}$ and an upper right triangular matrix $\boldsymbol{R} \in \mathbb{R}^{n \times n}$ is called

$$\text{LR decomposition or LR factorization.}$$

In the literature, the LR decomposition is also often referred to as the LU decomposition (lower, upper). Especially in the case of incomplete factorization, as described in Chapter 5, the term incomplete LU factorization is commonly used instead of incomplete LR factorization.

With the help of the matrices (3.1.1) and (3.1.2), the central part of the Gaussian elimination method can be formulated as follows:

Algorithm Gaussian Elimination I —

<table>
<tr><td>

$\boldsymbol{A}^{(1)} := \boldsymbol{A}$

</td></tr>
<tr><td>

For $k = 1, \dots, n-1$

<table>
<tr><td>

Choose from the k-th column of $\boldsymbol{A}^{(k)}$ any element $a_{jk}^{(k)} \neq 0$ with $j \geq k$.

</td></tr>
<tr><td>

Define $\boldsymbol{P}_{kj}$ with the above j and k according to (3.1.1).

</td></tr>
<tr><td>

$\widetilde{\boldsymbol{A}}^{(k)} := \boldsymbol{P}_{kj} \boldsymbol{A}^{(k)}$

</td></tr>
<tr><td>

Define $\boldsymbol{L}_k$ according to (3.1.2) with $l_{ik} = \widetilde{a}_{ik}^{(k)} / \widetilde{a}_{kk}^{(k)}$, $i = k+1, \dots, n$.

</td></tr>
<tr><td>

$\boldsymbol{A}^{(k+1)} := \boldsymbol{L}_k \widetilde{\boldsymbol{A}}^{(k)}$

</td></tr>
</table>

</td></tr>
</table>

With $\boldsymbol{A}^{(n)}$, we thus obtain the upper right triangular matrix $\boldsymbol{R}$, which, as already described, is used for the straightforward solution of the linear system of equations. In general, due to the special construction of the matrices $\boldsymbol{L}_k$, $k = 1, \dots, i-1$, the matrix $\boldsymbol{A}^{(i)}$ contains only zero elements below the diagonal up to and including the $(i-1)$-th column.

We now turn to the question of the existence and uniqueness of LU decompositions. Simple examples such as the matrix

$$A = \begin{pmatrix} 0 & 1 \\ 1 & 1 \end{pmatrix}$$

show that not every invertible matrix necessarily possesses an LU decomposition. Similarly, the proof of uniqueness of LU decompositions is only possible under an additional condition on one of the two triangular matrices. Before we address the actual statements regarding existence and uniqueness, it is useful to first introduce the concepts of leading principal submatrix and principal minor, and to prove some helpful lemmas.

Definition 3.2 Let $\boldsymbol{A} \in \mathbb{R}^{n \times n}$ be given, then

$$A[k] := \begin{pmatrix} a_{11} & \dots & a_{1k} \\ \vdots & \ddots & \vdots \\ a_{k1} & \dots & a_{kk} \end{pmatrix} \in \mathbb{R}^{k \times k} \text{ for } k \in \{1, \dots, n\}$$

is called the leading $k \times k$ principal submatrix of $\boldsymbol{A}$, and $\det A[k]$ the leading $k \times k$ principal minor of $\boldsymbol{A}$.

Lemma 3.3 *Let $\boldsymbol{\ell}_i = (0,\ldots,0,\ell_{i+1,i},\ldots,\ell_{n,i})^T \in \mathbb{R}^n$ and $\boldsymbol{e}_i \in \mathbb{R}^n$ be the i-th unit vector, then for $\boldsymbol{L}_i = \boldsymbol{I} - \boldsymbol{\ell}_i \boldsymbol{e}_i^T \in \mathbb{R}^{n\times n}$ the following holds:*

(a) $\boldsymbol{L}_i^{-1} = \boldsymbol{I} + \boldsymbol{\ell}_i \boldsymbol{e}_i^T$,

(b) $\boldsymbol{L}_1^{-1}\boldsymbol{L}_2^{-1}\ldots\boldsymbol{L}_k^{-1} = \boldsymbol{I} + \sum\limits_{i=1}^{k} \boldsymbol{\ell}_i \boldsymbol{e}_i^T$ *for* $k = 1,\ldots,n-1$.

Proof:
For (a):
Since $\boldsymbol{L}_i$ is a lower triangular matrix with unit diagonal, there exists exactly one matrix $\boldsymbol{L}_i^{-1}$ with $\boldsymbol{L}_i^{-1}\boldsymbol{L}_i = \boldsymbol{L}_i\boldsymbol{L}_i^{-1} = \boldsymbol{I}$. From this, statement (a) follows by

$$(\boldsymbol{I} - \boldsymbol{\ell}_i\boldsymbol{e}_i^T)(\boldsymbol{I} + \boldsymbol{\ell}_i\boldsymbol{e}_i^T) = \boldsymbol{I} - \boldsymbol{\ell}_i\boldsymbol{e}_i^T + \boldsymbol{\ell}_i\boldsymbol{e}_i^T - \boldsymbol{\ell}_i \underbrace{\boldsymbol{e}_i^T \boldsymbol{\ell}_i}_{=0} \boldsymbol{e}_i^T = \boldsymbol{I}.$$

For (b):
We prove this by induction on k . For $k = 1$, (a) gives the statement. Assume the statement holds for $j = 1,\ldots,k < n-1$, then

$$
\begin{aligned}
\boldsymbol{L}_1^{-1}\ldots\boldsymbol{L}_k^{-1}\boldsymbol{L}_{k+1}^{-1} &= \left(\boldsymbol{I} + \sum_{i=1}^{k}\boldsymbol{\ell}_i\boldsymbol{e}_i^T\right)\left(\boldsymbol{I} + \boldsymbol{\ell}_{k+1}\boldsymbol{e}_{k+1}^T\right)\\
&= \boldsymbol{I} + \boldsymbol{\ell}_{k+1}\boldsymbol{e}_{k+1}^T + \sum_{i=1}^{k}\boldsymbol{\ell}_i\boldsymbol{e}_i^T + \sum_{i=1}^{k}\boldsymbol{\ell}_i\underbrace{\boldsymbol{e}_i^T\boldsymbol{\ell}_{k+1}}_{=0}\boldsymbol{e}_{k+1}^T\\
&= \boldsymbol{I} + \sum_{i=1}^{k+1}\boldsymbol{\ell}_i\boldsymbol{e}_i^T.
\end{aligned}
$$

$\square$

Lemma 3.4 *Let $\boldsymbol{L} \in \mathbb{R}^{n\times n}$ be a non-singular lower left triangular matrix, then $\boldsymbol{L}^{-1} \in \mathbb{R}^{n\times n}$ is also a lower left triangular matrix. For non-singular upper right triangular matrices $\boldsymbol{R} \in \mathbb{R}^{n\times n}$, the analogous statement holds.*

Proof:
Define $\boldsymbol{D} = \mathrm{diag}\,\{\ell_{11},\ldots,\ell_{nn}\}$ using the diagonal entries of the matrix $\boldsymbol{L}$, then $\det\boldsymbol{D} \neq 0$ and

$$\widetilde{\boldsymbol{L}} := \boldsymbol{D}^{-1}\boldsymbol{L}$$

according to Lemma 2.26 is also a lower triangular matrix, which additionally has a unit diagonal. Thus, $\widetilde{\boldsymbol{L}}$ has the form $\widetilde{\boldsymbol{L}} = \boldsymbol{I} + \sum\limits_{i=1}^{n-1} \widetilde{\boldsymbol{\ell}}_i\boldsymbol{e}_i^T$ with

$$\widetilde{\boldsymbol{\ell}}_i = (0,\ldots,0,\widetilde{\ell}_{i+1,i},\ldots,\widetilde{\ell}_{n,i})^T$$

and can, using the matrices $\widetilde{\boldsymbol{L}}_i = \boldsymbol{I} + \widetilde{\boldsymbol{\ell}}_i\boldsymbol{e}_i^T$ ($i = 1,\ldots,n-1$), be represented as a product

$$\widetilde{\boldsymbol{L}} = \widetilde{\boldsymbol{L}}_1 \cdot \ldots \cdot \widetilde{\boldsymbol{L}}_{n-1}.$$

For the inverse, we obtain $\widetilde{\boldsymbol{L}}^{-1} = \widetilde{\boldsymbol{L}}_{n-1}^{-1} \cdot \ldots \cdot \widetilde{\boldsymbol{L}}_1^{-1}$ with $\widetilde{\boldsymbol{L}}_i^{-1} = \boldsymbol{I} - \widetilde{\boldsymbol{\ell}}_i \boldsymbol{e}_i^T$ ($i = 1,\ldots,n-1$) according to Lemma 3.3. The matrix $\boldsymbol{L}^{-1}$ can therefore also be written as a product of lower triangular matrices in the form $\boldsymbol{L}^{-1} = \widetilde{\boldsymbol{L}}_{n-1}^{-1} \cdot \ldots \cdot \widehat{\boldsymbol{L}}_1^{-1} \boldsymbol{D}^{-1}$ and thus, by Lemma 2.26, is also a left lower triangular matrix.

With $\boldsymbol{R}^T (\boldsymbol{R}^{-1})^T = (\boldsymbol{R}^{-1}\boldsymbol{R})^T = \boldsymbol{I} = \boldsymbol{R}^T (\boldsymbol{R}^T)^{-1}$ it follows that $(\boldsymbol{R}^T)^{-1} = (\boldsymbol{R}^{-1})^T$. Using the above part of the proof, with $\boldsymbol{L} = \boldsymbol{R}^T$ also $\boldsymbol{L}^{-1} = (\boldsymbol{R}^T)^{-1}$ is a left lower triangular matrix, so that $\boldsymbol{R}^{-1} = \left((\boldsymbol{R}^T)^{-1} \right)^T$ represents a right upper triangular matrix.
$\square$

Lemma 3.5 Let $\boldsymbol{A} = (a_{ij})_{i,j=1,\ldots,n} \in \mathbb{R}^{n\times n}$ and let $\boldsymbol{L} = (\ell_{ij})_{i,j=1,\ldots,n} \in \mathbb{R}^{n\times n}$ be a lower triangular matrix, then

$$(\boldsymbol{LA})[k] = \boldsymbol{L}[k]\boldsymbol{A}[k] \ \ for \ k = 1,\ldots,n.$$

Proof:
Let $k \in \{1,\ldots,n\}$. For $i,j \in \{1,\ldots,k\}$ it follows with $\ell_{im} = 0$ for $m > k \geq i$

$$((\boldsymbol{LA})[k])_{ij} = \sum_{m=1}^{n} \ell_{im} a_{mj} = \sum_{m=1}^{k} \ell_{im} a_{mj} + \sum_{m=k+1}^{n} \ell_{im} a_{mj} = \sum_{m=1}^{k} \ell_{im} a_{mj} = (\boldsymbol{L}[k]\boldsymbol{A}[k])_{ij}.$$

$\square$

Theorem 3.6 (Existence of an LR Factorization I)
Let $\boldsymbol{A} \in \mathbb{R}^{n\times n}$ be a non-singular matrix, then there exists a permutation matrix $\boldsymbol{P} \in \mathbb{R}^{n\times n}$ such that $\boldsymbol{PA}$ admits an LR factorization.

Proof:
For all matrices $\boldsymbol{A}^{(k)}$ computed in Algorithm I, it holds that

$$a_{i\ell}^{(k)} = 0 \quad \text{for} \quad \ell = 1,\ldots,k-1, \ \ i > \ell. \tag{3.1.3}$$

Since $\det \boldsymbol{L}_\ell \neq 0 \neq \det \boldsymbol{P}_{\ell,j_\ell}$ for all $\ell = 1,\ldots,k-1$, $j_\ell \geq \ell$, it follows that

$$\det \boldsymbol{A}^{(k)} = \det \left(\boldsymbol{L}_{k-1}\boldsymbol{P}_{k-1,j_{k-1}} \ldots \boldsymbol{L}_1 \boldsymbol{P}_{1,j_1} \boldsymbol{A} \right) \neq 0.$$

Thus, there exists an $i \in \{k,\ldots,n\}$ with $a_{ik}^{(k)} \neq 0$ and the Algorithm I does not terminate before the computation of $\boldsymbol{A}^{(n)}$. Furthermore,

$$\boldsymbol{R} := \boldsymbol{A}^{(n)} = \boldsymbol{L}_{n-1}\boldsymbol{P}_{n-1,j_{n-1}} \ldots \boldsymbol{L}_1 \boldsymbol{P}_{1,j_1} \boldsymbol{A} \tag{3.1.4}$$

with (3.1.3) is an upper triangular matrix. All $\boldsymbol{L}_i$ can be written in the form

$$\boldsymbol{L}_i = \boldsymbol{I} - \boldsymbol{\ell}_i \boldsymbol{e}_i^T, \ \boldsymbol{\ell}_i = (0,\ldots,0,\ell_{i+1,i},\ldots,\ell_{n,i})^T$$

and it holds that

$$\boldsymbol{P}_{k,j_k} \boldsymbol{e}_i = \boldsymbol{e}_i \text{ and } \boldsymbol{P}_{k,j_k} \boldsymbol{\ell}_i = \widehat{\boldsymbol{\ell}}_i = (0,\ldots,0,\widehat{\ell}_{i+1,i},\ldots,\widehat{\ell}_{n,i})^T \text{ for all } i < k \leq j_k.$$

For $i < k \le j_k$, it follows, taking into account that $\boldsymbol{P}_{k,j_k} = \boldsymbol{P}_{k,j_k}^T$, the equation

$$
\begin{aligned}
\boldsymbol{P}_{k,j_k}\boldsymbol{L}_i &= \boldsymbol{P}_{k,j_k}(\boldsymbol{I} - \boldsymbol{\ell}_i e_i^T) = \boldsymbol{P}_{k,j_k} - \hat{\boldsymbol{\ell}}_i e_i^T \\
&= \boldsymbol{P}_{k,j_k} - \hat{\boldsymbol{\ell}}_i(\boldsymbol{P}_{k,j_k}e_i)^T = \boldsymbol{P}_{k,j_k} - \hat{\boldsymbol{\ell}}_i e_i^T \boldsymbol{P}_{k,j_k} = \widehat{\boldsymbol{L}}_i \boldsymbol{P}_{k,j_k}
\end{aligned}
$$

with a lower triangular matrix $\widehat{\boldsymbol{L}}_i = \boldsymbol{I} - \hat{\boldsymbol{\ell}}_i e_i^T$. Using equation (3.1.4) yields

$$
\boldsymbol{R} = \underbrace{\boldsymbol{L}_{n-1}\widehat{\boldsymbol{L}}_{n-2}\ldots\widehat{\boldsymbol{L}}_1}_{\tilde{\boldsymbol{L}}:=} \underbrace{\boldsymbol{P}_{n-1,j_{n-1}}\ldots\boldsymbol{P}_{1,j_1}}_{\boldsymbol{P}:=}\boldsymbol{A}
$$

with a permutation matrix $\boldsymbol{P}$. Since $\tilde{\boldsymbol{L}}$ is a lower triangular matrix, it follows from Lemma 3.4 that

$$
\boldsymbol{L} := \tilde{\boldsymbol{L}}^{-1}
$$

represents a lower triangular matrix, and thus we obtain the claimed representation

$$
\boldsymbol{L}\boldsymbol{R} = \boldsymbol{P}\boldsymbol{A}.
$$

$\square$

We now want to illustrate the statement of Theorem 3.6 with a small example. For this, we consider the matrix

$$
\boldsymbol{A} = \begin{pmatrix} 1 & 2 & 0 & 2 \\ 2 & 4 & 0 & 9 \\ 3 & 8 & 6 & 14 \\ 4 & 9 & 12 & 14 \end{pmatrix} \in \mathbb{R}^{4\times 4} \tag{3.1.5}
$$

and follow the Gaussian Elimination I as presented on page 43. We write $\boldsymbol{A}^{(1)} = \boldsymbol{A}$ and, since $a_{11}^{(1)} \ne 0$, we can initially perform the first loop without row exchange, which corresponds to using $\boldsymbol{P}_{11} = \boldsymbol{I}$. Thus, $\tilde{\boldsymbol{A}}^{(1)} = \boldsymbol{A}^{(1)}$ holds, and we obtain

$$
\boldsymbol{L}_1 = \begin{pmatrix} 1 & 0 & 0 & 0 \\ -2 & 1 & 0 & 0 \\ -3 & 0 & 1 & 0 \\ -4 & 0 & 0 & 1 \end{pmatrix}
$$

as well as

$$
\boldsymbol{A}^{(2)} = \boldsymbol{L}_1\tilde{\boldsymbol{A}}^{(1)} = \begin{pmatrix} 1 & 2 & 0 & 2 \\ 0 & 0 & 0 & 5 \\ 0 & 2 & 6 & 8 \\ 0 & 1 & 12 & 6 \end{pmatrix}.
$$

The continuation of the procedure requires, since $a_{22}^{(2)} = 0$, a permutation, which we realize by $\boldsymbol{P}_{23}$, and thus

$$
\tilde{\boldsymbol{A}}^{(2)} = \boldsymbol{P}_{23}\boldsymbol{A}^{(2)} = \begin{pmatrix} 1 & 2 & 0 & 2 \\ 0 & 2 & 6 & 8 \\ 0 & 0 & 0 & 5 \\ 0 & 1 & 12 & 6 \end{pmatrix}
$$

is obtained. Using

$$L_2 = \begin{pmatrix} 1 & 0 & 0 & 0 \\ 0 & 1 & 0 & 0 \\ 0 & 0 & 1 & 0 \\ 0 & -\frac{1}{2} & 0 & 1 \end{pmatrix}$$

it follows that

$$A^{(3)} = L_2 \widetilde{A}^{(2)} = \begin{pmatrix} 1 & 2 & 0 & 2 \\ 0 & 2 & 6 & 8 \\ 0 & 0 & 0 & 5 \\ 0 & 0 & 9 & 2 \end{pmatrix}.$$

By means of the permutation matrix P_{34}, we obtain the upper right triangular matrix

$$R = P_{34} A^{(3)} = \begin{pmatrix} 1 & 2 & 0 & 2 \\ 0 & 2 & 6 & 8 \\ 0 & 0 & 9 & 2 \\ 0 & 0 & 0 & 5 \end{pmatrix},$$

for which the representation

$$R = P_{34} L_2 P_{23} L_1 A = \widehat{L}_2 \widehat{L}_1 P_{34} P_{23} A$$

holds. Here, the Frobenius matrices $\widehat{L}_1$ and $\widehat{L}_2$ have, according to the proof of Theorem 3.6, the form

$$\widehat{L}_1 = \begin{pmatrix} 1 & 0 & 0 & 0 \\ -3 & 1 & 0 & 0 \\ -4 & 0 & 1 & 0 \\ -2 & 0 & 0 & 1 \end{pmatrix} \quad \text{and} \quad \widehat{L}_2 = \begin{pmatrix} 1 & 0 & 0 & 0 \\ 0 & 1 & 0 & 0 \\ 0 & -\frac{1}{2} & 1 & 0 \\ 0 & 0 & 0 & 1 \end{pmatrix}.$$

With the permutation matrix

$$P = P_{34} P_{23} = \begin{pmatrix} 1 & 0 & 0 & 0 \\ 0 & 0 & 1 & 0 \\ 0 & 0 & 0 & 1 \\ 0 & 1 & 0 & 0 \end{pmatrix}$$

and the lower left triangular matrix

$$L = \left(\widehat{L}_2 \widehat{L}_1 \right)^{-1} = \begin{pmatrix} 1 & 0 & 0 & 0 \\ 3 & 1 & 0 & 0 \\ 4 & \frac{1}{2} & 1 & 0 \\ 2 & 0 & 0 & 1 \end{pmatrix}$$

we obtain the decomposition of the form given in Theorem 3.6:

$$PA = LR \tag{3.1.6}$$

as required.

Theorem 3.7 (Existence of an LR Factorization II)
Let $A \in \mathbb{R}^{n \times n}$ be invertible, then A has an LR factorization if and only if

$$\det A[k] \neq 0 \quad \forall k = 1, \dots, n$$

holds.

Proof:

" $\Rightarrow$ ":   Assume $A = LR$.

Due to the invertibility of the matrix A, the determinant multiplication theorem yields

$$\det L[n] \cdot \det R[n] = \det A[n] \neq 0$$

and consequently

$$\det L[n] \neq 0 \neq \det R[n].$$

Since L and R are triangular matrices, it follows that

$$\det L[k] \neq 0 \neq \det R[k]$$

for $k = 1, \ldots, n$, and by Lemma 3.5 we obtain

$$\det A[k] = \det(LR)[k] = \det L[k] \cdot \det R[k] \neq 0.$$

" $\Leftarrow$ ":   Assume $\det A[k] \neq 0$ for all $k = 1, \ldots, n$.

A has an LR decomposition if Algorithm I can be carried out with $P_{kk} = I$, $k = 1, \ldots, n-1$, that is, if $a_{kk}^{(k)} \neq 0$ for $k = 1, \ldots, n-1$ holds.

For $k = 1$ we have $a_{11}^{(1)} = \det A[k] \neq 0$, so $P_{11} = I$ can be chosen.

Assume $a_{kk}^{(k)} \neq 0$ for $k < n-1$, then it follows that

$$A^{(k+1)} = L_k \ldots L_1 A,$$

and again by Lemma 3.5 we obtain

$$\det A^{(k+1)}[k+1] = \det L_k[k+1] \cdot \ldots \cdot \det L_1[k+1] \cdot \det A[k+1] \neq 0.$$

Since $A^{(k+1)}[k+1]$ is an upper triangular matrix, it follows that

$$a_{k+1,k+1}^{(k+1)} \neq 0.$$

$\square$

For the matrix A given in equation (3.1.5), it was already clear that no LR decomposition exists, but rather a suitable permutation must first be performed. This fact is also confirmed by Theorem 3.7 due to

$$\det A[2] = \det \begin{pmatrix} 1 & 2 \\ 2 & 4 \end{pmatrix} = 0.$$

If, on the other hand, we consider the matrix

$$B = PA = \begin{pmatrix} 1 & 0 & 0 & 0 \\ 0 & 0 & 1 & 0 \\ 0 & 0 & 0 & 1 \\ 0 & 1 & 0 & 0 \end{pmatrix} \begin{pmatrix} 1 & 2 & 0 & 2 \\ 2 & 4 & 0 & 9 \\ 3 & 8 & 6 & 14 \\ 4 & 9 & 12 & 14 \end{pmatrix} = \begin{pmatrix} 1 & 2 & 0 & 2 \\ 3 & 8 & 6 & 14 \\ 4 & 9 & 12 & 14 \\ 2 & 4 & 0 & 9 \end{pmatrix},$$

in view of (3.1.6), the already computationally established existence of an LR decomposition is in agreement with the statement of Theorem 3.7 because of the properties $\det B[1] = 1$, $\det B[2] = 2$, $\det B[3] = 18$, and $\det B[4] = \det B = 90$.

Theorem 3.8 (Uniqueness of the LR Factorization)

Let $A \in \mathbb{R}^{n \times n}$ be invertible with $\det A[k] \neq 0$ for $k = 1, \ldots, n$, then there exists exactly one LR factorization of A such that L has a unit diagonal.

Proof:

By Theorem 3.7, at least one LR decomposition of the matrix A exists. Suppose two LR factorizations of the matrix A are given by $L_1 R_1 = A = L_2 R_2$, where L_1 and L_2 have unit diagonals, then it follows that $R_2 R_1^{-1} = L_2^{-1} L_1$. By Lemma 2.26 and Lemma 3.4, $L_2^{-1} L_1$ is both a lower left and upper right triangular matrix with a unit diagonal. Consequently, $L_2^{-1} L_1 = I$ and we obtain $L_1 = L_2$ and $R_1 = R_2$. $\qquad\qquad\square$

Remark:

Without the requirement that L has a unit diagonal, it follows that $L_1 = L_2 D$ and $R_1 = D^{-1} R_2$ with a non-singular diagonal matrix D. Thus, LR decompositions can be transformed into each other by multiplication with a diagonal matrix.

To solve $Ax = b$, the following explicit form of the Gaussian algorithm results.

Algorithm Gaussian Elimination without Pivoting —

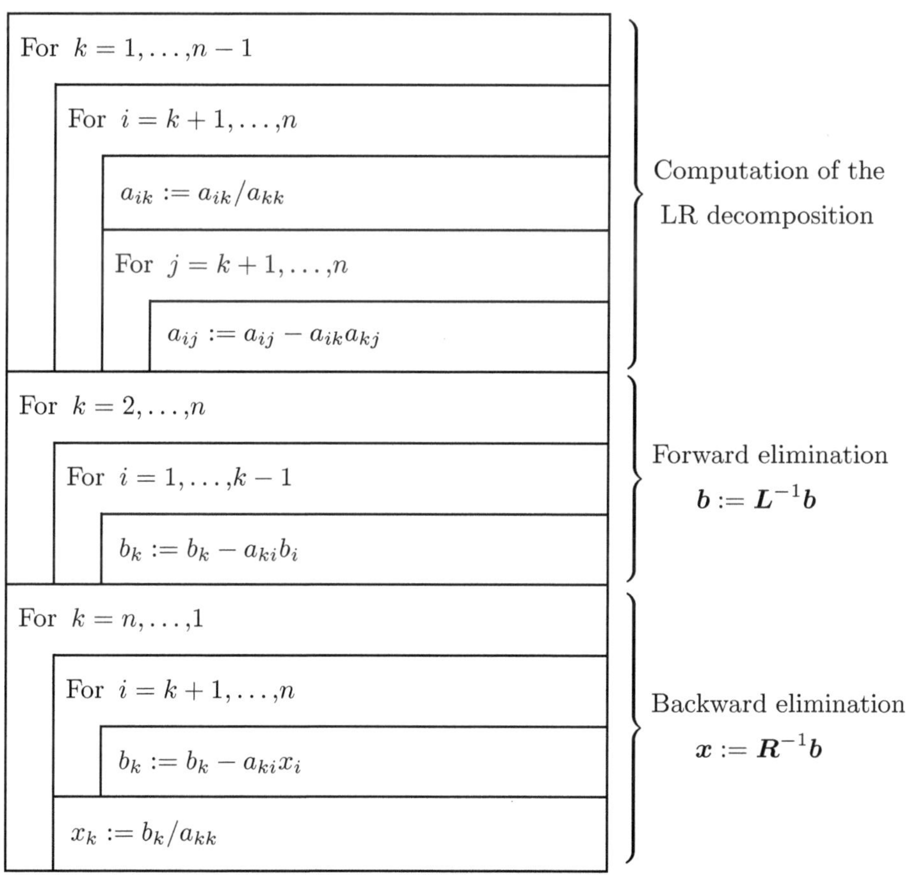

To analyze the computational effort, we consider only the time-consuming multiplications and divisions:

$$\# \text{ Divisions} \quad = \quad \sum_{k=1}^{n-1}(n-k)+n = \sum_{i=1}^{n-1}i+n = \sum_{i=1}^{n}i = \frac{n(n+1)}{2}$$

$$\# \text{ Multiplications} \quad = \quad \sum_{k=1}^{n-1}(n-k)^2 + \frac{n(n-1)}{2} + \frac{n(n-1)}{2}$$

$$= \quad \sum_{i=1}^{n-1}i^2 + n(n-1) = \frac{(n-1)n(2n-1)}{6} + n(n-1)$$

$$= \quad \frac{n^3}{3} + \frac{n^2}{2} - \frac{5n}{6}.$$

Thus, we obtain $\# \text{ divisions} + \# \text{ multiplications} = \dfrac{n^3}{3} + n^2 - \dfrac{n}{3}$.

If the computation time for a multiplication or division is $0.1\mu\,\mathrm{sec} = 10^{-7}\,\mathrm{sec}$, then the Gaussian algorithm requires the computation times listed in the following table for a real $n \times n$ matrix $\boldsymbol{A}$:

n	Computing time
10^3	≈ 33 seconds
10^4	≈ 9 hours and 15 minutes
10^5	≈ 1 year and 20 days

Systems of equations with 10^5 unknowns already occur in implicit finite volume and finite difference methods for the two-dimensional Euler equations of gas dynamics when a quite common discretization of the flow domain is used with 25,000 control volumes or grid points. In such applications, the Gaussian algorithm proves to be impractical for reasons of computation time alone, even if rounding errors and the immense memory requirements are neglected. On the other hand, if we consider systems of equations with a small number of unknowns, the Gaussian elimination method provides a direct method that is often superior to iterative algorithms. The following example is intended to illustrate the advantages of pivoting with regard to the accuracy of the computed solution vector.

Example 3.9 Let $\varepsilon \ll 1$ be such that, at machine precision[1]

$$1 \pm \varepsilon \doteq 1 \text{ respectively } 1 \pm \frac{1}{\varepsilon} \doteq \pm \frac{1}{\varepsilon}$$

holds. We consider the system of equations $\boldsymbol{Ax} = \boldsymbol{b}$ in the form

$$\begin{aligned} \varepsilon x_1 &+ 2x_2 &= 1 \\ x_1 &+ x_2 &= 1, \end{aligned}$$

which has the exact solution

---

[1] With the symbol $\doteq$ we indicate equality taking into account machine rounding errors.

$$x_1 = \frac{1}{2-\varepsilon} \approx 0.5, \ x_2 = \frac{1-\varepsilon}{2-\varepsilon} \approx 0.5$$

With the Gaussian elimination method we obtain

$$\varepsilon x_1 + 2x_2 = 1$$

$$\left(1 - \frac{2}{\varepsilon}\right) x_2 = 1 - \frac{1}{\varepsilon}$$

and thus, due to the given computational precision, $x_2 \doteq \frac{1}{\varepsilon} \cdot \frac{\varepsilon}{2} = 0.5$. Substituting into the first equation yields $\varepsilon x_1 + 1 = 1$, so that $\varepsilon x_1 = 0$ and thus $x_1 = 0$.

A prior row exchange, on the other hand, leads to the system of equations

$$x_1 \ + \ x_2 = 1$$

$$\varepsilon x_1 \ + \ 2x_2 = 1.$$

From this, using the Gaussian elimination method, we obtain

$$x_1 \ + \ x_2 = 1$$

$$(2-\varepsilon)x_2 = 1-\varepsilon,$$

so that the numerical solution reads $x_2 = \frac{1-\varepsilon}{2-\varepsilon} \doteq 0.5$ and $x_1 = 1 - x_2 = 0.5$. Using this system of equations, we want to examine the advantages of pivoting in more detail. Let us first consider the Gaussian elimination method without pivoting. The element $a_{11}^{(1)} = \varepsilon$ of the matrix

$$A = \begin{pmatrix} a_{11}^{(1)} & a_{12}^{(1)} \\ a_{21}^{(1)} & a_{22}^{(1)} \end{pmatrix} = \begin{pmatrix} \varepsilon & 2 \\ 1 & 1 \end{pmatrix}$$

is clearly smaller than $a_{21}^{(1)} = 1$. As a result, the algorithm forms a very large quotient $\frac{a_{21}^{(1)}}{a_{11}^{(1)}} = \frac{1}{\varepsilon}$. Since the other matrix elements are also of the order of $a_{21}^{(1)}$, we obtain

$$a_{22}^{(2)} = \underbrace{a_{22}^{(1)}}_{=\mathcal{O}(1) \text{ for } \varepsilon \to 0} - \underbrace{\frac{a_{12}^{(1)} a_{21}^{(1)}}{a_{11}^{(1)}}}_{=\mathcal{O}(\frac{1}{\varepsilon}) \text{ for } \varepsilon \to 0} = 1 - \frac{2}{\varepsilon} = \mathcal{O}\left(\frac{1}{\varepsilon}\right) \quad \text{for } \varepsilon \to 0.$$

Within the matrix $A^{(2)}$, all remaining nonzero elements thus exhibit a significantly different magnitude. Analogously, we obtain

$$b = \begin{pmatrix} 1 \\ 1 - \frac{1}{\varepsilon} \end{pmatrix} = \begin{pmatrix} \mathcal{O}(1) \\ \mathcal{O}\left(\frac{1}{\varepsilon}\right) \end{pmatrix} \quad \text{for } \varepsilon \to 0.$$

If a very small matrix coefficient $a_{11}^{(1)} = \varepsilon$ is present, then machine precision leads to $a_{22}^{(2)} = -\frac{2}{\varepsilon}$ and $b_2^{(2)} = -\frac{1}{\varepsilon}$. Consequently, in this case, the values $a_{22}^{(1)}$ and $b_2^{(1)}$ have no influence on the solution of the system of equations. Therefore, for sufficiently small $\varepsilon \neq 0$, the second equation is numerically equivalent to the first equation for $\varepsilon = 0$. Thus, we always obtain the final result

$$x_2 = \frac{b_1^{(1)}}{a_{12}^{(1)}} \quad \text{and} \quad x_1 = 0.$$

Due to the row interchange presented, a small quotient $\frac{\tilde{a}_{21}^{(1)}}{\tilde{a}_{11}^{(1)}} = \varepsilon$ arises, so that all relevant elements of the matrix

$$A^{(2)} = \begin{pmatrix} 1 & 1 \\ 0 & 2 - \varepsilon \end{pmatrix}$$

for $\varepsilon \to 0$ are of the same order of magnitude. Consequently, there are no problems due to the given computational accuracy. When using pivoting based on the largest row or column element, the quotient always lies in the interval $[-1,1]$, whereas with direct use of the Gaussian elimination method, no general bound for the quotient can be given. For the model case discussed, we obtain the numerical results given in the following tables, where the left table is without and the right table is with row interchange. The results obtained brilliantly confirm the previously derived theoretical statements.

ε	x_1	x_2
10^{-1}	0.5263	0.4737
10^{-2}	0.5025	0.4975
10^{-3}	0.5003	0.4997
10^{-5}	0.5000	0.5000
10^{-10}	0.5000	0.5000
10^{-14}	0.4996	0.5000
10^{-15}	0.5551	0.5000
10^{-16}	0.0000	0.5000
10^{-20}	0.0000	0.5000

ε	x_1	x_2
10^{-1}	0.5263	0.4737
10^{-2}	0.5025	0.4975
10^{-3}	0.5003	0.4997
10^{-5}	0.5000	0.5000
10^{-10}	0.5000	0.5000
10^{-14}	0.5000	0.5000
10^{-15}	0.5000	0.5000
10^{-16}	0.5000	0.5000
10^{-20}	0.5000	0.5000

The permutation of rows or columns thus proves useful not only in the case $a_{kk}^{(k)} = 0$. We distinguish three types of pivoting:

(a) Column pivoting:
    Define P_{kj} according to (3.1.1) with

$$j = \text{index} \max_{j=k,\ldots,n} |a_{jk}^{(k)}|$$

    and consider the system

$$A^{(k)} x = b,$$

    which is equivalent to

$$P_{kj} A^{(k)} x = P_{kj} b.$$

(b) Row pivoting:
    Define P_{kj} according to (3.1.1) with

$$j = \text{index} \max_{j=k,\ldots,n} |a_{kj}^{(k)}|$$

and consider the system

$$A^{(k)} P_{kj} y \;=\; b$$
$$x \;=\; P_{kj} y.$$

(c) Complete pivoting:
Define P_{k,j_1} and P_{k,j_2} according to (3.1.1) with

$$j_1 = \text{index} \max_{j=k,\ldots,n} \left(\max_{i=k,\ldots,n} |a_{ji}^{(k)}| \right),$$

$$j_2 = \text{index} \max_{j=k,\ldots,n} \left(\max_{i=k,\ldots,n} |a_{ij}^{(k)}| \right)$$

and consider the system

$$P_{k,j_1} A^{(k)} P_{k,j_2} y \;=\; P_{k,j_1} b$$
$$x \;=\; P_{k,j_2} y.$$

Remark:
If the Gaussian elimination method is applied to a system of equations

$$Ax = b$$

with a singular matrix using complete pivoting, then there exists a $k \le n$ such that

$$\max_{i,j=k,\ldots,n} |a_{ij}^{(k)}| = 0.$$

In this case, according to Lemma 2.25, the system is solvable if and only if, after performing all multiplications with L_i and P_{i,j_i}, $i = 1,\ldots,k-1$ applied to the augmented coefficient matrix

$$(A^{(1)}, b^{(1)}) = (A, b)$$

the result is of the form

$$(A^{(k)}, b^{(k)}) \text{ with } b^{(k)} = (b_1^{(k)}, \ldots, b_{k-1}^{(k)}, 0, \ldots 0)^T.$$

3.2 Cholesky Factorization

For symmetric, positive definite matrices, the computational effort required for calculating an LR decomposition in the context of the Gaussian elimination principle can be reduced, and it can also be shown that such a factorization always exists.

Definition 3.10 The decomposition of a matrix $A \in \mathbb{R}^{n \times n}$ into a product

$$A = LL^T$$

with a lower left triangular matrix $L \in \mathbb{R}^{n \times n}$ is called

$$\text{Cholesky decomposition or Cholesky factorization.}$$

As we can easily see from the above definition, a Cholesky decomposition is a special case of an LR decomposition, in which the right upper triangular matrix $\boldsymbol{R}$ is represented by the transpose of the lower left triangular matrix $\boldsymbol{L}$.

Theorem 3.11 (Existence and Uniqueness of the Cholesky Decomposition)
For every symmetric, positive definite matrix $\boldsymbol{A} \in \mathbb{R}^{n \times n}$, there exists exactly one lower left triangular matrix $\boldsymbol{L} \in \mathbb{R}^{n \times n}$ with $\ell_{ii} > 0, i = 1, \ldots, n$ such that

$$\boldsymbol{A} = \boldsymbol{L}\boldsymbol{L}^T$$

holds.

Proof:
Let $\boldsymbol{x} \in \mathbb{R}^k \setminus \{\boldsymbol{0}\}$, $k < n$, then we define $\boldsymbol{y} = (\boldsymbol{x}, 0, \ldots, 0)^T \in \mathbb{R}^n \setminus \{\boldsymbol{0}\}$. Thus, it follows that

$$\boldsymbol{x}^T \boldsymbol{A}[k] \boldsymbol{x} = \boldsymbol{y}^T \boldsymbol{A} \boldsymbol{y} > 0,$$

so that all $\boldsymbol{A}[k]$ for $k = 1, \ldots, n$ are positive definite.

An induction on n now yields the claim: For $n = 1$ we have $\boldsymbol{A} = (a_{11}) > 0$, so that $\ell_{11} := \sqrt{a_{11}} > 0$ gives the representation $\boldsymbol{A} = \boldsymbol{L}\boldsymbol{L}^T$ with $\boldsymbol{L} = (\ell_{11})$. Assume the claim holds for $n = 1, \ldots, j$, then for $n = j + 1$ it follows that

$$\boldsymbol{A} = \begin{pmatrix} \boldsymbol{A}[j] & \boldsymbol{c} \\ \boldsymbol{c}^T & a_{nn} \end{pmatrix},$$

where $\boldsymbol{A}[j]$ is positive definite. Thus, there exists exactly one lower left triangular matrix $\boldsymbol{L}_j \in \mathbb{R}^{j \times j}$ with $\ell_{ii} > 0$, $i = 1, \ldots, j$ and

$$\boldsymbol{A}[j] = \boldsymbol{L}_j \boldsymbol{L}_j^T.$$

We now make the ansatz

$$\boldsymbol{L}_n := \begin{pmatrix} \boldsymbol{L}_j & \boldsymbol{0} \\ \boldsymbol{d}^T & \alpha \end{pmatrix} \tag{3.2.1}$$

with $\boldsymbol{d} \in \mathbb{R}^j$, $\alpha \in \mathbb{R}$ such that

$$\begin{pmatrix} \boldsymbol{A}[j] & \boldsymbol{c} \\ \boldsymbol{c}^T & a_{nn} \end{pmatrix} = \boldsymbol{A}[n] = \boldsymbol{L}_n \boldsymbol{L}_n^T = \begin{pmatrix} \boldsymbol{A}[j] & \boldsymbol{L}_j \boldsymbol{d} \\ \boldsymbol{d}^T \boldsymbol{L}_j^T & \boldsymbol{d}^T \boldsymbol{d} + \alpha^2 \end{pmatrix}$$

should hold. Since $0 \neq \det \boldsymbol{A}[j] = (\det \boldsymbol{L}_j)^2$ we can conclude that the matrix $\boldsymbol{L}_j$ is invertible and thus $\boldsymbol{d} = \boldsymbol{L}_j^{-1} \boldsymbol{c}$ is uniquely determined. Furthermore, $\alpha^2 = a_{nn} - \boldsymbol{d}^T \boldsymbol{d}$. Due to the positive definiteness of $\boldsymbol{A} = \boldsymbol{A}[n]$ we have $\det \boldsymbol{A}[n] > 0$, so that

$$0 < \frac{\det(\boldsymbol{A}[n])}{(\det(\boldsymbol{L}_j))^2} = \alpha^2$$

results. Thus, we obtain

$$\alpha = \sqrt{a_{nn} - \boldsymbol{d}^T \boldsymbol{d}} \in \mathbb{R}^+,$$

which, together with (3.2.1), gives the desired matrix. $\square$

We perform a column-wise computation of the matrix coefficients. In deriving the algorithm, we thus assume that all l_{ij} for $i = 1,\ldots,n$ and $j \le k - 1$ are known. Then, from

$$A = \begin{pmatrix} \ell_{11} & & \\ \vdots & \ddots & \\ \ell_{n1} & \cdots & \ell_{nn} \end{pmatrix} \begin{pmatrix} \ell_{11} & \cdots & \ell_{n1} \\ & \ddots & \vdots \\ & & \ell_{nn} \end{pmatrix}$$

we obtain the relation

$$a_{kk} = \ell_{k1}^2 + \ldots + \ell_{kk}^2 = \sum_{j=1}^{k} \ell_{kj}^2,$$

which allows us to compute l_{kk} according to

$$\ell_{kk} = \sqrt{a_{kk} - \sum_{j=1}^{k-1} \ell_{kj}^2}. \tag{3.2.2}$$

From

$$a_{ik} = \ell_{i1}\ell_{k1} + \ldots + \ell_{ik}\ell_{kk} = \sum_{j=1}^{k} \ell_{ij}\ell_{kj} \text{ for } i = k+1,\ldots,n$$

we obtain the computation rule for the elements below the diagonal in the k-th column in the form

$$\ell_{ik} = \frac{1}{\ell_{kk}} \left(a_{ik} - \sum_{j=1}^{k-1} \ell_{ij}\ell_{kj} \right) \text{ for } i = k+1,\ldots,n. \tag{3.2.3}$$

Example 3.12 We will illustrate the described procedure for computing a Cholesky decomposition using the matrix

$$A = \begin{pmatrix} 4 & 2 & 6 \\ 2 & 10 & 3 \\ 6 & 3 & 10 \end{pmatrix}$$

as an example. The symmetry of the matrix A is obvious, while we will conclude its positive definiteness after the computation. According to (3.2.2), we first write

$$l_{11} = \sqrt{a_{11}} = \sqrt{4} = 2,$$

so that the rule (3.2.3) yields the coefficients

$$l_{21} = \frac{1}{l_{11}} a_{21} = 1 \quad \text{and} \quad l_{31} = \frac{1}{l_{11}} a_{31} = 3.$$

For the unknown values in the second column, we have

$$l_{22} = \sqrt{a_{22} - l_{21}^2} = \sqrt{10 - 1^2} = 3$$

and

$$l_{32} = \frac{1}{l_{22}}\left(a_{32} - l_{31}l_{21}\right) = \frac{1}{3}\left(3 - 3 \cdot 1\right) = 0.$$

The remaining coefficient is

$$l_{33} = \sqrt{a_{33} - l_{31}^2 - l_{32}^2} = \sqrt{10 - 3^2 - 0^2} = 1,$$

so that we obtain the desired lower triangular matrix

$$\boldsymbol{L} = \begin{pmatrix} 2 & 0 & 0 \\ 1 & 3 & 0 \\ 3 & 0 & 1 \end{pmatrix}$$

and thus the Cholesky factorization can be written as

$$\boldsymbol{A} = \underbrace{\begin{pmatrix} 2 & 0 & 0 \\ 1 & 3 & 0 \\ 3 & 0 & 1 \end{pmatrix}}_{=\boldsymbol{L}} \underbrace{\begin{pmatrix} 2 & 1 & 3 \\ 0 & 3 & 0 \\ 0 & 0 & 1 \end{pmatrix}}_{=\boldsymbol{L}^T}.$$

It should also be noted at this point that, due to

$$\boldsymbol{A}^T = \left(\boldsymbol{L}\boldsymbol{L}^T\right)^T = (\boldsymbol{L}^T)^T\boldsymbol{L}^T = \boldsymbol{L}\boldsymbol{L}^T = \boldsymbol{A}$$

the symmetry of the matrix $\boldsymbol{A}$ is a necessary condition for the existence of a Cholesky decomposition. If $\boldsymbol{A}$ is a non-singular matrix that admits a Cholesky decomposition, then the invertibility of the matrices $\boldsymbol{L}$ and $\boldsymbol{L}^T$ follows immediately. For any $\boldsymbol{x} \in \mathbb{R}^n \setminus \{\boldsymbol{0}\}$, it follows that $\boldsymbol{L}^T\boldsymbol{x} \neq \boldsymbol{0}$ and we obtain

$$\begin{aligned} (\boldsymbol{A}\boldsymbol{x},\boldsymbol{x})_2 &= (\boldsymbol{L}\boldsymbol{L}^T\boldsymbol{x},\boldsymbol{x})_2 = \left(\boldsymbol{L}\boldsymbol{L}^T\boldsymbol{x}\right)^T\boldsymbol{x} = \boldsymbol{x}^T(\boldsymbol{L}^T)^T\boldsymbol{L}^T\boldsymbol{x} \\ &= \left(\boldsymbol{L}^T\boldsymbol{x}\right)^T\boldsymbol{L}^T\boldsymbol{x} = (\boldsymbol{L}^T\boldsymbol{x},\boldsymbol{L}^T\boldsymbol{x})_2 = \|\underbrace{\boldsymbol{L}^T\boldsymbol{x}}_{\neq\boldsymbol{0}}\|_2^2 > 0, \end{aligned}$$

which means that the matrix $\boldsymbol{A}$ is necessarily positive definite. For a non-singular matrix, the existence of a Cholesky decomposition is therefore equivalent to the symmetry and positive definiteness of the matrix.

Cholesky Decomposition Algorithm —

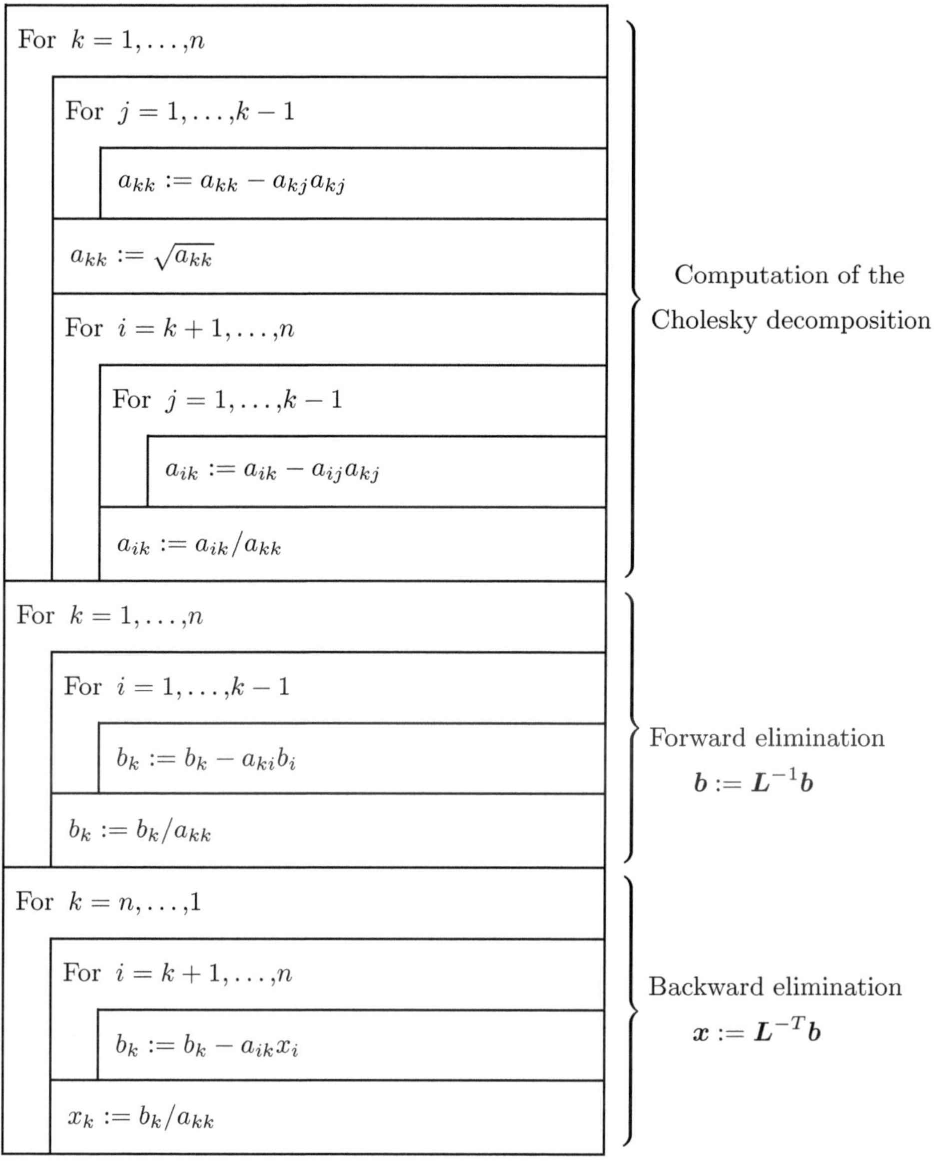

For the analysis of the computational effort we consider only the first part since it is decisive for large n. Here, we obtain

\# Multiplications + \# Divisions + \# Square roots

$$
= \underbrace{\sum_{k=1}^{n}(k-1)}_{\text{Multiplications}} + \underbrace{\sum_{k=1}^{n}1}_{\text{Square roots}} + \underbrace{\sum_{k=1}^{n}(n-k)(k-1)}_{\text{Multiplications}} + \underbrace{\sum_{k=1}^{n}(n-k)}_{\text{Divisions}}
$$

$$
= \sum_{k=1}^{n}k + n\sum_{k=1}^{n}k - \sum_{k=1}^{n}k^2
$$

$$
= \frac{n(n+1)}{2} + n\frac{n(n+1)}{2} - \frac{n(n+1)(2n+1)}{6}
$$

$$
= \frac{n^3}{6} + \frac{n^2}{2} + \frac{n}{3}.
$$

For large n, the Cholesky decomposition thus requires only about half as many costly operations as the Gaussian algorithm.

3.3   QR Factorization

An essential foundation for the definition of the GMRES method considered in Section 4.3.2.4, as well as for many methods for solving eigenvalue problems and linear least squares problems, is the QR factorization of a matrix. The advantage of the QR decomposition compared to the LR decomposition lies in the norm preservation of the unitary transformation with respect to the Euclidean norm. In addition, the system of equations $\boldsymbol{Ax} = \boldsymbol{b}$ can be easily solved due to $\boldsymbol{Q}^* = \boldsymbol{Q}^{-1}$ by

$$
\boldsymbol{Ax} = \boldsymbol{b} \Leftrightarrow \boldsymbol{QRx} = \boldsymbol{b} \Leftrightarrow \boldsymbol{Rx} = \boldsymbol{Q}^*\boldsymbol{b}.
$$

The three most well-known methods for computing this decomposition are the Gram-Schmidt scheme and the algorithms by Givens and Householder. In the following chapters, we will use only the first two methods, but for the sake of completeness, we will nevertheless derive all three algorithms in detail. Further descriptions of QR decompositions can be found, for example, in [16, 33, 55, 21].

Definition 3.13 The decomposition of a matrix $\boldsymbol{A} \in \mathbb{C}^{n \times n}$ into a product

$$
\boldsymbol{A} = \boldsymbol{QR}
$$

of a unitary matrix $\boldsymbol{Q}$ and an upper right triangular matrix $\boldsymbol{R}$ is called

QR decomposition or QR factorization.

3.3.1   The Gram-Schmidt Scheme

The derivation of the Gram-Schmidt scheme follows directly from the following constructive existence proof of the QR decomposition.

Theorem 3.14 (Existence of the QR Factorization)

Let $A \in \mathbb{C}^{n \times n}$ be an invertible matrix, then there exist a unitary matrix $Q \in \mathbb{C}^{n \times n}$ and an upper right triangular matrix $R \in \mathbb{C}^{n \times n}$ such that

$$A = QR$$

holds.

Proof:

We prove this by successively showing for $k = 1,\ldots,n$ the existence of vectors $q_1,\ldots,q_k \in \mathbb{C}^n$ with

$$(q_i,q_j)_2 = \delta_{ij} \quad \text{for } i,j = 1,\ldots,k \tag{3.3.1}$$

and

$$\operatorname{span}\{q_1,\ldots,q_k\} = \operatorname{span}\{a_1,\ldots,a_k\}, \tag{3.3.2}$$

where a_j $(1 \le j \le k)$ denotes the j-th column vector of the matrix A.

For $k = 1$, because $a_1 \in \mathbb{C}^n \setminus \{0\}$ both conditions are satisfied with

$$q_1 = \frac{a_1}{\|a_1\|_2}. \tag{3.3.3}$$

Now suppose $q_1,\ldots,q_k$ with $k \in \{1,\ldots,n-1\}$ are given, which satisfy conditions (3.3.1) and (3.3.2), then every vector

$$q_{k+1} \in \operatorname{span}\{a_1,\ldots,a_{k+1}\} \setminus \operatorname{span}\{q_1,\ldots,q_k\} \tag{3.3.4}$$

can be written in the form

$$q_{k+1} = c_{k+1}\left(a_{k+1} - \sum_{i=1}^{k} c_i q_i\right) \tag{3.3.5}$$

with $c_{k+1} \neq 0$. Motivated by

$$(q_{k+1},q_j)_2 = c_{k+1}\left[(a_{k+1},q_j)_2 - c_j\right] \quad \text{for } j = 1,\ldots,k$$

and the aim of orthogonality, we set in (3.3.5) $c_i := (a_{k+1},q_i)_2$ for $i = 1,\ldots,k$ and thus obtain the equation

$$(q_{k+1},q_j)_2 = c_{k+1}\left[(a_{k+1},q_j)_2 - \sum_{i=1}^{k}(a_{k+1},q_i)_2(q_i,q_j)_2\right]$$

$$= c_{k+1}[(a_{k+1},q_j)_2 - (a_{k+1},q_j)_2] = 0 \quad \text{for } j = 1,\ldots,k. \tag{3.3.6}$$

Since A is non-singular, it holds that $a_{k+1} \notin \operatorname{span}\{q_1,\ldots,q_k\}$, so that

$$\tilde{q}_{k+1} := a_{k+1} - \sum_{i=1}^{k}(a_{k+1},q_i)_2 q_i \neq 0$$

follows. With

$$c_{k+1} := \frac{1}{\|\tilde{q}_{k+1}\|_2}$$

it is easily seen that $\|q_{k+1}\|_2 = 1$, so that by

$$q_{k+1} = \frac{1}{\|\tilde{q}_{k+1}\|_2} \tilde{q}_{k+1} = \frac{1}{\|\tilde{q}_{k+1}\|_2} \left(a_{k+1} - \sum_{i=1}^{k} (a_{k+1}, q_i)_2 \, q_i \right) \tag{3.3.7}$$

due to (3.3.6) the desired vector is obtained. The definition

$$Q = (q_1 \cdots q_n)$$

with $q_1, \ldots, q_n$ according to (3.3.3) respectively (3.3.7) and

$$R := \begin{pmatrix} r_{11} & \cdots & r_{1n} \\ & \ddots & \vdots \\ & & r_{nn} \end{pmatrix} \quad \text{with} \quad r_{ik} = \begin{cases} \|\tilde{q}_i\|_2 & \text{for} \ k = i, \\ (a_k, q_i)_2 & \text{for} \ k > i \end{cases}$$

thus yields

$$A = QR.$$

$\square$

Based on a QR decomposition, we obtain the solution to the system of equations

$$Ax = b$$

by backward elimination of the system

$$Rx = \tilde{b} \quad \text{with} \quad \tilde{b} = Q^* b.$$

Example 3.15 With this example, we want to illustrate both the computation of a QR decomposition using the Gram-Schmidt method and, at the same time, the simple solution of a system of equations using it. We consider the system

$$\underbrace{\begin{pmatrix} 8 & 2 \\ 6 & 3 \end{pmatrix}}_{=A} \underbrace{\begin{pmatrix} x_1 \\ x_2 \end{pmatrix}}_{=x} = \underbrace{\begin{pmatrix} 2 \\ -3 \end{pmatrix}}_{=b}.$$

First, we obtain the first column vector q_1 of the orthogonal matrix Q by

$$q_1 = \frac{a_1}{\|a_1\|_2} = \frac{1}{10} \begin{pmatrix} 8 \\ 6 \end{pmatrix} = \frac{1}{5} \begin{pmatrix} 4 \\ 3 \end{pmatrix}$$

as well as the diagonal entry of the upper right triangular matrix R by

$$r_{11} = \|a_1\|_2 = 10.$$

Next, we have

$$r_{12} = (\boldsymbol{a}_2, \boldsymbol{q}_1)_2 = \left(\begin{pmatrix} 2 \\ 3 \end{pmatrix}, \frac{1}{5} \begin{pmatrix} 4 \\ 3 \end{pmatrix} \right)_2 = \frac{17}{5},$$

which gives

$$\tilde{\boldsymbol{q}}_2 = \boldsymbol{a}_2 - r_{12}\boldsymbol{q}_1 = \begin{pmatrix} 2 \\ 3 \end{pmatrix} - \frac{17}{25} \begin{pmatrix} 4 \\ 3 \end{pmatrix} = \frac{1}{25} \begin{pmatrix} -18 \\ 24 \end{pmatrix}$$

as well as

$$r_{22} = \|\tilde{\boldsymbol{q}}_2\|_2 = \frac{6}{5}$$

and

$$\boldsymbol{q}_2 = \frac{1}{r_{22}}\tilde{\boldsymbol{q}}_2 = \frac{1}{30} \begin{pmatrix} -18 \\ 24 \end{pmatrix} = \frac{1}{5} \begin{pmatrix} -3 \\ 4 \end{pmatrix}.$$

The QR factorization is therefore

$$\boldsymbol{A} = \underbrace{\begin{pmatrix} \frac{4}{5} & -\frac{3}{5} \\ \frac{3}{5} & \frac{4}{5} \end{pmatrix}}_{=\boldsymbol{Q}} \underbrace{\begin{pmatrix} 10 & \frac{17}{5} \\ 0 & \frac{6}{5} \end{pmatrix}}_{=\boldsymbol{R}}$$

and for the solution of the corresponding system of equations, we can use

$$\tilde{\boldsymbol{b}} = \boldsymbol{Q}^T \boldsymbol{b} = \begin{pmatrix} \frac{4}{5} & \frac{3}{5} \\ -\frac{3}{5} & \frac{4}{5} \end{pmatrix} \begin{pmatrix} 2 \\ -3 \end{pmatrix} = \begin{pmatrix} -\frac{1}{5} \\ -\frac{18}{5} \end{pmatrix}$$

and by the well-known backward elimination with respect to

$$\underbrace{\begin{pmatrix} 10 & \frac{17}{5} \\ 0 & \frac{6}{5} \end{pmatrix}}_{=\boldsymbol{R}} \underbrace{\begin{pmatrix} x_1 \\ x_2 \end{pmatrix}}_{=\boldsymbol{x}} = \underbrace{\begin{pmatrix} -\frac{1}{5} \\ -\frac{18}{5} \end{pmatrix}}_{=\tilde{\boldsymbol{b}}}$$

determine the solution vector

$$\boldsymbol{x} = \begin{pmatrix} 1 \\ -3 \end{pmatrix}.$$

If the columns of the matrix $\boldsymbol{A} \in \mathbb{C}^{m \times n}$, $m \geq n$, are linearly independent, then from the above proof follows the existence of a matrix $\widetilde{\boldsymbol{Q}} \in \mathbb{C}^{m \times n}$, whose columns are pairwise orthonormal, and a non-singular upper right triangular matrix $\boldsymbol{R} \in \mathbb{C}^{n \times n}$ with $\boldsymbol{A} = \widetilde{\boldsymbol{Q}}\boldsymbol{R}$. If we extend $\widetilde{\boldsymbol{Q}}$ to a unitary matrix $\boldsymbol{Q} \in \mathbb{C}^{m \times m}$, then one can write

$$\boldsymbol{A} = \boldsymbol{Q} \begin{pmatrix} \boldsymbol{R} \\ \boldsymbol{0} \end{pmatrix}.$$

In summary, from the proof of the above theorem, we obtain the Gram-Schmidt scheme for a non-singular matrix $\boldsymbol{A} \in \mathbb{R}^{n \times n}$ in the following form:

Algorithm Gram-Schmidt —

For $k = 1, \ldots, n$

    For $i = 1, \ldots, k-1$

        $r_{ik} := 0$

        For $j = 1, \ldots, n$

            $r_{ik} := r_{ik} + a_{ji} a_{jk}$

    For $i = 1, \ldots, k-1$

        For $j = 1, \ldots, n$

            $a_{jk} := a_{jk} - r_{ik} a_{ji}$

    $r_{kk} := 0$

    For $j = 1, \ldots, n$

        $r_{kk} := r_{kk} + a_{jk} a_{jk}$

    $r_{kk} := \sqrt{r_{kk}}$

    For $j = 1, \ldots, n$

        $a_{jk} := a_{jk} / r_{kk}$

                 Computation of the QR decomposition

For $k = 1, \ldots, n$

    $\tilde{b}_k := 0$

    For $i = 1, \ldots, n$

        $\tilde{b}_k := \tilde{b}_k + a_{ik} b_i$

                 Matrix multiplication $\tilde{b} := Q^T b$

For $k = n, \ldots, 1$

    For $i = k+1, \ldots, n$

        $\tilde{b}_k := \tilde{b}_k - r_{ki} x_i$

    $x_k := \tilde{b}_k / r_{kk}$

                 Back substitution $x := R^{-1} \tilde{b}$

In the computational cost analysis of the QR decomposition via Gram-Schmidt, we consider only the computationally intensive first part and again neglect the explicit solution of the system of equations. Thus, we obtain

\# multiplications + \# divisions + \# square roots

$$
= \underbrace{2n \sum_{k=1}^{n}(k-1) + n \sum_{k=1}^{n}1}_{\text{multiplications}} + \underbrace{n \sum_{k=1}^{n}1}_{\text{divisions}} + \underbrace{\sum_{k=1}^{n}1}_{\text{square roots}}
$$

$$
= 2n \sum_{k=1}^{n} k + n = 2\frac{n^2(n+1)}{2} + n = n^3 + n^2 + n.
$$

For large n, the computational cost is therefore approximately three times as high as for the Gaussian algorithm.

The numerical issue with the Gram-Schmidt scheme lies in the occurrence of rounding errors, which can lead to cancellation effects and consequently to insufficient orthogonality of the vectors $q_1, \ldots, q_n$. To improve this, we consider two variants of the algorithm.

Variant I: The Modified Gram-Schmidt Process

Basic idea: After computing the k-th column vector q_k of the unitary matrix Q, the vectors $a_{k+1}, \ldots, a_n$ of the matrix A are modified in such a way that they are orthogonal to all column vectors $q_1, \ldots, q_k$.

Procedure: Let $A = (a_1 \ldots a_n) \in \mathbb{R}^{n \times n}$

Step 1: Set $a_k^{(1)} = a_k$ for $k = 1, \ldots, n$.

Step 2: For $k = 1, \ldots, n$

$$
q_k := \frac{a_k^{(k)}}{\|a_k^{(k)}\|_2}
$$

$$
a_j^{(k+1)} = a_j^{(k)} - (a_j^{(k)}, q_k)_2 q_k \quad \text{for } j = k+1, \ldots, n.
$$

The computational effort of the modified Gram-Schmidt scheme is identical to the original algorithm. If we compare both methods, a difference can occur at the earliest for $n = 3$.

For $A = (a_1, a_2, a_3) = (a_1^{(1)}, a_2^{(1)}, a_3^{(1)}) \in \mathbb{R}^{3 \times 3}$ it follows:

Gram-Schmidt Process	Modified Gram-Schmidt Process
$$q_1 = \frac{a_1}{\|a_1\|_2}$$	$$q_1 = \frac{a_1^{(1)}}{\|a_1^{(1)}\|_2}$$ $$a_2^{(2)} = a_2^{(1)} - \left(a_2^{(1)},q_1\right)_2 q_1$$ $$a_3^{(2)} = a_3^{(1)} - \left(a_3^{(1)},q_1\right)_2 q_1$$
$$\tilde{a}_2 = a_2 - (a_2,q_1)_2\, q_1$$ $$q_2 = \frac{\tilde{a}_2}{\|\tilde{a}_2\|_2}$$	$$q_2 = \frac{a_2^{(2)}}{\|a_2^{(2)}\|_2}$$ $$a_3^{(3)} = a_3^{(2)} - \left(a_3^{(2)},q_2\right)_2 q_2$$
$$\tilde{a}_3 = a_3 - (a_3,q_1)_2\, q_1$$ $$\qquad - (a_3,q_2)_2\, q_2$$ $$q_3 = \frac{\tilde{a}_3}{\|\tilde{a}_3\|_2}$$	$$q_3 = \frac{a_3^{(3)}}{\|a_3^{(3)}\|_2}$$

From this, the relationship follows

$$
\begin{aligned}
a_3^{(3)} &= a_3^{(1)} - \left(a_3^{(1)},q_1\right)_2 q_1 - \left(a_3^{(1)},q_2\right)_2 q_2 + \left(\left(a_3^{(1)},q_1\right)_2 q_1, q_2\right)_2 q_2 \\
&= \tilde{a}_3 + ((a_3,q_1)_2 q_1, q_2)_2\, q_2,
\end{aligned}
$$

so that the difference lies in the vector $((a_3,q_1)_2 q_1, q_2)_2\, q_2$, which vanishes in exact arithmetic.

Variant II: Gram-Schmidt Process with Reorthogonalization

Basic idea: After computing the unnormalized k-th column vector q_k, it is orthogonalized with respect to $q_1,\ldots,q_{k-1}$ and then normalized.

Procedure: Let $\tilde{q}_k$ be the unnormalized vector, then set

$$
\bar{q}_k = \tilde{q}_k - \sum_{i=1}^{k-1} (\tilde{q}_k,q_i)_2 q_i
$$

$$
q_k = \frac{\bar{q}_k}{\|\bar{q}_k\|_2}.
$$

Additionally, the coefficients of the matrix $\boldsymbol{R}$ must be modified by

$$r_{i,k}^{\text{new}} = r_{i,k}^{\text{old}} + (\tilde{\boldsymbol{q}}_k, \boldsymbol{q}_i)_2 \text{ for } i = 1,\dots,k-1.$$

The computational effort is about twice as high compared to the original Gram-Schmidt process.

Theorem 3.16 (Uniqueness of the QR Factorization)
Let $\boldsymbol{A} \in \mathbb{C}^{n \times n}$ be invertible, then for any two QR decompositions

$$\boldsymbol{Q}_1 \boldsymbol{R}_1 = \boldsymbol{A} = \boldsymbol{Q}_2 \boldsymbol{R}_2 \tag{3.3.8}$$

there exists a unitary diagonal matrix $\boldsymbol{D} \in \mathbb{C}^{n \times n}$ such that

$$\boldsymbol{Q}_1 = \boldsymbol{Q}_2 \boldsymbol{D} \quad and \quad \boldsymbol{R}_2 = \boldsymbol{D} \boldsymbol{R}_1. \tag{3.3.9}$$

Proof:
With $\boldsymbol{D} = \boldsymbol{Q}_2^* \boldsymbol{Q}_1$, we have, by Lemma 2.27, a unitary matrix, and it holds that

$$\boldsymbol{Q}_1 = \boldsymbol{Q}_2 \boldsymbol{D}.$$

Since $\boldsymbol{A}$ is invertible, both $\boldsymbol{R}_1$ and $\boldsymbol{R}_2$ are also invertible, and we obtain from (3.3.8)

$$\boldsymbol{D} = \boldsymbol{Q}_2^* \boldsymbol{Q}_1 = \boldsymbol{R}_2 \boldsymbol{R}_1^{-1}.$$

Lemma 3.4 in combination with Lemma 2.26 and Lemma 2.27 states that $\boldsymbol{D}$ is an upper triangular matrix. Since $\boldsymbol{D}$ is also unitary, $\boldsymbol{D}$ is in fact a diagonal matrix. $\qquad\square$

Corollary 3.17 *Let $\boldsymbol{A} \in \mathbb{C}^{n \times n}$ be non-singular, then there exists exactly one QR decomposition of the matrix $\boldsymbol{A}$ such that the diagonal elements of the matrix $\boldsymbol{R}$ are real and positive.*

3.3.2　The QR Factorization via Givens Rotations

We restrict ourselves to the case of a matrix $\boldsymbol{A} \in \mathbb{R}^{n \times n}$. The idea of the Givens method is a successive elimination of the subdiagonal elements. Starting with the first column, the subdiagonal elements of each column are eliminated in ascending order using orthogonal rotation matrices.

For example, let us consider the matrix

$$
\boldsymbol{A} \;=\;
\begin{pmatrix}
* & \cdots & \cdots & \cdots & \cdots & \cdots & \cdots & \cdots & \cdots & \cdots & * \\
0 & \ddots & & & & & & & & & \vdots \\
\vdots & \ddots & \ddots & & & & & & & & \vdots \\
\vdots & & 0 & * & & & & & & & \vdots \\
\vdots & & 0 & 0 & * & & & & & & \vdots \\
\vdots & & \vdots & \vdots & * & \ddots & & & & & \vdots \\
\vdots & & \vdots & \vdots & \vdots & \ddots & \ddots & & & & \vdots \\
\vdots & & \vdots & 0 & \vdots & & \ddots & \ddots & & & \vdots \\
\vdots & & \vdots & * & \vdots & & & \ddots & \ddots & & \vdots \\
\vdots & & \vdots & \vdots & \vdots & & & & \ddots & \ddots & \vdots \\
0 & \cdots & 0 & * & * & \cdots & \cdots & \cdots & \cdots & \cdots & *
\end{pmatrix}
\in \mathbb{R}^{n\times n}.
$$

$$\uparrow$$
$$i\text{-th column}$$

$\leftarrow j$-th row

Here, we have

$$
a_{k\ell} = 0 \quad \forall \ell \in \{1,\dots,i-1\} \ \text{with} \ \ell < k \in \{1,\dots,n\}, \tag{3.3.10}
$$

$$
a_{i+1,i} = \dots = a_{j-1,i} = 0 \tag{3.3.11}
$$

and $a_{ji} \neq 0$. Then we look for an orthogonal matrix

$$
\boldsymbol{G}_{ji} =
\begin{pmatrix}
1 & & & & & & & & & \\
 & \ddots & & & & & & & & \\
 & & 1 & & & & & & & \\
 & & & g_{ii} & 0 & \cdots & 0 & g_{ij} & & \\
 & & & 0 & 1 & & & 0 & & \\
 & & & \vdots & & \ddots & & \vdots & & \\
 & & & 0 & & & 1 & 0 & & \\
 & & & g_{ji} & 0 & \cdots & 0 & g_{jj} & & \\
 & & & & & & & & 1 & \\
 & & & & & & & & & \ddots & \\
 & & & & & & & & & & 1
\end{pmatrix}
\in \mathbb{R}^{n\times n}
$$

such that for

$$
\widetilde{\boldsymbol{A}} = \boldsymbol{G}_{ji}\boldsymbol{A}
$$

in addition to

$$
\tilde{a}_{k\ell} = 0 \quad \forall \ell \in \{1,\dots,i-1\} \ \text{with} \ \ell < k \in \{1,\dots,n\}, \tag{3.3.12}
$$

and

$$
\tilde{a}_{i+1,i} = \dots = \tilde{a}_{j-1,i} = 0 \tag{3.3.13}
$$

also

$$
\tilde{a}_{ji} = 0 \tag{3.3.14}
$$

holds.

Initially, $\widetilde{A}$ differs from A only in the i-th and j-th row, and for $\ell = 1,\ldots,n$ we have

$$\tilde{a}_{i\ell} = g_{ii}a_{i\ell} + g_{ij}a_{j\ell}$$

$$\tilde{a}_{j\ell} = g_{ji}a_{i\ell} + g_{jj}a_{j\ell}.$$

From (3.3.10) it follows that $a_{i\ell} = a_{j\ell} = 0$ for $\ell < i < j$, so that

$$\tilde{a}_{i\ell} = \tilde{a}_{j\ell} = 0 \quad \text{for} \quad \ell = 1,\ldots,i-1$$

holds and thus the requirements (3.3.12) and (3.3.13) are satisfied. Well-defined by the property $a_{ji} \neq 0$, we set

$$g_{ii} = g_{jj} = \frac{a_{ii}}{\sqrt{a_{ii}^2 + a_{ji}^2}} \quad \text{and} \quad g_{ij} = -g_{ji} = \frac{a_{ji}}{\sqrt{a_{ii}^2 + a_{ji}^2}}.$$

Thus, G_{ji} represents an orthogonal rotation matrix by the angle $\alpha = \arccos g_{ii}$, and we have

$$\tilde{a}_{ji} = -\frac{a_{ji}}{\sqrt{a_{ii}^2 + a_{ji}^2}}a_{ii} + \frac{a_{ii}}{\sqrt{a_{ii}^2 + a_{ji}^2}}a_{ji} = 0.$$

If we define $G_{ji} = I$ in the case of a matrix A that satisfies (3.3.10) and (3.3.11) and also contains $a_{ji} = 0$, then with

$$\widetilde{Q} := \prod_{i=n-1}^{1} \prod_{j=n}^{i+1} G_{ji} := G_{n,n-1} \cdot \ldots \cdot G_{3,2} \cdot G_{n,1} \cdot \ldots \cdot G_{3,1} \cdot G_{2,1}$$

we obtain an orthogonal matrix for which

$$R = \widetilde{Q}A$$

is an upper triangular matrix. With $Q = \widetilde{Q}^T$ it follows that $A = QR$.

Example 3.18 For the matrix already used in Example 3.15

$$A = \begin{pmatrix} 8 & 2 \\ 6 & 3 \end{pmatrix}$$

the Givens method yields the coefficients of the rotation matrix G_{21} as

$$g_{11} = g_{22} = \frac{a_{11}}{\sqrt{a_{11}^2 + a_{21}^2}} = \frac{8}{\sqrt{8^2 + 6^2}} = \frac{4}{5} \quad \text{and} \quad g_{12} = -g_{21} = \frac{a_{21}}{\sqrt{a_{11}^2 + a_{21}^2}} = \frac{3}{5},$$

so that

$$G_{21} = \frac{1}{5}\begin{pmatrix} 4 & 3 \\ -3 & 4 \end{pmatrix}.$$

Consequently, analogous to the Gram-Schmidt scheme, we have

$$R = G_{21}A = \frac{1}{5}\begin{pmatrix} 4 & 3 \\ -3 & 4 \end{pmatrix}\begin{pmatrix} 8 & 2 \\ 6 & 3 \end{pmatrix} = \begin{pmatrix} 10 & \frac{17}{5} \\ 0 & \frac{6}{5} \end{pmatrix}$$

and the QR decomposition is given by

$$A = G_{21}^T R = \underbrace{\frac{1}{5}\begin{pmatrix} 4 & -3 \\ 3 & 4 \end{pmatrix}}_{=Q}\underbrace{\begin{pmatrix} 10 & \frac{17}{5} \\ 0 & \frac{6}{5} \end{pmatrix}}_{=R}.$$

In the context of the Givens method, we have the possibility to solve the system of equations $\boldsymbol{Ax} = \boldsymbol{b}$ using a QR decomposition without explicitly storing the orthogonal matrix $\boldsymbol{Q}$. For this, the rotations must be applied not only to the columns of the matrix $\boldsymbol{A}$, but also to the right-hand side $\boldsymbol{b}$. Therefore, we augment the matrix $\boldsymbol{A}$ with the right-hand side $\boldsymbol{b}$ according to $\boldsymbol{a}_{n+1} = \boldsymbol{b}$ and obtain the algorithm shown below.

Algorithm Givens Method —

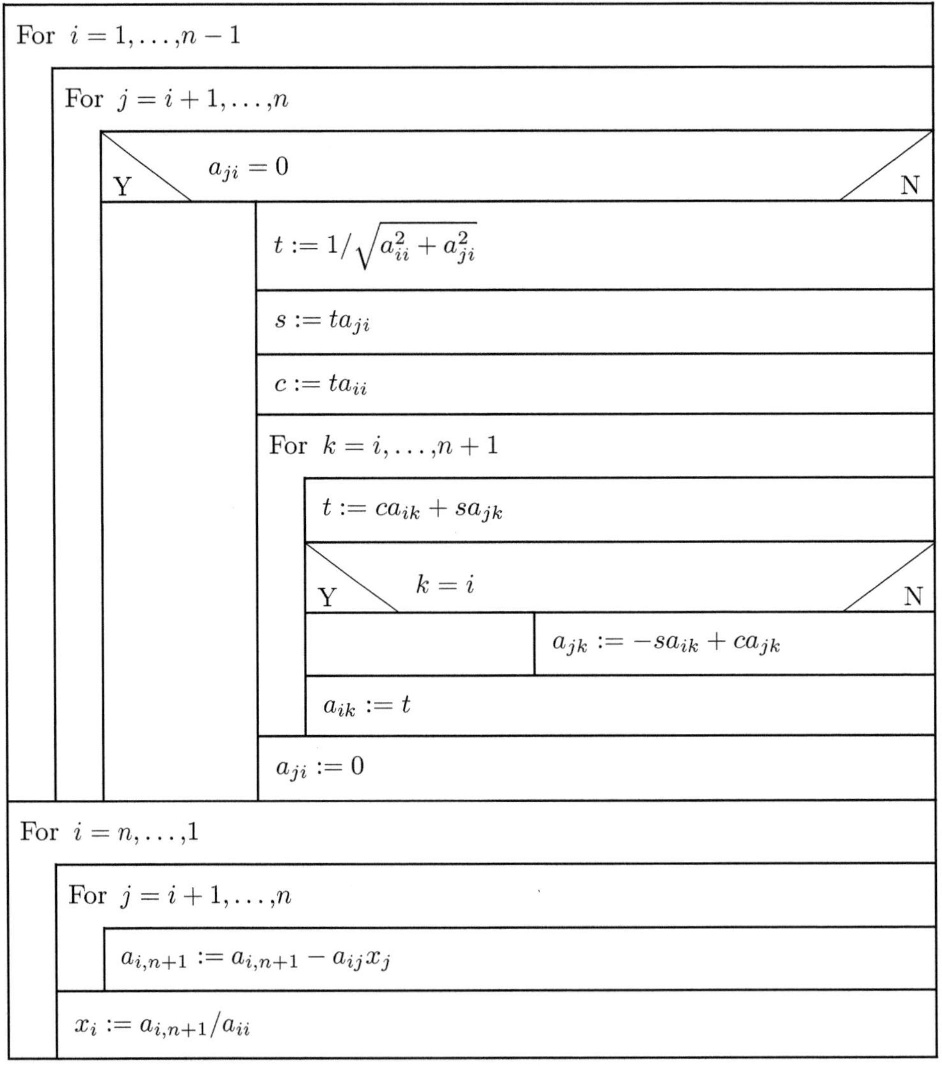

For the Givens method, disregarding the right-hand side $\boldsymbol{b}$ and the explicit solution of the system of equations by backward elimination, the following computational effort results for a dense matrix:

\# Multiplications + \# Divisions + \# Square roots

$$\leq \underbrace{\sum_{i=1}^{n-1}\{2(n-i)+4(n-i)(n-i+1)\}}_{\text{multiplications}}+\underbrace{\sum_{i=1}^{n-1}(n-i)}_{\text{divisions}}+\underbrace{\sum_{i=1}^{n-1}(n-i)}_{\text{square roots}}$$

$$= \frac{4}{3}n^3 + 2n^2 - \frac{10}{3}n.$$

In the general case, the Givens method requires more computational effort compared to the Gram-Schmidt process. However, due to the query performed in the algorithm, the effort of the Givens method can be significantly reduced if the matrix has a special structure. For example, if we consider an upper Hessenberg matrix[2], only $n-1$ Givens rotations are needed to transform it into an upper triangular form. In this special case, the computational effort is reduced from order n^3 to order n^2. We will exploit this property later when deriving the GMRES method.

3.3.3 The QR Factorization via Householder

With the Householder transformation, we now learn another way to compute a QR decomposition of a given matrix $\boldsymbol{A} \in \mathbb{R}^{n \times n}$. We will successively transform the matrix into an upper triangular form, whereby, in contrast to the Givens method, reflection matrices are always used instead of rotation matrices. While the goal of the Givens rotation is to eliminate exactly one matrix element in each step, the Householder transformation always performs reflections in such a way that all subdiagonal elements of a column become zero.

Let us consider an arbitrary vector $\boldsymbol{v} \in \mathbb{R}^s \setminus \{\boldsymbol{0}\}$, and seek an orthogonal, symmetric matrix[3] $\boldsymbol{H} \in \mathbb{R}^{s \times s}$ with $\det \boldsymbol{H} = -1$, for which

$$\boldsymbol{H}\boldsymbol{v} = c \cdot \boldsymbol{e}_1, \, c \in \mathbb{R} \setminus \{0\} \tag{3.3.15}$$

holds, where $\boldsymbol{e}_1$ denotes the first unit vector. Due to the desired orthogonality of the matrix, the mapping is length-preserving according to Theorem 2.28, so for the constant c we have

$$\|\boldsymbol{v}\|_2 = \|\boldsymbol{H}\boldsymbol{v}\|_2 = \|c \cdot \boldsymbol{e}_1\|_2 = |c| \cdot \|\boldsymbol{e}_1\|_2 = |c|$$

which yields

$$c = \pm\|\boldsymbol{v}\|_2. \tag{3.3.16}$$

Let $\boldsymbol{u} \in \mathbb{R}^s$ be the unit normal vector to the reflection hyperplane[4] S of the matrix $\boldsymbol{H}$ with the angle between the vectors $\boldsymbol{u}$ and $\boldsymbol{v}$ given by $\varphi = \angle(\boldsymbol{u},\boldsymbol{v})$. Then, the length ℓ of the orthogonal projection of $\boldsymbol{v}$ onto $\boldsymbol{u}$ is, according to Figure 3.1 (left), given by

---

[2] An upper Hessenberg matrix has nonzero entries only in the upper right triangular part and on the subdiagonal directly below the main diagonal.

[3] Real-valued reflection matrices are always orthogonal and symmetric and have determinant -1, see [17].

[4] In the case considered here, the reflection hyperplane is an $(n-1)$-dimensional subspace of $\mathbb{R}^n$.

$$\frac{\ell}{\|\boldsymbol{v}\|_2} = \cos\varphi = \frac{(\boldsymbol{u},\boldsymbol{v})_2}{\underbrace{\|\boldsymbol{u}\|_2}_{=1}\|\boldsymbol{v}\|_2} = \frac{\boldsymbol{u}^T\boldsymbol{v}}{\|\boldsymbol{v}\|_2}$$

so that $\ell = \boldsymbol{u}^T\boldsymbol{v}$.

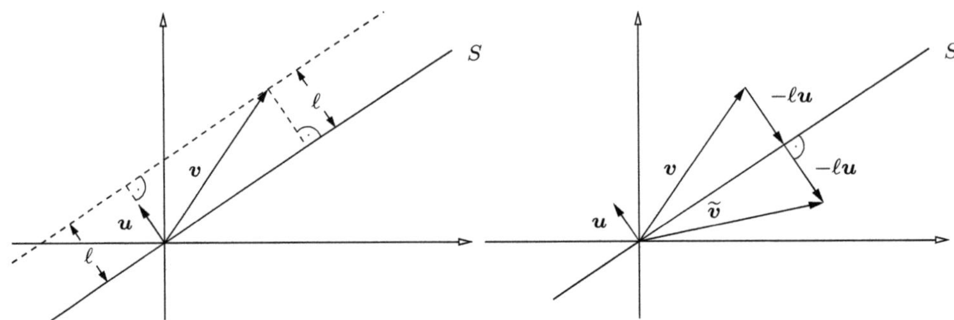

Figure 3.1 Length of the orthogonal projection (left) and effect of the matrix $\boldsymbol{H}$ on the vector $\boldsymbol{v}$ (right)

For given vectors $\boldsymbol{u}$ and $\boldsymbol{v}$, the vector reflected from $\boldsymbol{v}$ across S is given by

$$\tilde{\boldsymbol{v}} = \boldsymbol{v} - 2\boldsymbol{u}\ell = \boldsymbol{v} - 2\boldsymbol{u}\boldsymbol{u}^T\boldsymbol{v} = \boldsymbol{H}\boldsymbol{v}$$

with $\boldsymbol{H} = \boldsymbol{I} - 2\boldsymbol{u}\boldsymbol{u}^T$. The effect of the matrix $\boldsymbol{H}$ on the vector $\boldsymbol{v}$ is shown in Figure 3.1 (right). One can verify that the reflection property also holds if the vectors $\boldsymbol{u}$ and $\boldsymbol{v}$, contrary to Figure 3.1 (right), point into different half-spaces with respect to the reflection hyperplane S.

Before we turn to the determination of the vector $\boldsymbol{u}$, we will first analyze the properties of the resulting matrix $\boldsymbol{H}$.

Theorem and Definition 3.19 *Let $\boldsymbol{u} \in \mathbb{R}^s$ with $\|\boldsymbol{u}\|_2 = 1$, then*

$$\boldsymbol{H} = \boldsymbol{I} - 2\boldsymbol{u}\boldsymbol{u}^T \in \mathbb{R}^{s \times s}$$

is an orthogonal and symmetric matrix with $\det\boldsymbol{H} = -1$ *and is called Householder matrix.*

Proof:
From

$$\boldsymbol{H}^T = (\boldsymbol{I} - 2\boldsymbol{u}\boldsymbol{u}^T)^T = \boldsymbol{I} - 2(\boldsymbol{u}\boldsymbol{u}^T)^T = \boldsymbol{I} - 2\boldsymbol{u}^{T^T}\boldsymbol{u}^T = \boldsymbol{I} - 2\boldsymbol{u}\boldsymbol{u}^T = \boldsymbol{H}$$

the claimed symmetry follows, and taking into account the Euclidean norm condition $\|\boldsymbol{u}\|_2 = 1$, we obtain the confirmation of the orthogonality simply by

$$\boldsymbol{H}^T\boldsymbol{H} = \boldsymbol{H}^2 = (\boldsymbol{I} - 2\boldsymbol{u}\boldsymbol{u}^T)(\boldsymbol{I} - 2\boldsymbol{u}\boldsymbol{u}^T) = \boldsymbol{I} - 4\boldsymbol{u}\boldsymbol{u}^T + 4\boldsymbol{u}\underbrace{\boldsymbol{u}^T\boldsymbol{u}}_{=1}\boldsymbol{u}^T = \boldsymbol{I}.$$

If we extend the vector $\boldsymbol{u}$ by the vectors $\boldsymbol{y}_1,\ldots,\boldsymbol{y}_{s-1} \in \mathbb{R}^s$ to form an orthonormal basis of $\mathbb{R}^s$ and thus define the orthogonal matrix

$$Q = (u, y_1, \ldots, y_{s-1}) \in \mathbb{R}^{s \times s},$$

then $u^T Q = (1, 0, \ldots, 0) = e_1^T$ holds, and thus we obtain from

$$Q^{-1} H Q = Q^T H Q = Q^T (I - 2uu^T) Q$$

$$= Q^T Q - 2 Q^T u u^T Q = I - 2 e_1 e_1^T = \begin{pmatrix} -1 & & & \\ & 1 & & \\ & & \ddots & \\ & & & 1 \end{pmatrix}$$

directly

$$-1 = \det(Q^{-1} H Q) = \underbrace{\det Q^{-1}}_{=1/\det Q} \det H \det Q = \det H.$$

$\square$

Since the desired property $Hv = c\, e_1$ is only uniquely determined up to sign due to $c = \pm \|v\|_2$, there are two possible reflection planes $S_\pm$, each defined by a unit normal vector $u_\pm$. Geometrically, from Figure 3.2 (left), it is clear that reflecting v onto $\|v\|_2 e_1$ requires a reflection plane S_+ that contains the angle bisector between v and e_1. We therefore use

$$u_+ = \frac{v - \|v\|_2 e_1}{\|v - \|v\|_2 e_1\|_2}$$

as the vector in the direction of a diagonal of the parallelogram formed by v and $\|v\|_2 e_1$.

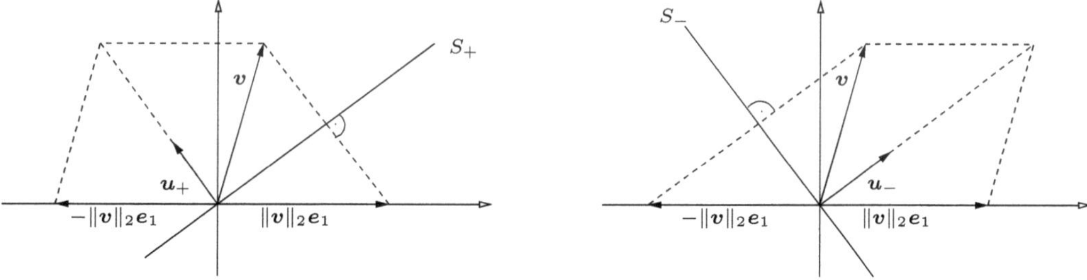

Figure 3.2 Definition of the unit normal vector u_+ or u_- for reflection onto $\|v\|_2 e_1$ (left) or $-\|v\|_2 e_1$ (right)

Figure 3.2 (right) also shows that the reflection of v onto $-\|v\|_2 e_1$ can be achieved by choosing

$$u_- = \frac{v + \|v\|_2 e_1}{\|v + \|v\|_2 e_1\|_2}$$

as expected. The correctness of this heuristic approach for determining the vectors u_+ and u_- can be verified by simply checking the corresponding mapping properties. Taking into account

$$\|v \mp \|v\|_2 e_1\|_2^2 = v^T v \mp 2\|v\|_2 e_1^T v + \underbrace{\|v\|_2^2}_{=v^T v} = 2(v^T \mp \|v\|_2 e_1^T) v \tag{3.3.17}$$

we obtain for the corresponding matrices $H_\pm = I - 2 u_\pm u_\pm^T$ from

$$
\boldsymbol{H}_{\pm}\boldsymbol{v} \;=\; \boldsymbol{v} - 2\boldsymbol{u}_{\pm}\boldsymbol{u}_{\pm}^{T}\boldsymbol{v} = \boldsymbol{v} - 2\frac{\boldsymbol{v}\mp\|\boldsymbol{v}\|_2\boldsymbol{e}_1}{\|\boldsymbol{v}\mp\|\boldsymbol{v}\|_2\boldsymbol{e}_1\|_2^2}\underbrace{(\boldsymbol{v}\mp\|\boldsymbol{v}\|_2\boldsymbol{e}_1)^{T}\boldsymbol{v}}_{\overset{(3.3.17)}{=}\frac{1}{2}\|\boldsymbol{v}\mp\|\boldsymbol{v}\|_2\boldsymbol{e}_1\|_2^2}
$$

$$
=\; \boldsymbol{v} - (\boldsymbol{v}\mp\|\boldsymbol{v}\|_2\boldsymbol{e}_1) = \pm\|\boldsymbol{v}\|_2\boldsymbol{e}_1 \tag{3.3.18}
$$

the desired result. To avoid unnecessarily amplifying a rounding error that may occur in the computation of $\boldsymbol{v}\mp\|\boldsymbol{v}\|_2\boldsymbol{e}_1$ by dividing by a small number, the sign is chosen so that

$$
\|\boldsymbol{v}\mp\|\boldsymbol{v}\|_2\boldsymbol{e}_1\|_2^2 = (v_1\mp\|\boldsymbol{v}\|_2)^2 + v_2^2 + \ldots v_s^2 \tag{3.3.19}
$$

is maximized. From (3.3.19) it is immediately clear that

$$
\boldsymbol{u} = \frac{\boldsymbol{v} + \operatorname{sgn}(v_1)\|\boldsymbol{v}\|_2\boldsymbol{e}_1}{\|\boldsymbol{v} + \operatorname{sgn}(v_1)\|\boldsymbol{v}\|_2\boldsymbol{e}_1\|_2} \quad\text{with}\quad \operatorname{sgn}(x) = \left\{\begin{array}{rl} 1, & \text{for } x \geq 0 \\ -1, & \text{for } x < 0 \end{array}\right. \tag{3.3.20}
$$

results. For the final construction of the matrices required for the Householder algorithm, we first consider the following auxiliary result, where the abbreviation $\boldsymbol{I}_k$ always denotes the identity matrix in $\mathbb{R}^{k\times k}$.

Lemma 3.20 *Let $\boldsymbol{u}_s \in \mathbb{R}^s$ with $\|\boldsymbol{u}_s\|_2 = 1$ be given and $\widetilde{\boldsymbol{H}}_s = \boldsymbol{I}_s - 2\boldsymbol{u}_s\boldsymbol{u}_s^{T} \in \mathbb{R}^{s\times s}$ the corresponding Householder matrix, then*

$$
\boldsymbol{H}_s = \left(\begin{array}{cc} \boldsymbol{I}_{n-s} & \boldsymbol{0} \\ \boldsymbol{0} & \widetilde{\boldsymbol{H}}_s \end{array}\right) \in \mathbb{R}^{n\times n}
$$

is an orthogonal matrix.

Proof:
The proof follows immediately from

$$
\boldsymbol{H}_s^{T}\boldsymbol{H}_s = \left(\begin{array}{cc} \boldsymbol{I}_{n-s} & \boldsymbol{0} \\ \boldsymbol{0} & \widetilde{\boldsymbol{H}}_s^{T}\widetilde{\boldsymbol{H}}_s \end{array}\right) = \left(\begin{array}{cc} \boldsymbol{I}_{n-s} & \boldsymbol{0} \\ \boldsymbol{0} & \boldsymbol{I}_s \end{array}\right) = \boldsymbol{I}_n\,.
$$

$\square$

Using the results obtained, we can successively transform a given matrix $\boldsymbol{A} \in \mathbb{R}^{n\times n}$ into an upper triangular form. For this, we set $\boldsymbol{A}^{(0)} = \boldsymbol{A}$ and write

$$
\boldsymbol{A}^{(0)} = \left(\begin{array}{ccc} a_{11}^{(0)} & \cdots & a_{1n}^{(0)} \\ \vdots & & \vdots \\ a_{n1}^{(0)} & \cdots & a_{nn}^{(0)} \end{array}\right).
$$

With $\boldsymbol{a}^{(0)} = (a_{11}^{(0)},\ldots,a_{n1}^{(0)})^{T}$ we define, using

$$
\boldsymbol{u}_n = \frac{\boldsymbol{a}^{(0)} + \operatorname{sgn}(a_{11}^{(0)})\|\boldsymbol{a}^{(0)}\|_2\boldsymbol{e}_1}{\|\boldsymbol{a}^{(0)} + \operatorname{sgn}(a_{11}^{(0)})\|\boldsymbol{a}^{(0)}\|_2\boldsymbol{e}_1\|_2}
$$

the Householder matrix

$$
\boldsymbol{H}_n = \boldsymbol{I}_n - 2\boldsymbol{u}_n\boldsymbol{u}_n^{T} \in \mathbb{R}^{n\times n}.
$$

Consequently, according to (3.3.18), $\boldsymbol{A}^{(1)} = \boldsymbol{H}_n \boldsymbol{A}^{(0)}$ has the form

$$\boldsymbol{A}^{(1)} = \begin{pmatrix} a_{11}^{(1)} & a_{12}^{(1)} & \cdots & a_{1n}^{(1)} \\ 0 & a_{22}^{(1)} & \cdots & a_{2n}^{(1)} \\ \vdots & \vdots & & \vdots \\ 0 & a_{n2}^{(1)} & \cdots & a_{nn}^{(1)} \end{pmatrix} .$$

We will apply this procedure to further columns, each time considering smaller lower right portions of the matrix. The goal is to eliminate the subdiagonal elements. After k transformation steps, if the matrix

$$\boldsymbol{A}^{(k)} = \left(\begin{array}{c|ccc} \boldsymbol{R}^{(k)} & & \boldsymbol{B}^{(k)} & \\ \hline & a_{k+1,k+1}^{(k)} & \cdots & a_{k+1,n}^{(k)} \\ \boldsymbol{0} & \vdots & & \vdots \\ & a_{n,k+1}^{(k)} & \cdots & a_{nn}^{(k)} \end{array} \right) \in \mathbb{R}^{n \times n}$$

is given, where $\boldsymbol{R}^{(k)} \in \mathbb{R}^{k \times k}$ is an upper right triangular matrix and $\boldsymbol{B}^{(k)} \in \mathbb{R}^{k \times (n-k)}$, we set $\boldsymbol{a}^{(k)} = (a_{k+1,k+1}^{(k)}, \ldots, a_{n,k+1}^{(k)})^T \in \mathbb{R}^{n-k}$. Analogous to the transition from $\boldsymbol{A}^{(0)}$ to $\boldsymbol{A}^{(1)}$, we construct with

$$\boldsymbol{u}_{n-k} = \frac{\boldsymbol{a}^{(k)} + \mathrm{sgn}(a_{k+1,k+1}^{(k)})\|\boldsymbol{a}^{(k)}\|_2 \boldsymbol{e}_1}{\|\boldsymbol{a}^{(k)} + \mathrm{sgn}(a_{k+1,k+1}^{(k)})\|\boldsymbol{a}^{(k)}\|_2 \boldsymbol{e}_1\|_2}$$

the Householder matrix $\widetilde{\boldsymbol{H}}_{n-k} = \boldsymbol{I}_{n-k} - 2\boldsymbol{u}_{n-k}\boldsymbol{u}_{n-k}^T \in \mathbb{R}^{(n-k) \times (n-k)}$. According to Lemma 3.20,

$$\boldsymbol{H}_{n-k} = \begin{pmatrix} \boldsymbol{I}_k & \boldsymbol{0} \\ \boldsymbol{0} & \widetilde{\boldsymbol{H}}_{n-k} \end{pmatrix} \in \mathbb{R}^{n \times n}$$

is an orthogonal matrix, and for $\boldsymbol{A}^{(k+1)} = \boldsymbol{H}_{n-k}\boldsymbol{A}^{(k)}$ we obtain the representation

$$\boldsymbol{A}^{(k+1)} = \left(\begin{array}{c|cccc} \boldsymbol{R}^{(k)} & & & \boldsymbol{B}^{(k)} & \\ \hline & a_{k+1,k+1}^{(k+1)} & a_{k+1,k+2}^{(k+1)} & \cdots & a_{k+1,n}^{(k+1)} \\ & 0 & a_{k+2,k+2}^{(k+1)} & \cdots & a_{k+2,n}^{(k+1)} \\ \boldsymbol{0} & \vdots & \vdots & & \vdots \\ & 0 & a_{n,k+2}^{(k+1)} & \cdots & a_{nn}^{(k+1)} \end{array} \right)$$

$$= \left(\begin{array}{c|ccc} \boldsymbol{R}^{(k+1)} & & \boldsymbol{B}^{(k+1)} & \\ \hline & a_{k+2,k+2}^{(k+1)} & \cdots & a_{k+2,n}^{(k+1)} \\ \boldsymbol{0} & \vdots & & \vdots \\ & a_{n,k+2}^{(k+1)} & \cdots & a_{nn}^{(k+1)} \end{array} \right) \in \mathbb{R}^{n \times n}$$

with an upper right triangular matrix $\boldsymbol{R}^{(k+1)} \in \mathbb{R}^{(k+1)\times(k+1)}$. After $n-1$ steps[5] we thus obtain the upper right triangular matrix

$$\boldsymbol{R} = \boldsymbol{R}^{(n)} = \boldsymbol{H}_2 \boldsymbol{H}_3 \cdots \boldsymbol{H}_n \boldsymbol{A}. \tag{3.3.21}$$

By Lemma 3.20 we know that all matrices used, $\boldsymbol{H}_2, \ldots, \boldsymbol{H}_n$, are orthogonal and consequently, with $\boldsymbol{Q} = \boldsymbol{H}_n^T \boldsymbol{H}_{n-1}^T \ldots \boldsymbol{H}_2^T$, we also have an orthogonal matrix, which, due to

$$\boldsymbol{A} = (\boldsymbol{H}_2 \boldsymbol{H}_3 \cdots \boldsymbol{H}_n)^T \boldsymbol{R} = \boldsymbol{H}_n^T \boldsymbol{H}_{n-1}^T \ldots \boldsymbol{H}_2^T \boldsymbol{R} = \boldsymbol{Q}\boldsymbol{R}$$

yields the desired QR decomposition.

Example 3.21 Analogous to the previously presented Gram-Schmidt and Givens methods, we want to study the effect of the Householder transformation using the matrix

$$\boldsymbol{A} = \begin{pmatrix} 8 & 2 \\ 6 & 3 \end{pmatrix} = (\boldsymbol{v}\ \boldsymbol{w}) \quad \text{with} \quad \boldsymbol{v} = \begin{pmatrix} 8 \\ 6 \end{pmatrix} \quad \text{and} \quad \boldsymbol{w} = \begin{pmatrix} 2 \\ 3 \end{pmatrix}.$$

First, according to the rule (3.3.20), we determine the vector

$$\boldsymbol{u} = \frac{\boldsymbol{v} + \text{sgn}(v_1)\|\boldsymbol{v}\|_2 \boldsymbol{e}_1}{\|\boldsymbol{v} + \text{sgn}(v_1)\|\boldsymbol{v}\|_2 \boldsymbol{e}_1\|_2} = \frac{\begin{pmatrix} 8 \\ 6 \end{pmatrix} + 10 \begin{pmatrix} 1 \\ 0 \end{pmatrix}}{\left\| \begin{pmatrix} 8 \\ 6 \end{pmatrix} + 10 \begin{pmatrix} 1 \\ 0 \end{pmatrix} \right\|_2} = \frac{1}{\sqrt{10}} \begin{pmatrix} 3 \\ 1 \end{pmatrix}.$$

Consequently, we obtain the Householder matrix

$$\boldsymbol{H} = \boldsymbol{I} - 2\boldsymbol{u}\boldsymbol{u}^T = \boldsymbol{I} - \frac{1}{5} \begin{pmatrix} 3 \\ 1 \end{pmatrix} (3\ 1) = \begin{pmatrix} 1 & 0 \\ 0 & 1 \end{pmatrix} - \frac{1}{5} \begin{pmatrix} 9 & 3 \\ 3 & 1 \end{pmatrix} = \frac{1}{5} \begin{pmatrix} -4 & -3 \\ -3 & 4 \end{pmatrix}$$

and hence

$$\boldsymbol{R} = \boldsymbol{H}\boldsymbol{A} = \frac{1}{5} \begin{pmatrix} -4 & -3 \\ -3 & 4 \end{pmatrix} \begin{pmatrix} 8 & 2 \\ 6 & 3 \end{pmatrix} = \begin{pmatrix} -10 & -\frac{17}{5} \\ 0 & \frac{6}{5} \end{pmatrix}.$$

The QR decomposition thus results in the form

$$\begin{pmatrix} 8 & 2 \\ 6 & 3 \end{pmatrix} = \boldsymbol{A} = \boldsymbol{H}^T \boldsymbol{R} = \underbrace{\frac{1}{5} \begin{pmatrix} -4 & -3 \\ -3 & 4 \end{pmatrix}}_{=\boldsymbol{Q}} \underbrace{\begin{pmatrix} -10 & -\frac{17}{5} \\ 0 & \frac{6}{5} \end{pmatrix}}_{=\boldsymbol{R}}.$$

Since the Householder method is based on orthogonal reflection matrices and the Givens method, on the other hand, is based on orthogonal rotation matrices, different QR decompositions must be computed in the examples 3.18 and 3.21. However, the respective matrices $\boldsymbol{Q}$ and $\boldsymbol{R}$ can be transformed into each other according to Theorem 3.16 by means of the diagonal matrix

$$\boldsymbol{D} = \begin{pmatrix} -1 & 0 \\ 0 & 1 \end{pmatrix}$$

---

[5] Consider why $n-1$ steps are already sufficient.

as in (3.3.9).

In summary, the QR decomposition using Householder transformations can be represented in the following form, where we augment the matrix A by the right-hand side b according to $a_{n+1} = b$, thus enabling a direct solution of the system of equations $Ax = b$. The matrix Q is given implicitly via the vectors u_{n-k+1}, $k = 1,\ldots,n$, which form the lower left triangular part of the resulting matrix. However, since the vectors u_{n-k+1} always have dimension $n - k + 1$, this leads to a double occupancy of the diagonal in combination with the upper right triangular matrix R. Therefore, the diagonal elements of the matrix R are stored in the vector d. For a better understanding of the algorithmic representation, some algebraic relationships should be pointed out. Let $a = (a_{kk},\ldots,a_{kn})^T \in \mathbb{R}^{n-k+1}$, then for the auxiliary quantities d_k and β occurring in the procedure, the representations $d_k = -\mathrm{sgn}(a_{kk})\|a\|_2$ and $\beta \overset{(3.3.19)}{=} 1/(\|a + \mathrm{sgn}(a_{kk})\|a\|_2 e_1\|_2^2)$ hold.

If, as in the previous methods, we neglect the right-hand side and the explicit solution of the system of equations by backward elimination, the following computational effort results:

\# multiplications + \# divisions + \# square roots

$$= \underbrace{\sum_{k=1}^{n}\{2(n-k+1)+(n-k)(2(n-k+1)+1)\}}_{\text{multiplications}} + \underbrace{\sum_{k=1}^{n}\{1+(n-k+1)\}}_{\text{divisions}}$$

$$+ \underbrace{\sum_{k=1}^{n}\{1+(n-k+1)\}}_{\text{square roots}}$$

$$= \frac{2}{3}n^3 + 2n^2 - \frac{16}{3}n.$$

For a large dimension n of the system of equations, in the case of a full matrix, the Householder method requires about half the number of arithmetic operations as the Givens method and about 2/3 of the operations of the Gram-Schmidt method. However, it should be noted again in this comparison that the Givens algorithm often has the lowest computational effort of all three methods for matrices with special structure.

Algorithm Householder Method —

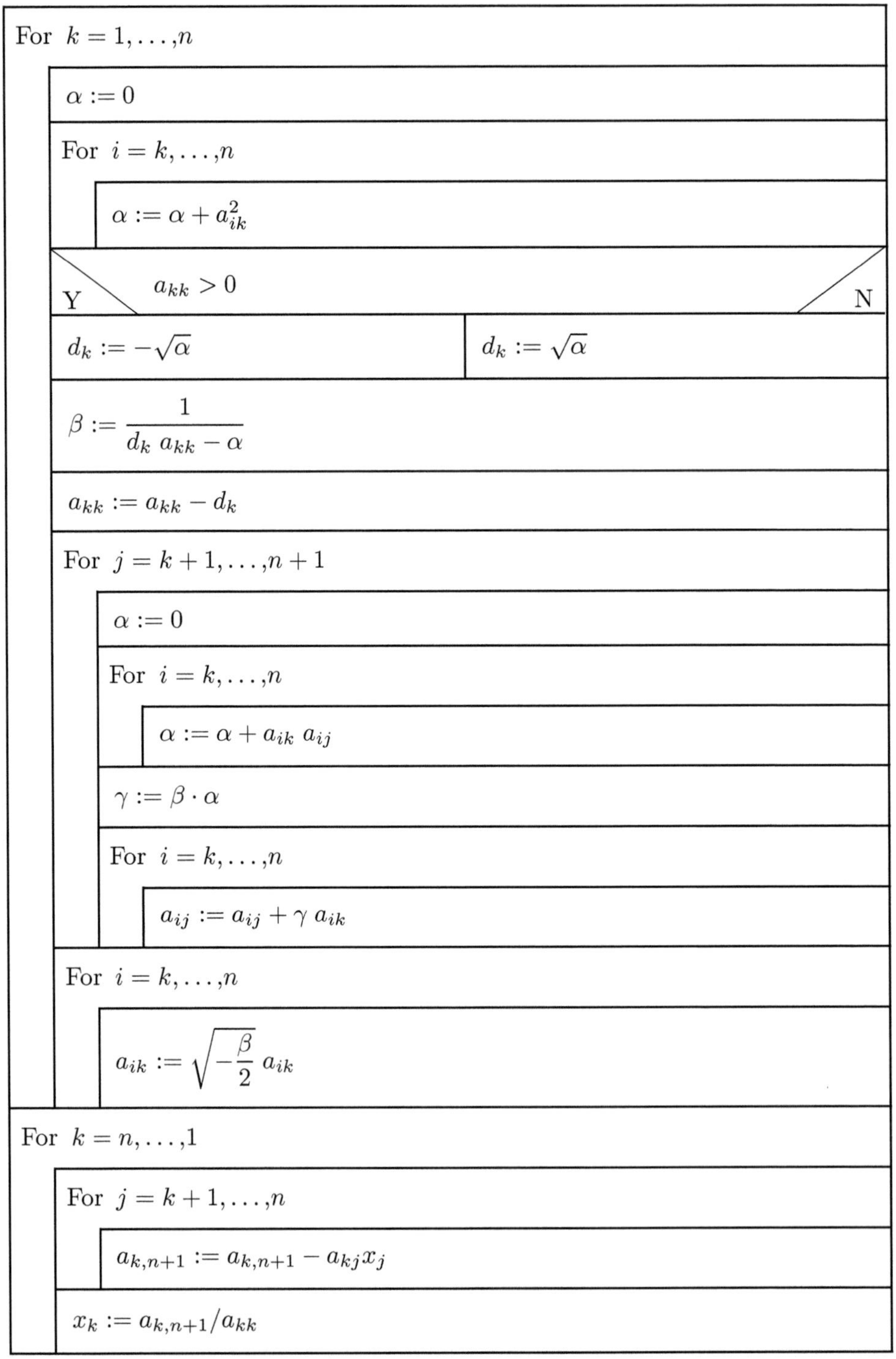

For $k = 1, \ldots, n$

> $\alpha := 0$
>
> For $i = k, \ldots, n$
>
> > $\alpha := \alpha + a_{ik}^2$
>
> $a_{kk} > 0$ Y / N
>
> | Y | N |
> |---|---|
> | $d_k := -\sqrt{\alpha}$ | $d_k := \sqrt{\alpha}$ |
>
> $\beta := \dfrac{1}{d_k\, a_{kk} - \alpha}$
>
> $a_{kk} := a_{kk} - d_k$
>
> For $j = k + 1, \ldots, n + 1$
>
> > $\alpha := 0$
> >
> > For $i = k, \ldots, n$
> >
> > > $\alpha := \alpha + a_{ik}\, a_{ij}$
> >
> > $\gamma := \beta \cdot \alpha$
> >
> > For $i = k, \ldots, n$
> >
> > > $a_{ij} := a_{ij} + \gamma\, a_{ik}$
>
> For $i = k, \ldots, n$
>
> > $a_{ik} := \sqrt{-\dfrac{\beta}{2}}\, a_{ik}$

For $k = n, \ldots, 1$

> For $j = k + 1, \ldots, n$
>
> > $a_{k,n+1} := a_{k,n+1} - a_{kj} x_j$
>
> $x_k := a_{k,n+1} / a_{kk}$

3.4   Exercises

Problem 1:
Compute the LR decomposition of the matrix

$$A = \begin{pmatrix} 1 & 4 & 5 \\ 1 & 6 & 11 \\ 2 & 14 & 31 \end{pmatrix}$$

and use it to solve the linear system $A\boldsymbol{x} = (17, 31, 82)^T$.

Problem 2:
Given is the linear system $A\,\boldsymbol{x} = \boldsymbol{b}$ with

$$A = \begin{pmatrix} 4 & \alpha \\ 6 & 15 \end{pmatrix}, \qquad b = \begin{pmatrix} 8 \\ \beta \end{pmatrix}.$$

(a) For which values of α and β does this system

  (1) have a unique solution,
  (2) have no solution,
  (3) have infinitely many solutions?

(b) In the case that $A\boldsymbol{x} = \boldsymbol{b}$ has a unique solution, formulate the solution in terms of the parameters α, β.

Problem 3:
Prove that for every symmetric, positive definite matrix $A \in \mathbb{R}^{n \times n}$ there exists exactly one lower triangular matrix $L \in \mathbb{R}^{n \times n}$ with positive diagonal elements such that $A = LL^T$.

Problem 4:
Compute the QR decomposition of the matrix

$$A = \begin{pmatrix} 1 & 3 \\ -1 & 1 \end{pmatrix}$$

and use it to solve the linear system $A\boldsymbol{x} = (16, 0)^T$.

Problem 5:
Given is the matrix

$$A = \begin{pmatrix} 1 & 3 & 4 & 1 \\ 2 & 7 & a & 4 \\ 1 & 4 & 6 & 1 \\ 3 & 4 & 9 & 0 \end{pmatrix}.$$

(a) For which value of a does A not have an LR decomposition?

(b) Compute an LR decomposition of A in the case it exists.

(c) In the case that A does not have an LR decomposition, give a permutation matrix P such that PA has an LR decomposition.

Problem 6:

Let $\boldsymbol{A} \in \mathbb{R}^{n \times n}$ be a non-singular matrix with bandwidth d, that is, $\boldsymbol{A} = (a_{ij})_{i,j=1,\ldots,n}$ with $a_{ij} = 0$ for $|i-j| \geq d$. Furthermore, all leading principal minors of $\boldsymbol{A}$ are nonzero.

Show: In the LR decomposition of $\boldsymbol{A}$, the band structure is preserved, i.e., $\boldsymbol{L}$ and $\boldsymbol{R}$ have bandwidth d.

Problem 7:

Given is the tridiagonal matrix

$$
\boldsymbol{A}_n = \begin{pmatrix} a_1 & c_2 & & \\ b_1 & a_2 & \ddots & \\ & \ddots & \ddots & c_n \\ & & b_{n-1} & a_n \end{pmatrix}, \text{ as well as the numbers } c_1 := 0, \quad b_n := 0.
$$

Show that under the assumptions

$$
|b_j| + |c_j| \leq |a_j| \quad \text{and} \quad |b_j| < |a_j|, \quad j = 1,\ldots,n
$$

an LR decomposition of $\boldsymbol{A}_n$ of the form

$$
\boldsymbol{L}_n = \begin{pmatrix} 1 & & & \\ \alpha_1 & \ddots & & \\ & \ddots & \ddots & \\ & & \alpha_{n-1} & 1 \end{pmatrix} \quad \text{and} \quad \boldsymbol{R}_n = \begin{pmatrix} \beta_1 & c_2 & & \\ & \ddots & \ddots & \\ & & \ddots & c_n \\ & & & \beta_n \end{pmatrix}
$$

exists. Give a recursion formula for the computation of α_j and β_j.

Hint: Show by induction: $|\alpha_j| < 1$, $j = 1,\ldots,n-1$ and $|\beta_j| > 0$, $j = 1,\ldots,n$.

Problem 8:

Let $\boldsymbol{A} \in \mathbb{C}^{n \times n}$ be invertible and suppose $\boldsymbol{A} = \boldsymbol{Q}\boldsymbol{R}$, where $\boldsymbol{Q} \in \mathbb{C}^{n \times n}$ is unitary and $\boldsymbol{R} \in \mathbb{C}^{n \times n}$ is an upper right triangular matrix. Let $\boldsymbol{a}_j \in \mathbb{C}^n$ and $\boldsymbol{q}_j \in \mathbb{C}^n$ denote the j-th column vector of $\boldsymbol{A}$ and $\boldsymbol{Q}$, respectively, $j = 1,\ldots,n$. Show:

$$
\text{span}\{\boldsymbol{a}_1,\ldots,\boldsymbol{a}_j\} = \text{span}\{\boldsymbol{q}_1,\ldots,\boldsymbol{q}_j\} \quad \text{for all} \quad j = 1,\ldots,n.
$$

Problem 9:

Compute a Cholesky decomposition of the matrix

$$
\boldsymbol{A} = \begin{pmatrix} 9 & 3 & 9 \\ 3 & 9 & 11 \\ 9 & 11 & 17 \end{pmatrix}
$$

and use it to solve the linear system

$$
\boldsymbol{A}\boldsymbol{x} = \begin{pmatrix} 24 \\ 16 \\ 32 \end{pmatrix}.
$$

Problem 10:

(a) Given the matrix

$$A = \left(\begin{array}{cc} A_{11} & A_{12} \\ 0 & A_{22} \end{array} \right) \in \mathbb{R}^{n \times n}$$

with $A_{ii} \in \mathbb{R}^{m_i \times m_i}$, $i = 1,2$, where $m_1 + m_2 = n$ and $\det A_{ii} \neq 0$, $i = 1,2$ hold. Show that A is invertible and determine a block representation of the inverse A^{-1}.

(b) Compute the inverse of the matrix

$$B = \left(\begin{array}{ccccc} 1 & 2 & 3 & 1 & 0 \\ 2 & 5 & 7 & 1 & 1 \\ 0 & 1 & 2 & 1 & 1 \\ 0 & 0 & 0 & 1 & 0 \\ 0 & 0 & 0 & 0 & 1 \end{array} \right)$$

by using a suitable block structure.

Problem 11:

Let $A_{n-1} \in \mathbb{R}^{(n-1) \times (n-1)}$ be a symmetric matrix and $L_{n-1} \in \mathbb{R}^{(n-1) \times (n-1)}$ a lower triangular matrix with positive diagonal elements, such that $L_{n-1} L_{n-1}^T = A_{n-1}$. Furthermore, let $b \in \mathbb{R}^{n-1}$ be a column vector, $\alpha \in \mathbb{R}$, and

$$A_n := \left(\begin{array}{cc} A_{n-1} & b \\ b^T & \alpha \end{array} \right) \in \mathbb{R}^{n \times n}$$

be symmetric and positive definite. Show that there exists a vector $c \in \mathbb{R}^{n-1}$ and a positive real number β such that

$$A_n = \left(\begin{array}{cc} L_{n-1} & 0 \\ c^T & \beta \end{array} \right) \left(\begin{array}{cc} L_{n-1}^T & c \\ 0 & \beta \end{array} \right)$$

holds.

Problem 12:

Given the matrix

$$A = \left(\begin{array}{ccc} 1 & 4 & 7 \\ 2 & \alpha & \beta \\ 0 & 1 & 1 \end{array} \right).$$

Under which conditions on the values $\alpha, \beta \in \mathbb{R}$ is the matrix non-singular (invertible) and also possesses an LU decomposition? Also, give a pair of parameters (α, β) such that the matrix A is non-singular but does not have an LU decomposition.

Problem 13:

Show that the matrices H_n defined in Lemma 3.20 are not only orthogonal as previously shown, but also symmetric and have determinant -1, and thus represent a reflection.

4   Iterative Methods

The direct methods presented often provide an efficient approach when the number of unknowns is small. However, practical problems (see Examples 1.1 and 1.4) frequently lead to large, sparse systems of equations. The storage of such systems is usually only made possible by neglecting the zero elements of the matrix, which sometimes account for more than 99% of the matrix coefficients. Considering the discretization of the convection-diffusion equation explained earlier with $N = 500$, and assuming a memory requirement of 8 bytes for each real number, we need about 10 megabytes to store the nonzero matrix coefficients. In contrast, storing the entire matrix would require 500 gigabytes.

Direct methods generally cannot exploit the special structure of the systems of equations, resulting in the generation of dense intermediate matrices, which on the one hand exceed the available memory and on the other hand lead to unacceptable computation times (see table on page 50). Furthermore, such systems of equations usually arise from a discretization of the underlying problem, so that even the exact solution of the system of equations only represents an approximation to the desired solution. Consequently, an approximate solution to the system of equations with an error on the order of the discretization error is sufficient. Iterative methods are particularly well suited for this purpose.

We consider a linear system of equations of the form

$$\boldsymbol{Ax} = \boldsymbol{b} \tag{4.0.1}$$

with given right-hand side $\boldsymbol{b} \in \mathbb{C}^n$ and a non-singular matrix $\boldsymbol{A} \in \mathbb{C}^{n \times n}$. Iterative methods successively compute approximations $\boldsymbol{x}_m$ to the exact solution $\boldsymbol{A}^{-1}\boldsymbol{b}$ by repeatedly applying a fixed computational rule

$$\boldsymbol{x}_{m+1} = \phi(\boldsymbol{x}_m, \boldsymbol{b}) \text{ for } m = 0, 1, \ldots$$

with a chosen initial vector $\boldsymbol{x}_0 \in \mathbb{C}^n$.

Before we address specific numerical methods, we will first describe in this section some useful properties of iterative methods.

Definition 4.1 An iterative method is given by a mapping

$$\phi : \mathbb{C}^n \times \mathbb{C}^n \to \mathbb{C}^n$$

and is called linear if there exist matrices $\boldsymbol{M}, \boldsymbol{N} \in \mathbb{C}^{n \times n}$ such that

$$\phi(\boldsymbol{x}, \boldsymbol{b}) = \boldsymbol{Mx} + \boldsymbol{Nb}$$

holds. The matrix $\boldsymbol{M}$ is called the iteration matrix of the iteration ϕ.

The matrices $\boldsymbol{M}$ and $\boldsymbol{N}$ are uniquely determined by the iteration rule.

Supplementary Information The online version contains supplementary material available at https://doi.org/10.1007/978-3-658-50260-7_4.

A. Meister, *Numerical Methods for Linear Systems of Equations*,
Mathematics Study Resources 26, https://doi.org/10.1007/978-3-658-50260-7_4

Definition 4.2 A vector $\tilde{x} \in \mathbb{C}^n$ is called a fixed point of the iterative method ϕ : $\mathbb{C}^n \times \mathbb{C}^n \to \mathbb{C}^n$ for $b \in \mathbb{C}^n$ if

$$\tilde{x} = \phi(\tilde{x}, b)$$

holds.

Definition 4.3 An iterative method ϕ is called consistent with the matrix A if for all $b \in \mathbb{C}^n$ the solution $A^{-1}b$ is a fixed point of ϕ for b. An iterative method ϕ is called convergent if for all $b \in \mathbb{C}^n$ and all initial values $x_0 \in \mathbb{C}^n$ there exists a limit

$$\hat{x} = \lim_{m \to \infty} x_m = \lim_{m \to \infty} \phi(x_{m-1}, b)$$

that is independent of the initial value.

Consistency is a necessary condition for any iterative method, as it ensures a meaningful connection between the numerical method and the system of equations. For an inconsistent algorithm, the user would have to stop the iteration at an appropriate moment, since the exact solution is not a stationary point of the iteration and the sequence of approximate solutions would necessarily move away from the solution vector again after reaching it. For a linear iterative method, the consistency of the method can be determined directly from the matrices M and N used. This essential property is established by the following theorem.

Theorem 4.4 *A linear iterative method is consistent with the matrix A if and only if*

$$M = I - NA$$

holds.

Proof:
Let $\tilde{x} = A^{-1}b$.

" $\Rightarrow$ " ϕ is consistent with the matrix A.

Thus, we obtain

$$\tilde{x} = \phi(\tilde{x}, b) = M\tilde{x} + Nb = M\tilde{x} + NA\tilde{x}. = (M + NA)\tilde{x}.$$

Since consistency holds for all $b \in \mathbb{C}^n$, taking into account the invertibility of the matrix A, the validity of the above equation follows for all $\tilde{x} \in \mathbb{C}^n$, which yields

$$M = I - NA$$

as a result.

" $\Leftarrow$ " Assume $M = I - NA$.

Then we have

$$\tilde{x} = M\tilde{x} + NA\tilde{x} = M\tilde{x} + Nb = \phi(\tilde{x}, b),$$

which shows that the consistency of the iteration method ϕ with respect to the matrix A follows. $\qquad\square$

The choice $M = I$ and $N = 0$ already clearly shows that the requirement of consistency of the method is a necessary, but not a sufficient condition for defining practical linear iteration methods. We therefore need an additional requirement to ensure the convergence of the method to the desired solution.

Theorem 4.5 *A linear iteration method ϕ is convergent if and only if the spectral radius of the iteration matrix M satisfies the condition*

$$\rho(M) < 1$$

Proof:
" $\Rightarrow$ " Let ϕ be convergent.

Let λ be an eigenvalue of M with $|\lambda| = \rho(M)$ and $x \in \mathbb{C}^n \setminus \{0\}$ the corresponding eigenvector. If we choose $b = 0 \in \mathbb{C}^n$, then for $x_0 = cx$ with arbitrary $c \in \mathbb{R} \setminus \{0\}$, the iteration sequence is

$$x_m = \phi(x_{m-1}, b) = M x_{m-1} = \ldots = M^m x_0 = \lambda^m x_0.$$

In the case $|\lambda| > 1$, it follows from $\|x_m\| = |\lambda|^m \|x_0\|$ that the sequence $\{x_m\}_{m \in \mathbb{N}}$ diverges.

For $|\lambda| = 1$, M represents a rotation for the eigenvector. The sequence $\{x_m\}_{m \in \mathbb{N}}$ therefore only converges in the case $\lambda = 1$. In this case, we have $x_m = x_0$ for all $m \in \mathbb{N}$ regardless of the chosen scaling parameter c, so that the limit

$$\hat{x} = \lim_{m \to \infty} x_m = x_0$$

is dependent on the initial vector x_0, and therefore the iteration method is not convergent.

The condition $|\lambda| < 1$ and thus $\rho(M) < 1$ therefore represents a necessary criterion for the convergence of the iteration method.

" $\Leftarrow$ " Assume $\rho(M) < 1$.

Since, according to Theorem 2.11, all norms on $\mathbb{C}^n$ are equivalent, the convergence can be shown in any norm.

Let $\varepsilon := \frac{1}{2}(1 - \rho(M)) > 0$, then by Theorem 2.37 there exists a norm on $\mathbb{C}^{n \times n}$ such that

$$q := \|M\| \leq \rho(M) + \varepsilon < 1$$

holds. For given $b \in \mathbb{C}^n$ we define

$$\begin{array}{rcl} F: \ \mathbb{C}^n & \to & \mathbb{C}^n \\ x & \overset{F}{\mapsto} & F(x) = Mx + Nb. \end{array}$$

With this we obtain

$$\|F(x) - F(y)\| = \|Mx - My\| \leq \|M\| \|x - y\| = q\|x - y\|,$$

so that, due to the Banach fixed point theorem, the sequence defined by

$$x_{m+1} = F(x_m)$$

$\{x_m\}_{m \in \mathbb{N}}$ for any starting element $x_0 \in \mathbb{C}^n$ converges to the uniquely determined fixed point

$$\hat{x} = \lim_{m \to \infty} x_{m+1} = \lim_{m \to \infty} F(x_m) = \lim_{m \to \infty} \phi(x_m, b)$$

and thus ϕ is a convergent iteration method. $\square$

Of course, as with consistency, matrices M and N can also be specified that generate a convergent linear iteration method without being in a meaningful relationship to the system of equations (e.g., $M = 0$, $N = I$). Only the combination of consistency and convergence yields a suitable iteration scheme.

Theorem 4.6 *Let ϕ be a linear iterative method that is consistent with the matrix A and convergent, then the limit element $\widetilde{x}$ of the sequence*

$$x_m = \phi(x_{m-1}, b) \text{ for } m = 1, 2, \ldots$$

satisfies the system of equations (4.0.1) for every initial guess $x_0 \in \mathbb{C}^n$ given.

Proof:
By Theorem 4.5, the sequence $x_m = \phi(x_{m-1}, b)$ converges to the uniquely determined fixed point

$$\widetilde{x} = \phi(\widetilde{x}, b),$$

which, due to the consistency of the iterative method, represents the solution of the equation $Ax = b$. $\square$

4.1   Splitting Methods

Splitting methods for solving the system of equations (4.0.1) are based on a decomposition of the matrix A in the form

$$A = B + (A - B), \quad B \in \mathbb{C}^{n \times n}, \tag{4.1.1}$$

so that from

$$Ax = b$$

we obtain the equivalent system

$$Bx = (B - A)x + b.$$

If B is also invertible, then we have

$$x = B^{-1}(B - A)x + B^{-1}b$$

and thus define the linear iterative method

$$x_{m+1} = \phi(x_m, b) = Mx_m + Nb \text{ for } m = 0, 1, \ldots$$

with

$$M := B^{-1}(B - A)$$

and

$$N := B^{-1}.$$

Before we analyze splitting methods with respect to their consistency and convergence behavior, we will introduce the concept of the symmetric splitting method with the following definition. Such methods prove to be advantageous in the preconditioning of symmetric, positive definite matrices described in Chapter 5, since with the help of the iteration matrix of symmetric splitting methods, these properties of the matrix can be preserved even after the equivalent transformation. Consequently, methods for positive definite, symmetric matrices, such as the conjugate gradient method derived in Section 4.3.1.3, can also be applied after preconditioning.

Definition 4.7 The splitting method

$$\boldsymbol{x}_{m+1} = \boldsymbol{B}^{-1}(\boldsymbol{B} - \boldsymbol{A})\boldsymbol{x}_m + \boldsymbol{B}^{-1}\boldsymbol{b}$$

for solving equation (4.0.1) is called symmetric if for every positive definite and symmetric matrix $\boldsymbol{A}$, the matrix $\boldsymbol{B}$ is also positive definite and symmetric.

Theorem 4.8 *Let $\boldsymbol{B} \in \mathbb{C}^{n \times n}$ be non-singular, then the linear iterative method*

$$\boldsymbol{x}_{m+1} = \phi(\boldsymbol{x}_m, \boldsymbol{b}) = \boldsymbol{B}^{-1}(\boldsymbol{B} - \boldsymbol{A})\boldsymbol{x}_m + \boldsymbol{B}^{-1}\boldsymbol{b}$$

is consistent with the matrix $\boldsymbol{A}$.

Proof:
With $\boldsymbol{M} = \boldsymbol{B}^{-1}(\boldsymbol{B} - \boldsymbol{A}) = \boldsymbol{I} - \boldsymbol{B}^{-1}\boldsymbol{A} = \boldsymbol{I} - \boldsymbol{N}\boldsymbol{A}$, the claim follows by application of Theorem 4.4. $\qquad\qquad\qquad\qquad\qquad\qquad\qquad\qquad\qquad\qquad\qquad\qquad\quad\square$

If $\rho(\boldsymbol{M}) < 1$ holds for a splitting method, then the uniquely determined limit element of the sequence

$$\boldsymbol{x}_m = \phi(\boldsymbol{x}_{m-1}, \boldsymbol{b}) \text{ for } m = 1, 2, \dots$$

for any initial vector $\boldsymbol{x}_0 \in \mathbb{C}^n$ is the solution of the corresponding equation $\boldsymbol{A}\boldsymbol{x} = \boldsymbol{b}$. Let us consider a linear iterative method

$$\boldsymbol{x}_m = \boldsymbol{M}\boldsymbol{x}_{m-1} + \boldsymbol{N}\boldsymbol{b} \text{ for } m = 1, 2, \dots$$

with $\rho(\boldsymbol{M}) < 1$, then according to Theorem 2.37, for every ε with $0 < \varepsilon < 1 - \rho(\boldsymbol{M})$ there exists a norm such that

$$\rho(\boldsymbol{M}) \leq \underbrace{\|\boldsymbol{M}\|}_{q:=} \leq \rho(\boldsymbol{M}) + \varepsilon < 1$$

holds. From the Banach fixed point theorem, the a priori error estimate

$$\|\boldsymbol{x}_m - \boldsymbol{A}^{-1}\boldsymbol{b}\| \leq \frac{q^m}{1 - q}\|\boldsymbol{x}_1 - \boldsymbol{x}_0\| \text{ for } m = 1, 2, \dots.$$

follows. For any other norm $\|\cdot\|_a$ we have

$$\|\boldsymbol{x}_m - \boldsymbol{A}^{-1}\boldsymbol{b}\|_a \leq \frac{q^m}{1 - q}C_a\|\boldsymbol{x}_1 - \boldsymbol{x}_0\|_a$$

with a constant $C_a > 0$ that depends only on the norms. Thus, the spectral radius is a measure of the convergence rate in any norm.

Theorem 4.9 *Let ϕ be a linear iterative method consistent with the matrix $\boldsymbol{A}$, for whose associated iteration matrix $\boldsymbol{M}$ there exists a norm such that $q := \|\boldsymbol{M}\| < 1$, then for given $\varepsilon > 0$ the estimate*

$$\|\boldsymbol{x}_m - \boldsymbol{A}^{-1}\boldsymbol{b}\| \le \varepsilon$$

holds for all $m \in \mathbb{N}$ with

$$m \ge \frac{\ln \dfrac{\varepsilon(1-q)}{\|\boldsymbol{x}_1 - \boldsymbol{x}_0\|}}{\ln q}$$

and $\boldsymbol{x}_1 = \phi(\boldsymbol{x}_0,\boldsymbol{b}) \ne \boldsymbol{x}_0$.

Proof:
With $\|\phi(\boldsymbol{x},\boldsymbol{b}) - \phi(\boldsymbol{y},\boldsymbol{b})\| \le q\|\boldsymbol{x} - \boldsymbol{y}\|$, the inequality

$$\|\boldsymbol{x}_m - \boldsymbol{A}^{-1}\boldsymbol{b}\| \le \frac{q^m}{1-q}\|\boldsymbol{x}_1 - \boldsymbol{x}_0\|$$

follows from the a priori error estimate of Banach's fixed point theorem. For a given $\varepsilon > 0$, using $\boldsymbol{x}_1 \ne \boldsymbol{x}_0$, we obtain for

$$m \ge \frac{\ln \dfrac{\varepsilon(1-q)}{\|\boldsymbol{x}_1 - \boldsymbol{x}_0\|}}{\ln q}.$$

thus the estimate

$$\|\boldsymbol{x}_m - \boldsymbol{A}^{-1}\boldsymbol{b}\| \le \frac{q^m}{1-q}\|\boldsymbol{x}_1 - \boldsymbol{x}_0\| \le \frac{\dfrac{\varepsilon(1-q)}{\|\boldsymbol{x}_1 - \boldsymbol{x}_0\|}}{1-q}\|\boldsymbol{x}_1 - \boldsymbol{x}_0\| = \varepsilon.$$

$\square$

If $0 < q < 1$ and $\tilde{q} = q^2$, then it follows that

$$\frac{\tilde{q}^m}{1-\tilde{q}} = \frac{q^{2m}}{1-q^2} < \frac{q^{2m}}{1-q}.$$

If we consider two convergent linear iterative methods ϕ_1 and ϕ_2 whose associated iteration matrices $\boldsymbol{M}_1$ and $\boldsymbol{M}_2$ satisfy the property

$$\rho(\boldsymbol{M}_1) = \rho(\boldsymbol{M}_2)^2$$

then, according to Theorem 4.9, a guaranteed accuracy statement for the method ϕ_1 is usually achieved after half the number of iterations required for the method ϕ_2. Within an iterative method of this class, one can therefore expect a halving of the required number of iterations if the spectral radius, for example, is reduced from 0.9 to 0.81.

Example 4.10 Trivial method

We consider the model problem

$$\underbrace{\begin{pmatrix} 0.7 & -0.4 \\ -0.2 & 0.5 \end{pmatrix}}_{A:=} \underbrace{\begin{pmatrix} x_1 \\ x_2 \end{pmatrix}}_{x:=} = \underbrace{\begin{pmatrix} 0.3 \\ 0.3 \end{pmatrix}}_{b:=}. \tag{4.1.2}$$

The exact solution is $A^{-1}b = (1,1)^T$. Of course, for this system of equations there is no need to use an iterative method. However, this example is very well suited to illustrate the efficiency of the individual splitting methods. With $A = I - (I - A)$, it follows that there is an equivalence between $Ax = b$ and

$$x = (I - A)x + b.$$

This yields the simple, consistent, and linear iterative method

$$x_{m+1} = \phi(x_m,b) = \underbrace{(I - A)}_{M:=} x_m + \underbrace{I}_{N:=} b. \tag{4.1.3}$$

We have $\det(M - \lambda I) = (\lambda - 0.4)^2 - 0.09$, so that the eigenvalues of the iteration matrix are $\lambda_1 = 0.1$ and $\lambda_2 = 0.7$, and thus the method (4.1.3) converges with $\rho(M) = 0.7 < 1$. Let $x_0 = (21, -19)^T$, then we obtain the convergence history listed in the following table.

Trivial method				
m	$x_{m,1}$	$x_{m,2}$	$\varepsilon_m := \|x_m - A^{-1}b\|_\infty$	$\varepsilon_m/\varepsilon_{m-1}$
0	2.100000e+01	-1.900000e+01	2.000000e+01	
10	8.116832e-01	8.116832e-01	1.883168e-01	7.000000e-01
40	9.999958e-01	9.999958e-01	4.244537e-06	7.000000e-01
70	1.000000e-00	1.000000e-00	9.566903e-11	7.000002e-01
96	1.000000e-00	1.000000e-00	8.881784e-15	6.956522e-01

The convergence history shows that such a primitive choice of the iteration matrix generally does not lead to a satisfactory algorithm. The development of effective splitting methods correlates with the targeted selection of the matrix B, which on the one hand should provide a good approximation of the matrix A with the goal $\rho(M) \ll 1$, and on the other hand must either be easily invertible or at least allow the computation of the matrix-vector product $B^{-1}z$ for an arbitrary vector z to be carried out efficiently.

4.1.1 Jacobi Method

The Jacobi method for the iterative solution of the linear system of equations $Ax = b$ with a non-singular matrix $A \in \mathbb{C}^{n\times n}$ requires nonzero diagonal elements $a_{ii} \neq 0$, $i = 1,\ldots,n$ of the matrix A, so that with $D = \mathrm{diag}\{a_{11},\ldots,a_{nn}\}$ a non-singular diagonal matrix is present. Following the basic idea of splitting methods, we write the linear system $Ax = b$ in the equivalent form

$$x = \underbrace{D^{-1}(D - A)}_{M_J:=} x + \underbrace{D^{-1}}_{N_J:=} b.$$

According to Theorem 4.8, the linear iteration method defined in this way

$$\boldsymbol{x}_{m+1} = \boldsymbol{D}^{-1}(\boldsymbol{D} - \boldsymbol{A})\boldsymbol{x}_m + \boldsymbol{D}^{-1}\boldsymbol{b} \text{ for } m = 0,1,2,\dots \qquad (4.1.4)$$

is consistent with the matrix $\boldsymbol{A}$, and we obtain the component-wise form

$$x_{m+1,i} = \frac{1}{a_{ii}} \left(b_i - \sum_{\substack{j=1 \\ j \neq i}}^{n} a_{ij} x_{m,j} \right) \text{ for } i = 1,\dots,n \text{ and } m = 0,1,2,\dots.$$

In the Jacobi method, the new iterate $\boldsymbol{x}_{m+1}$ is thus determined exclusively by the previous iterate $\boldsymbol{x}_m$. For this reason, the method is also called a simultaneous update method and is therefore independent of the chosen ordering of the unknowns $\boldsymbol{x} = (x_1,\dots,x_n)^T$. Since it is a splitting method, the convergence of the Jacobi method depends solely on the spectral radius of the iteration matrix $\boldsymbol{M}_J = \boldsymbol{D}^{-1}(\boldsymbol{D} - \boldsymbol{A})$. According to Theorem 2.36, this value can be estimated by any matrix norm, which allows the convergence of the method to be checked based on the size of the matrix coefficients without computing the eigenvalues.

Theorem 4.11 *If the invertible matrix $\boldsymbol{A} \in \mathbb{C}^{n \times n}$ with $a_{ii} \neq 0$, $i = 1,\dots,n$ satisfies the strong row sum criterion*

$$q_\infty := \max_{i=1,\dots,n} \sum_{\substack{k=1 \\ k \neq i}}^{n} \frac{|a_{ik}|}{|a_{ii}|} < 1$$

or the strong column sum criterion

$$q_1 := \max_{k=1,\dots,n} \sum_{\substack{i=1 \\ i \neq k}}^{n} \frac{|a_{ik}|}{|a_{ii}|} < 1$$

or the square sum criterion

$$q_2 := \sum_{\substack{i,k=1 \\ i \neq k}}^{n} \left(\frac{|a_{ik}|}{|a_{ii}|} \right)^2 < 1,$$

then the Jacobi method converges for any initial vector $\boldsymbol{x}_0 \in \mathbb{C}^n$ and for any right-hand side $\boldsymbol{b} \in \mathbb{C}^n$ to $\boldsymbol{A}^{-1}\boldsymbol{b}$.

Proof:
Because

$$\boldsymbol{M}_J = \boldsymbol{D}^{-1}(\boldsymbol{D} - \boldsymbol{A})$$

it follows that

$$q_\infty = \|\boldsymbol{M}_J\|_\infty, \quad q_1 = \|\boldsymbol{M}_J\|_1 \quad \text{and} \quad 1 > \sqrt{q_2} \geq \|\boldsymbol{M}_J\|_2,$$

which proves the statement by applying Theorem 4.5 on the basis of

$$\rho(\boldsymbol{M}_J) \leq \min\{\|\boldsymbol{M}_J\|_\infty, \|\boldsymbol{M}_J\|_1, \|\boldsymbol{M}_J\|_2\} < 1.$$

$\square$

Remark:
A matrix that satisfies the strong row sum criterion is called strictly diagonally dominant.

In practical applications, matrices often occur that do not satisfy any of the three convergence criteria listed in Theorem 4.11. A well-known example is the matrix derived in Example 1.1 from the discretization of the Poisson equation. Nevertheless, for such systems of equations, the convergence of the Jacobi method can be proven. As we will see, the decisive property in this case is the irreducibility of the matrix.

Definition 4.12 A matrix $\boldsymbol{A} \in \mathbb{C}^{n \times n}$ is called reducible or decomposable if there exists a permutation matrix $\boldsymbol{P} \in \mathbb{R}^{n \times n}$ such that

$$\boldsymbol{P}\boldsymbol{A}\boldsymbol{P}^T = \begin{pmatrix} \tilde{\boldsymbol{A}}_{11} & \tilde{\boldsymbol{A}}_{12} \\ \boldsymbol{0} & \tilde{\boldsymbol{A}}_{22} \end{pmatrix}$$

with $\tilde{\boldsymbol{A}}_{ii} \in \mathbb{C}^{n_i \times n_i}$, $n_i > 0$, $i = 1,2$, $n_1 + n_2 = n$. Otherwise, $\boldsymbol{A}$ is called irreducible or indecomposable.

Example 4.13 Let us consider the matrix

$$\boldsymbol{A} = \begin{pmatrix} 6 & 8 & 5 & 7 \\ 0 & 12 & 0 & 11 \\ 2 & 4 & 1 & 3 \\ 0 & 10 & 0 & 9 \end{pmatrix}.$$

Due to the arrangement of the zero entries, the reducibility of the matrix can already be anticipated. The proof is provided by using the permutation matrix

$$\boldsymbol{P} = \begin{pmatrix} 0 & 0 & 1 & 0 \\ 1 & 0 & 0 & 0 \\ 0 & 0 & 0 & 1 \\ 0 & 1 & 0 & 0 \end{pmatrix}.$$

With this, we obtain

$$\boldsymbol{P}\boldsymbol{A}\boldsymbol{P}^T = \begin{pmatrix} 0 & 0 & 1 & 0 \\ 1 & 0 & 0 & 0 \\ 0 & 0 & 0 & 1 \\ 0 & 1 & 0 & 0 \end{pmatrix} \begin{pmatrix} 6 & 8 & 5 & 7 \\ 0 & 12 & 0 & 11 \\ 2 & 4 & 1 & 3 \\ 0 & 10 & 0 & 9 \end{pmatrix} \begin{pmatrix} 0 & 1 & 0 & 0 \\ 0 & 0 & 0 & 1 \\ 1 & 0 & 0 & 0 \\ 0 & 0 & 1 & 0 \end{pmatrix}$$

$$= \begin{pmatrix} 0 & 0 & 1 & 0 \\ 1 & 0 & 0 & 0 \\ 0 & 0 & 0 & 1 \\ 0 & 1 & 0 & 0 \end{pmatrix} \begin{pmatrix} 5 & 6 & 7 & 8 \\ 0 & 0 & 11 & 12 \\ 1 & 2 & 3 & 4 \\ 0 & 0 & 9 & 10 \end{pmatrix} = \begin{pmatrix} 1 & 2 & 3 & 4 \\ 5 & 6 & 7 & 8 \\ 0 & 0 & 9 & 10 \\ 0 & 0 & 11 & 12 \end{pmatrix},$$

which proves the expected reducibility.

In the case of an irreducible matrix, we will show in the following that even a weakened form of the strong row sum criterion is sufficient for the convergence of the Jacobi method.

Definition 4.14 A matrix $A \in \mathbb{C}^{n \times n}$ is called irreducibly diagonally dominant if it is irreducible, satisfies the inequalities

$$|a_{ii}| \geq \sum_{\substack{j=1 \\ j \neq i}}^{n} |a_{ij}| \tag{4.1.5}$$

for $i = 1, \ldots, n$, and in addition, there exists a $k \in \{1, \ldots, n\}$ with

$$|a_{kk}| > \sum_{\substack{j=1 \\ j \neq k}}^{n} |a_{kj}|. \tag{4.1.6}$$

Remark:
A matrix that satisfies the row sum criterion (4.1.5) is called diagonally dominant.

As we can easily see from the example of the zero matrix A, the property of diagonal dominance is neither sufficient for the invertibility of the matrix nor for ensuring the applicability of the Jacobi method. Even if we additionally require irreducibility, the invertibility of the matrix A does not necessarily follow, as can be seen, for example, from the matrix

$$A = \begin{pmatrix} 1 & 1 \\ 1 & 1 \end{pmatrix}.$$

Only the condition (4.1.6) will provide us with the invertibility of the matrix as well as the applicability and convergence of the Jacobi method.

Theorem 4.15 *An irreducibly diagonally dominant matrix* $A \in \mathbb{C}^{n \times n}$ *is always invertible, and the Jacobi method converges for any initial vector* $x_0 \in \mathbb{C}^n$ *and any right-hand side* $b \in \mathbb{C}^n$ *to* $A^{-1}b$.

The basic idea of the convergence statement lies in a graph-theoretical consideration: Let us denote by $G^A := \{V^A, E^A\}$ the directed graph of the matrix A, which consists of the nodes $V^A := \{1, \ldots, n\}$ and the set of edges of ordered pairs $E^A := \{(i,j) \in V^A \times V^A \mid a_{ij} \neq 0\}$. A matrix A is irreducible if and only if the corresponding directed graph G^A is connected, that is, if for any two indices $i_0 = i, i_\ell = j \in V^A$ there exists a directed path of length $\ell \in \mathbb{N}$ in the form

$$(i_0, i_1)(i_1, i_2) \ldots (i_{\ell-1}, i_\ell)$$

with $(i_k, i_{k+1}) \in E^A$ for $k = 0, \ldots, \ell - 1$. Reducibility thus depends solely on the sparsity structure of the matrix and not on the absolute values of the matrix coefficients. Moreover, the diagonal elements have no influence on reducibility.

If

$$e_m := x_m - A^{-1}b$$

is the error vector of the m-th iterate of the Jacobi method, then from (4.1.5) it follows that $|e_{m+1,i}| \leq \|e_m\|_\infty$ for $i = 1,\ldots,n$, and from (4.1.6) there exists a $k \in \{1,\ldots,n\}$ such that

$$|e_{m+1,k}| \leq \gamma \|e_m\|_\infty \text{ with } \gamma < 1.$$

From this it follows that

$$|e_{m+2,j}| \leq \gamma \|e_m\|_\infty \text{ with } \gamma < 1$$

for all $j \in \{1,\ldots,n\}$ with $(j,k) \in E^{\boldsymbol{A}}$, and due to the connectedness of the graph, we have

$$\|e_{m+n}\|_\infty \leq \gamma \|e_m\|_\infty \tag{4.1.7}$$

with $\gamma < 1$, which establishes the convergence of the Jacobi method. The irreducibility of the matrix thus ensures that property (4.1.6) for diagonally dominant matrices leads, after at most n iterations, to a reduction in the absolute value of all components within the error vector. The number of required iterations is determined by the maximal value of the length of the shortest directed path between rows that satisfy inequality (4.1.6) and rows $i \in \{1,\ldots,n\}$ that satisfy the equation

$$|a_{ii}| = \sum_{\substack{j=1 \\ j \neq i}}^{n} |a_{ij}|$$

If a dense matrix is given that satisfies the conditions of Theorem 4.15, then inequality (4.1.7) holds after at most two iterations.

We now turn to the proof of Theorem 4.15.

Proof:
We first address the invertibility of the matrix $\boldsymbol{A} \in \mathbb{C}^{n\times n}$ and proceed by contradiction by assuming that $\boldsymbol{A}$ is singular. Consequently, there exists a vector $\boldsymbol{x} \neq \boldsymbol{0}$ with $\boldsymbol{A}\boldsymbol{x} = \boldsymbol{0}$, which gives

$$a_{ii}x_i = -\sum_{\substack{j=1 \\ j \neq i}}^{n} a_{ij}x_j$$

and therefore also

$$|a_{ii}||x_i| \leq \sum_{\substack{j=1 \\ j \neq i}}^{n} |a_{ij}||x_j| \tag{4.1.8}$$

for all $i = 1,\ldots,n$. If all components of the vector $\boldsymbol{x}$ had the same absolute value, then with $0 < \|\boldsymbol{x}\|_\infty = |x_1| = \ldots = |x_n|$, from (4.1.8) we would obtain for all $i = 1,\ldots,n$ the inequality

$$|a_{ii}| \leq \sum_{\substack{j=1 \\ j \neq i}}^{n} |a_{ij}| \underbrace{\frac{|x_j|}{|x_i|}}_{=1} = \sum_{\substack{j=1 \\ j \neq i}}^{n} |a_{ij}|.$$

However, this contradicts property (4.1.6). Accordingly, we can partition the index set $\mathcal{M} = \{1,\ldots,n\}$ disjointly into the two nonempty sets

$$\mathcal{I} = \{i \in \mathcal{M} \mid |x_i| = \|\boldsymbol{x}\|_\infty\} \text{ and } \mathcal{J} = \{j \in \mathcal{M} \mid |x_j| < \|\boldsymbol{x}\|_\infty\}.$$

Starting from any $i \in \mathcal{I}$ and $j \in \mathcal{J}$, we always have $i \neq j$, and due to the given irreducibility, there exists a directed path

$$(i,i_1)(i_1,i_2)\ldots(i_\ell,j)$$

from i to j. Thus, the path contains an index pair (r,s) with $r \in \mathcal{I}$ and $s \in \mathcal{J}$. From this, taking into account $|x_r| > 0$, it follows directly from (4.1.8) that

$$|a_{rr}| \leq \sum_{\substack{j=1\\j\neq r}}^{n} |a_{rj}|\frac{|x_j|}{|x_r|} = \underbrace{|a_{rs}|}_{>0}\underbrace{\frac{|x_s|}{|x_r|}}_{<1} + \sum_{\substack{j=1\\j\neq r,s}}^{n} |a_{rj}|\underbrace{\frac{|x_j|}{|x_r|}}_{\leq 1} < \sum_{\substack{j=1\\j\neq r}}^{n} |a_{rj}|.$$

However, this inequality contradicts (4.1.6), and thus the invertibility of the matrix is proven. Since an invertible matrix cannot contain a zero row, it must hold by (4.1.6) that $|a_{ii}| > 0$ for all $i = 1,\ldots,n$, so that the Jacobi method can be applied.

Let us consider the iteration matrix $\boldsymbol{M}_J = (m_{ij})_{i,j=1,\ldots,n}$ of the Jacobi method and define $|\boldsymbol{M}_J| := (|m_{ij}|)_{i,j=1,\ldots,n}$ as well as $\boldsymbol{y} := (1,\ldots,1)^T \in \mathbb{R}^n$. Then, using (4.1.5), one can write

$$0 \leq (|\boldsymbol{M}_J|\boldsymbol{y})_i \leq y_i \text{ for } i = 1,\ldots,n, \tag{4.1.9}$$

and with (4.1.6) there exists a $k \in \{1,\ldots,n\}$ such that

$$0 \leq (|\boldsymbol{M}_J|\boldsymbol{y})_k < y_k. \tag{4.1.10}$$

From (4.1.9) it follows that $(|\boldsymbol{M}_J|^m\boldsymbol{y})_i \leq y_i$ for $i = 1,\ldots,n$ and all $m \in \mathbb{N}$. If there exists a $\tilde{m} \in \mathbb{N}$ such that $(|\boldsymbol{M}_J|^{\tilde{m}}\boldsymbol{y})_j < y_j$, then for all $m \geq \tilde{m}$, using (4.1.5), we obtain the inequality

$$(|\boldsymbol{M}_J|^m\boldsymbol{y})_j < y_j. \tag{4.1.11}$$

With

$$\boldsymbol{t}^m := \boldsymbol{y} - |\boldsymbol{M}_J|^m\boldsymbol{y}$$

and

$$\tau^m := \#\{t_i^m \,|\, t_i^m \neq 0,\, i = 1,\ldots,n\}$$

it follows, using (4.1.10) and (4.1.11), the chain of inequalities

$$0 < \tau^1 \leq \tau^2 \leq \ldots \leq \tau^m \leq \tau^{m+1} \leq \ldots\,.$$

Now, let us assume that there exists an $m \in \{1,\ldots,n-1\}$ with $\tau^m = \tau^{m+1} < n$ and derive a contradiction.

Without loss of generality, suppose the vector $\boldsymbol{t}^m$ has the form

$$\boldsymbol{t}^m = \begin{pmatrix} \boldsymbol{u} \\ \boldsymbol{0} \end{pmatrix}, \quad \boldsymbol{u} \in \mathbb{R}^p \text{ for } 1 \leq p < n, \text{ and } u_i > 0 \text{ for } i = 1,\ldots,p$$

then, using (4.1.11) and $\tau^m = \tau^{m+1}$, it follows that

$$\boldsymbol{t}^{m+1} = \begin{pmatrix} \boldsymbol{v} \\ \boldsymbol{0} \end{pmatrix}, \quad \boldsymbol{v} \in \mathbb{R}^p, \text{ and } v_i > 0 \text{ for } i = 1,\ldots,p.$$

Let

$$|\boldsymbol{M}_J| = \begin{pmatrix} |\boldsymbol{M}_{11}| & |\boldsymbol{M}_{12}| \\ |\boldsymbol{M}_{21}| & |\boldsymbol{M}_{22}| \end{pmatrix}$$

with $|\boldsymbol{M}_{11}| \in \mathbb{R}^{p \times p}$, then using the inequality (4.1.9) we obtain

$$\begin{pmatrix} \boldsymbol{v} \\ \boldsymbol{0} \end{pmatrix} = \boldsymbol{t}^{m+1} = \boldsymbol{y} - |\boldsymbol{M}_J|^{m+1}\boldsymbol{y} \geq |\boldsymbol{M}_J|\boldsymbol{y} - |\boldsymbol{M}_J|^{m+1}\boldsymbol{y} = |\boldsymbol{M}_J|\boldsymbol{t}^m = \begin{pmatrix} |\boldsymbol{M}_{11}|\boldsymbol{u} \\ |\boldsymbol{M}_{21}|\boldsymbol{u} \end{pmatrix}.$$

With $u_i > 0$, $i = 1,\dots,p$, we obtain, due to the non-negativity of the elements of the matrix $|\boldsymbol{M}_{21}|$, the equation

$$|\boldsymbol{M}_{21}| = \boldsymbol{0},$$

which means that $\boldsymbol{M}_J$ is reducible. Since the sparsity patterns of $\boldsymbol{M}_J$ and $\boldsymbol{A}$ are the same except for the diagonal elements, and these have no influence on the reducibility, we obtain a contradiction with respect to the irreducibility of the matrix $\boldsymbol{A}$. Thus, in the case $\tau^m < n$, it follows directly that

$$0 < \tau^1 < \tau^2 < \dots < \tau^m < \tau^{m+1}$$

for $m \in \{1,\dots,n-1\}$, which implies the existence of an $m \in \{1,\dots,n\}$ with

$$0 \leq (|\boldsymbol{M}_J|^m\boldsymbol{y})_i < y_i$$

for $i = 1,\dots,n$. From this, we obtain

$$\rho(\boldsymbol{M}_J)^m \leq \rho(\boldsymbol{M}_J^m) \leq \|\boldsymbol{M}_J^m\|_\infty \leq \||\boldsymbol{M}_J|^m\|_\infty < 1$$

and consequently $\rho(\boldsymbol{M}_J) < 1$. $\square$

Example 4.16 For the system of equations $\boldsymbol{Au} = \boldsymbol{g}$ with the matrix $\boldsymbol{A}$ resulting from the discretization of the Poisson problem (see (1.0.3)), the Jacobi method converges for any $\boldsymbol{g} \in \mathbb{R}^N$ and $\boldsymbol{u}_0 \in \mathbb{R}^N$. We prove this statement by verifying the assumptions of Theorem 4.15. First, $\boldsymbol{A}$ satisfies the weak row sum criterion, and we have

$$\sum_{i=2}^{n} \frac{|a_{1i}|}{|a_{11}|} = \frac{1}{4} + \frac{1}{4} < 1,$$

so it remains to show the irreducibility of the matrix. As a first step, we prove the irreducibility of the submatrices $\boldsymbol{B} \in \mathbb{R}^{N \times N}$ contained in the matrix $\boldsymbol{A} \in \mathbb{R}^{N^2 \times N^2}$. Let $i,j \in V^{\boldsymbol{B}}$ be given. Due to the symmetry of the matrix, we may assume without loss of generality that $i < j$. Since $(\ell,\ell+1) \in E^{\boldsymbol{B}}$, $\ell = 1,\dots,N-1$, we obtain

$$(i,i+1)(i+1,i+2)\dots(i+k-1,j)$$

with $k = j - i$ as a directed path from i to j, which proves the irreducibility of the matrix $\boldsymbol{B}$.

Now we turn to the matrix $\boldsymbol{A}$: Let $i,j \in V^{\boldsymbol{A}}$ be given, where, analogously to the matrix $\boldsymbol{B}$, due to the symmetry of the matrix $\boldsymbol{A}$, we again assume $i < j$ without loss of generality. Then there always exists a unique representation of the indices in the form

$$i = i_1 N + i_2 \quad \text{and} \quad j = j_1 N + j_2$$

with $0 \leq i_1 \leq j_1 \leq N - 1$ and $i_2, j_2 \in \{1, \ldots, N\}$. If $i_1 = j_1$, then, due to the irreducibility of the matrix $\boldsymbol{B}$, there exists a directed path from i to j. In the case $i_1 < j_1$, there first exists a directed path from i to $\tilde{j} = i_1 N + j_2 \leq N^2 - N$. Since $(\ell, \ell + N) \in E^{\boldsymbol{A}}$, for $l = 1, \ldots, N^2 - N$, we obtain

$$(\tilde{j}, \tilde{j} + N)(\tilde{j} + N, \tilde{j} + 2N) \ldots (\tilde{j} + (k-1)N, j)$$

with $k = j_1 - i_1$ as the completion of the directed path from i to j.

A reducible matrix arises, for example, in the discretization of the convection-diffusion equation in the limiting case of a vanishing diffusion parameter. In this case, however, the corresponding system of equations can be solved by simple forward elimination.

Example 4.17 For the model problem (4.1.2) with

$$\boldsymbol{A} = \begin{pmatrix} 0.7 & -0.4 \\ -0.2 & 0.5 \end{pmatrix}, \boldsymbol{b} = \begin{pmatrix} 0.3 \\ 0.3 \end{pmatrix}$$

we have

$$q_\infty := \max_{i=1,2} \sum_{\substack{j=1 \\ j \neq i}}^{2} \frac{|a_{ij}|}{|a_{ii}|} = \max \left\{ \frac{4}{7}, \frac{2}{5} \right\} = \frac{4}{7} < 1,$$

which proves the convergence of the Jacobi method. The eigenvalues of the iteration matrix

$$\boldsymbol{M}_J = \boldsymbol{D}^{-1}(\boldsymbol{D} - \boldsymbol{A}) = \begin{pmatrix} 0 & \frac{4}{7} \\ \frac{2}{5} & 0 \end{pmatrix}$$

are

$$\lambda_{1,2} = \pm \sqrt{\frac{8}{35}},$$

so that

$$\rho(\boldsymbol{M}_J) = \sqrt{\frac{8}{35}} \approx 0.4781$$

follows. Using the initial vector $\boldsymbol{x}_0 = (21, -19)^T$, we obtain the following sequence of approximate solutions, which also, compared to the convergence of the trivial method (see Example 4.10) is confirmed due to $\rho(\boldsymbol{M}_J) \approx (\rho(\boldsymbol{I} - \boldsymbol{A}))^2$, with an expected halving of the number of iterations.

Jacobi Method				
m	$x_{m,1}$	$x_{m,2}$	$\varepsilon_m := \|\boldsymbol{x}_m - \boldsymbol{A}^{-1}\boldsymbol{b}\|_\infty$	$\varepsilon_m / \varepsilon_{m-1}$
0	2.100000e+01	-1.900000e+01	2.000000e+01	
1	-1.042857e+01	9.000000e+00	1.142857e+01	5.714286e-01
2	5.571429e+00	-3.571429e+00	4.571429e+00	4.000000e-01
3	-1.612245e+00	2.828571e+00	2.612245e+00	5.714286e-01
4	2.044898e+00	-4.489796e-02	1.044898e+00	4.000000e-01
5	4.029155e-01	1.417959e+00	5.970845e-01	5.714286e-01
6	1.238834e+00	7.611662e-01	2.388338e-01	4.000000e-01
15	9.996275e-01	1.000261e+00	3.725165e-04	5.714286e-01
30	1.000000e+00	1.000000e-00	4.856900e-09	4.000000e-01
45	1.000000e-00	1.000000e+00	9.037215e-14	5.700280e-01
48	1.000000e+00	1.000000e-00	8.437695e-15	4.086022e-01

From the error reduction in the table above, we can see that the convergence behavior is indeed monotonic, but the same factor between ε_m and ε_{m-1} does not always occur. The convergence of the method is directly related to the spectral radius of the iteration matrix $\boldsymbol{M}_J$, and this also represents, asymptotically, the geometric mean of the error reductions. However, within the table, we are only considering individual iteration steps, so the convergence behavior is directly linked to the vector norm used. Taking into account the consistency of the Jacobi method, the following estimate holds for the maximum norm chosen here:

$$
\begin{aligned}
\varepsilon_m &= \|\boldsymbol{x}_m - \boldsymbol{A}^{-1}\boldsymbol{b}\|_\infty = \|\boldsymbol{M}_J\boldsymbol{x}_{m-1} + \boldsymbol{N}_J\boldsymbol{b} - \left(\boldsymbol{M}_J\boldsymbol{A}^{-1}\boldsymbol{b} + \boldsymbol{N}_J\boldsymbol{b}\right)\|_\infty \\
&= \|\boldsymbol{M}_J\left(\boldsymbol{x}_{m-1} - \boldsymbol{A}^{-1}\boldsymbol{b}\right)\|_\infty \\
&\leq \|\boldsymbol{M}_J\|_\infty\|\left(\boldsymbol{x}_{m-1} - \boldsymbol{A}^{-1}\boldsymbol{b}\right)\|_\infty = \underbrace{\|\boldsymbol{M}_J\|_\infty}_{=q_\infty=\frac{4}{7}}\varepsilon_{m-1} = \frac{4}{7}\,\varepsilon_{m-1}.
\end{aligned}
$$

This proves that the convergence behavior based on the maximum norm is monotonic, but the spectral radius does not have to appear as the reduction factor in every single step. However, from the values of $\varepsilon_m/\varepsilon_{m-1}$, one can see that the geometric mean of the error reduction over two consecutive iteration steps corresponds to the spectral radius of the iteration matrix.

4.1.2 Gauss-Seidel Method

Analogous to the previously introduced Jacobi method, the Gauss-Seidel method also considers a linear system of equations $\boldsymbol{A}\boldsymbol{x} = \boldsymbol{b}$ with a non-singular matrix $\boldsymbol{A} \in \mathbb{C}^{n\times n}$ that has a non-singular diagonal part $\boldsymbol{D} = \mathrm{diag}\{a_{11},\ldots,a_{nn}\}$. We define the strict lower left triangular matrix

$$\boldsymbol{L} = (\ell_{ij})_{i,j=1,\ldots,n} \quad \text{with} \quad \ell_{ij} = \begin{cases} a_{ij}, & i > j \\ 0, & \text{otherwise} \end{cases} \tag{4.1.12}$$

and the strict upper right triangular matrix

$$\boldsymbol{R} = (r_{ij})_{i,j=1,\ldots,n} \quad \text{with} \quad r_{ij} = \begin{cases} a_{ij}, & i < j \\ 0, & \text{otherwise.} \end{cases} \tag{4.1.13}$$

In this way, we obtain the system of equations equivalent to $\boldsymbol{A}\boldsymbol{x} = \boldsymbol{b}$ in the form

$$(\boldsymbol{D} + \boldsymbol{L})\boldsymbol{x} = -\boldsymbol{R}\boldsymbol{x} + \boldsymbol{b} \tag{4.1.14}$$

and

$$\boldsymbol{x} = \underbrace{-(\boldsymbol{D} + \boldsymbol{L})^{-1}\boldsymbol{R}}_{\boldsymbol{M}_{GS}:=}\boldsymbol{x} + \underbrace{(\boldsymbol{D} + \boldsymbol{L})^{-1}}_{\boldsymbol{N}_{GS}:=}\boldsymbol{b}.$$

Thus, we have

$$\boldsymbol{M}_{GS} = (\boldsymbol{D} + \boldsymbol{L})^{-1}(\boldsymbol{D} + \boldsymbol{L} - \boldsymbol{A}) = \boldsymbol{I} - \boldsymbol{N}_{GS}\boldsymbol{A},$$

and the linear iterative method

$$\boldsymbol{x}_{m+1} = -(\boldsymbol{D} + \boldsymbol{L})^{-1}\boldsymbol{R}\boldsymbol{x}_m + (\boldsymbol{D} + \boldsymbol{L})^{-1}\boldsymbol{b} \text{ for } m = 0,1,\ldots \tag{4.1.15}$$

is consistent with the matrix $\boldsymbol{A}$.

To derive the component-wise notation, we consider the i-th row of the iterative method (4.1.15) in the form according to the system of equations (4.1.14)

$$\sum_{j=1}^{i} a_{ij} x_{m+1,j} = - \sum_{j=i+1}^{n} a_{ij} x_{m,j} + b_i.$$

Suppose $x_{m+1,j}$ for $j = 1, \ldots, i-1$ are known, then $x_{m+1,i}$ can be determined by

$$x_{m+1,i} = \frac{1}{a_{ii}} \left(b_i - \sum_{j=1}^{i-1} a_{ij} x_{m+1,j} - \sum_{j=i+1}^{n} a_{ij} x_{m,j} \right) \text{ for } i = 1, \ldots, n \text{ and } m = 0,1,2,\ldots.$$

$$(4.1.16)$$

From this representation, it becomes clear that in the Gauss-Seidel method, to compute the i-th component of the $(m+1)$-th iterate, in addition to the components of the previous m-th iterate $\boldsymbol{x}_m$, the already known first $i-1$ components of the $(m+1)$-th iterate $\boldsymbol{x}_{m+1}$ are used. The method is therefore also called a single-step method and depends on the chosen ordering of the unknowns $(x_1, \ldots, x_n)^T$. Compared to the Jacobi method, the Gauss-Seidel algorithm uses $\boldsymbol{D} + \boldsymbol{L}$, which is usually a better approximation of the matrix $\boldsymbol{A}$ than $\boldsymbol{D}$. Hence, both a smaller spectral radius of the iteration matrix and thus faster convergence can be expected.

Theorem 4.18 *Let the invertible matrix $\boldsymbol{A} \in \mathbb{C}^{n \times n}$ with $a_{ii} \neq 0$ for $i = 1, \ldots, n$ be given. If the numbers $p_1, \ldots, p_n$ defined recursively by*

$$p_i = \sum_{j=1}^{i-1} \frac{|a_{ij}|}{|a_{ii}|} p_j + \sum_{j=i+1}^{n} \frac{|a_{ij}|}{|a_{ii}|} \text{ for } i = 1,2,\ldots,n$$

satisfy the condition

$$p := \max_{i=1,\ldots,n} p_i < 1,$$

then the Gauss-Seidel method converges for any starting vector $\boldsymbol{x_0}$ and for any right-hand side $\boldsymbol{b}$ to $\boldsymbol{A}^{-1}\boldsymbol{b}$.

Proof:
Our goal is to show $\|\boldsymbol{M}_{GS}\|_\infty < 1$. Let $\boldsymbol{x} \in \mathbb{C}^n$ with $\|\boldsymbol{x}\|_\infty = 1$. For

$$\boldsymbol{z} := \boldsymbol{M}_{GS}\boldsymbol{x} = -(\boldsymbol{D} + \boldsymbol{L})^{-1} \boldsymbol{R}\boldsymbol{x}$$

it holds that

$$z_i = -\sum_{j=1}^{i-1} \frac{a_{ij}}{a_{ii}} z_j - \sum_{j=i+1}^{n} \frac{a_{ij}}{a_{ii}} x_j. \tag{4.1.17}$$

Thus, using $\|\boldsymbol{x}\|_\infty = 1$, we obtain the estimate

$$|z_1| \leq \sum_{j=2}^{n} \frac{|a_{ij}|}{|a_{ii}|} = p_1 < 1.$$

Suppose $z_1,\ldots,z_{i-1}$ are given with $|z_j| \le p_j$, $j = 1,\ldots,i-1 < n$, then for the i-th component of the vector z, using (4.1.17), it follows that

$$|z_i| \le \sum_{j=1}^{i-1} \frac{|a_{ij}|}{|a_{ii}|} p_j + \sum_{j=i+1}^{n} \frac{|a_{ij}|}{|a_{ii}|} = p_i < 1.$$

From this, we obtain $\|z\|_\infty < 1$ and thus, due to the compactness of the unit circle, the estimate

$$\|M_{GS}\|_\infty = \sup_{\substack{x \in \mathbb{C}^n \\ \|x\|_\infty = 1}} \|M_{GS}x\|_\infty < 1$$

is proven, which implies $\rho(M_{GS}) < 1$ and thus the convergence of the Gauss-Seidel method. $\qquad\square$

Example 4.19 As already in Example 4.16, we consider the matrix A according to Example 1.1. With $p_j = 0$ for all $j \le 0$, we obtain

$$
\begin{aligned}
p_1 &= \tfrac{1}{4} + \tfrac{1}{4} < 1, \\[4pt]
p_i &\le \tfrac{1}{4} p_{i-1} + \tfrac{1}{4} + \tfrac{1}{4} < 1 \text{ for } i = 2,\ldots,N, \\[4pt]
p_i &\le \tfrac{1}{4} p_{i-N} + \tfrac{1}{4} p_{i-1} + \tfrac{1}{4} + \tfrac{1}{4} < 1 \text{ for } i = N+1,\ldots,N^2,
\end{aligned}
$$

and consequently the convergence of the Gauss-Seidel method.

Strictly diagonally dominant matrices inherently satisfy the conditions of Theorem 4.18 because $\displaystyle\max_{i=1,\ldots,n} \sum_{\substack{j=1 \\ j \ne i}}^{n} \frac{|a_{ij}|}{|a_{ii}|} < 1$, which leads to the following corollary:

Corollary 4.20 *Let the invertible matrix $A \in \mathbb{C}^{n \times n}$ be strictly diagonally dominant, then the Gauss-Seidel method converges for any starting vector x_0 and for any right-hand side b to $A^{-1}b$.*

Example 4.21 The matrix

$$A = \begin{pmatrix} 0.7 & -0.4 \\ -0.2 & 0.5 \end{pmatrix}$$

is strictly diagonally dominant, which ensures the convergence of the Gauss-Seidel method. The corresponding iteration matrix

$$M_{GS} = -(D+L)^{-1}R = \begin{pmatrix} 0 & \frac{4}{7} \\ 0 & \frac{8}{35} \end{pmatrix}$$

has the eigenvalues $\lambda_1 = 0$ and $\lambda_2 = \frac{8}{35}$, so that

$$\rho(M_{GS}) = \rho(M_J)^2 = \frac{8}{35} \approx 0.22857$$

holds, and thus convergence can be expected to be about twice as fast as with the Jacobi method. For the starting vector $x_0 = (21, -19)^T$ and the right-hand side $b = (0.3, 0.3)^T$ this expectation is confirmed by the convergence history listed in the following table.

Gauss-Seidel Method				
m	$x_{m,1}$	$x_{m,2}$	$\varepsilon_m := \|x_m - A^{-1}b\|_\infty$	$\varepsilon_m/\varepsilon_{m-1}$
0	2.100000e+01	-1.900000e+01	2.000000e+01	
5	9.688054e-01	9.875222e-01	3.119462e-02	2.285714e-01
10	9.999805e-01	9.999922e-01	1.946209e-05	2.285714e-01
15	1.000000e-00	1.000000e-00	1.214225e-08	2.285714e-01
20	1.000000e-00	1.000000e-00	7.575385e-12	2.285702e-01
25	1.000000e-00	1.000000e-00	4.551914e-15	2.204301e-01

4.1.3 Relaxation Methods

We write the linear iteration method

$$x_{m+1} = B^{-1}(B - A)x_m + B^{-1}b$$

in the form

$$x_{m+1} = x_m + \underbrace{B^{-1}(b - Ax_m)}_{r_m :=}. \tag{4.1.18}$$

Thus, x_{m+1} can be interpreted as a correction of x_m using the vector r_m.

Let us first restrict our considerations to methods that have a global step character. The goal of relaxation methods is to improve the convergence rate of method (4.1.18) by weighting the correction vector r_m. Hence, we modify (4.1.18) to

$$x_{m+1} = x_m + \omega B^{-1}(b - Ax_m)$$

with $\omega \in \mathbb{R}^+$. Starting from x_m, we seek the optimal x_{m+1} in the direction of r_m.

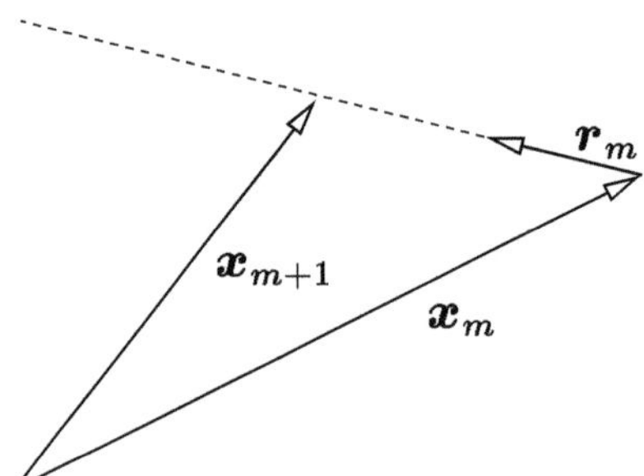

Figure 4.1   Computation of the iterates in the relaxation method

Optimal in the above sense means that the spectral radius of the iteration matrix is minimized. With

$$\begin{aligned}
x_{m+1} &= x_m + \omega B^{-1}(b - Ax_m) \\
&= \underbrace{(I - \omega B^{-1}A)}_{M(\omega):=} x_m + \underbrace{\omega B^{-1}}_{N(\omega):=} b
\end{aligned} \tag{4.1.19}$$

$\omega \in \mathbb{R}^+$ must therefore be chosen such that $\rho(\boldsymbol{M}(\omega))$ is minimized, that is,

$$\omega = \arg \min_{\alpha \in \mathbb{R}^+} \rho(\boldsymbol{M}(\alpha)).$$

The weighting factor ω is called the relaxation parameter, and the method (4.1.19) is called an underrelaxation method for $\omega < 1$ and an overrelaxation method for $\omega > 1$.

Let us now consider as underlying methods those based on a single-step approach. For these algorithms, the aforementioned procedure is also possible, but uncommon. In these methods, relaxation is taken into account during the iteration of each individual component. The exact procedure is explained using the example of the Gauss-Seidel method in Section 4.1.3.2.

4.1.3.1   Jacobi Relaxation Method

According to (4.1.18), we write the Jacobi method in the form

$$\boldsymbol{x}_{m+1} = \boldsymbol{x}_m + \boldsymbol{D}^{-1}(\boldsymbol{b} - \boldsymbol{A}\boldsymbol{x}_m) \text{ for } m = 0,1,\dots.$$

Thus, the Jacobi relaxation method has the representation

$$\begin{aligned}
\boldsymbol{x}_{m+1} &= \boldsymbol{x}_m + \omega \boldsymbol{D}^{-1}(\boldsymbol{b} - \boldsymbol{A}\boldsymbol{x}_m) \\
&= \underbrace{(\boldsymbol{I} - \omega \boldsymbol{D}^{-1}\boldsymbol{A})}_{\boldsymbol{M}_J(\omega):=} \boldsymbol{x}_m + \underbrace{\omega \boldsymbol{D}^{-1}}_{\boldsymbol{N}_J(\omega):=} \boldsymbol{b} \text{ for } m = 0,1,\dots,
\end{aligned}$$

and we obtain the component-wise notation

$$\begin{aligned}
x_{m+1,i} &= x_{m,i} + \frac{\omega}{a_{ii}}\left(b_i - \sum_{j=1}^{n} a_{ij}x_{m,j}\right) \\
&= (1-\omega)x_{m,i} + \frac{\omega}{a_{ii}}\left(b_i - \sum_{\substack{j=1 \\ j\neq i}}^{n} a_{ij}x_{m,j}\right) \text{ for } i = 1,\dots,n \text{ and } m = 0,1,\dots.
\end{aligned}$$

Theorem 4.22 *Let the iteration matrix of the Jacobi method $\boldsymbol{M}_J$ have only real eigenvalues $\lambda_1 \leq \dots \leq \lambda_n$ with the corresponding linearly independent eigenvectors $\boldsymbol{u}_1,\dots,\boldsymbol{u}_n$, and suppose $\rho(\boldsymbol{M}_J) < 1$. Then the iteration matrix $\boldsymbol{M}_J(\omega)$ of the Jacobi relaxation method has the eigenvalues*

$$\mu_i = 1 - \omega + \omega\lambda_i \text{ for } i = 1,\dots,n,$$

and it holds that

$$\omega_{opt} = \arg \min_{\omega \in \mathbb{R}^+} \rho(\boldsymbol{M}_J(\omega)) = \frac{2}{2 - \lambda_1 - \lambda_n}.$$

Proof:
With

$$-\boldsymbol{D}^{-1}(\boldsymbol{L} + \boldsymbol{R})\boldsymbol{u}_i = \boldsymbol{M}_J\boldsymbol{u}_i = \lambda_i\boldsymbol{u}_i \text{ for } i = 1,\dots,n$$

we obtain

$$\begin{aligned}
\boldsymbol{M}_J(\omega)\boldsymbol{u}_i &= (\boldsymbol{I} - \omega \boldsymbol{D}^{-1}\boldsymbol{A})\boldsymbol{u}_i \\[2mm]
&= \left((1-\omega)\boldsymbol{I} - \omega \boldsymbol{D}^{-1}(\boldsymbol{L}+\boldsymbol{R})\right)\boldsymbol{u}_i \\[2mm]
&= (1-\omega+\omega\lambda_i)\boldsymbol{u}_i \ \text{ for } i = 1,\ldots,n.
\end{aligned}$$

Since the eigenvectors $\boldsymbol{u}_1,\ldots,\boldsymbol{u}_n$ are linearly independent, there are no further eigenvalues. For the eigenvalues $\mu_i(\omega) = 1 - \omega + \omega\lambda_i$, $(i = 1,\ldots,n)$ of the iteration matrix of the Jacobi relaxation method, it holds for $\omega \geq 0$

$$\mu_1(\omega) \leq \ldots \leq \mu_n(\omega).$$

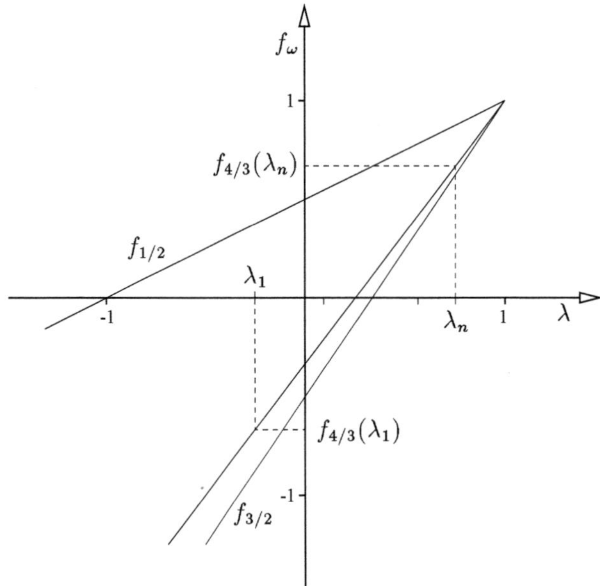

Figure 4.2　　Behavior of the relaxation functions f_ω depending on the chosen relaxation parameter ω

Let us consider the relaxation function

$$\begin{aligned}
f_\omega : \ \mathbb{R} &\to \mathbb{R} \\
\lambda &\xmapsto{f_\omega} f_\omega(\lambda) = 1 - \omega + \omega\lambda,
\end{aligned}$$

which is shown in Figure 4.2 for different relaxation parameters ω. The idea is to determine the optimal relaxation parameter by the condition

$$\mu_n(\omega^*) = f_{\omega^*}(\lambda_n) = -f_{\omega^*}(\lambda_1) = -\mu_1(\omega^*).$$

This yields, from

$$\mu_n(\omega^*) = 1 - \omega^* + \omega^*\lambda_n = -(1 - \omega^* + \omega^*\lambda_1) = -\mu_1(\omega^*)$$

using $\rho(\boldsymbol{M}_j) < 1$, the representation

$$\omega^* = \frac{2}{2 - \lambda_1 - \lambda_n} > 0.$$

Taking into account $f_\omega(1) = 1$ for all $\omega \in \mathbb{R}^+$, it always holds that $\mu_n(\omega^*) = -\mu_1(\omega^*) \geq 0$. To prove optimality, let $\omega > \omega^* > 0$. Thus, it follows that

$$\mu_1(\omega) = 1 - \omega \underbrace{(1 - \lambda_1)}_{>0} < 1 - \omega^*(1 - \lambda_1) = \mu_1(\omega^*) \leq 0,$$

which implies

$$\rho(\boldsymbol{M}_J(\omega)) \geq |\mu_1(\omega)| > |\mu_1(\omega^*)| = \rho(\boldsymbol{M}_J(\omega^*)).$$

An analogous statement is obtained in the case $\omega < \omega^*$ by considering the eigenvalues $\mu_n(\omega)$ and $\mu_n(\omega^*)$, so that

$$\omega^* = \omega_{\text{opt}} = \arg \min_{\omega \in \mathbb{R}^+} \rho(\boldsymbol{M}_J(\omega))$$

follows. $\qquad\square$

Systems of equations for which the iteration matrix of the Jacobi method has an eigenvalue distribution symmetric about the origin yield, according to Theorem 4.22, the optimal relaxation parameter $\omega_{\text{opt}} = 1$. Consequently, in such cases, the relaxation approach cannot accelerate the underlying global step method. The model equation considered in (4.1.2) exhibits this property, which is why a discussion of the method with respect to this issue is omitted.

4.1.3.2 *Gauss-Seidel Relaxation Method*

We consider the component form of the Gauss-Seidel method (4.1.16) with weighted correction vector component $r_{m,i}$ in the form

$$\begin{aligned}
x_{m+1,i} &= x_{m,i} + \frac{\omega}{a_{ii}} \left(b_i - \sum_{j=1}^{i-1} a_{ij} x_{m+1,j} - \sum_{j=i}^{n} a_{ij} x_{m,j} \right) \\
&= (1 - \omega) x_{m,i} + \frac{\omega}{a_{ii}} \left(b_i - \sum_{j=1}^{i-1} a_{ij} x_{m+1,j} - \sum_{j=i+1}^{n} a_{ij} x_{m,j} \right)
\end{aligned}$$

for $i = 1, \ldots, n$ and $m = 0, 1, \ldots$. From this we obtain

$$(\boldsymbol{I} + \omega \boldsymbol{D}^{-1} \boldsymbol{L}) x_{m+1} = \left[(1 - \omega) \boldsymbol{I} - \omega \boldsymbol{D}^{-1} \boldsymbol{R} \right] x_m + \omega \boldsymbol{D}^{-1} \boldsymbol{b},$$

which yields

$$\boldsymbol{D}^{-1}(\boldsymbol{D} + \omega \boldsymbol{L}) x_{m+1} = \boldsymbol{D}^{-1} \left[(1 - \omega) \boldsymbol{D} - \omega \boldsymbol{R} \right] x_m + \omega \boldsymbol{D}^{-1} \boldsymbol{b}$$

and thus the Gauss-Seidel relaxation method in the form

$$x_{m+1} = \underbrace{(\boldsymbol{D} + \omega \boldsymbol{L})^{-1} \left[(1 - \omega) \boldsymbol{D} - \omega \boldsymbol{R} \right]}_{\boldsymbol{M}_{GS}(\omega):=} x_m + \underbrace{\omega (\boldsymbol{D} + \omega \boldsymbol{L})^{-1}}_{\boldsymbol{N}_{GS}(\omega):=} \boldsymbol{b}$$

follows.

Theorem 4.23 *Let $A \in \mathbb{C}^{n \times n}$ with $a_{ii} \neq 0$ for $i = 1, \ldots, n$, then for $\omega \in \mathbb{R}$ it holds that*

$$\rho(M_{GS}(\omega)) \geq |\omega - 1|.$$

Proof:
Let $\lambda_1, \ldots, \lambda_n$ be the eigenvalues of $M_{GS}(\omega)$, then it follows that

$$
\begin{aligned}
\prod_{i=1}^{n} \lambda_i &= \det M_{GS}(\omega) = \det((D + \omega L)^{-1}) \det((1 - \omega)D - \omega R) \\
&= \det(D^{-1}) \det((1 - \omega)D) \\
&= \det(D)^{-1} (1 - \omega)^n \det D \\
&= (1 - \omega)^n.
\end{aligned}
$$

From this it follows that

$$\rho(M_{GS}(\omega)) = \max_{i=1,\ldots,n} |\lambda_i| \geq |1 - \omega|.$$

$\square$

The above theorem states that a choice of relaxation parameter $\omega \leq 0$ always leads to a divergent method, so that the initially arbitrary requirement of positivity for the relaxation parameter in the relaxation method finds its justification here in the context of the Gauss-Seidel relaxation method. Analogously, the theorem includes the requirement $\omega < 2$. We summarize both conditions in the following corollary.

Corollary 4.24 *The Gauss-Seidel relaxation method converges at most for a relaxation parameter $\omega \in (0,2)$.*

Theorem 4.25 *Let A be Hermitian and positive definite, then the Gauss-Seidel relaxation method converges if and only if $\omega \in (0,2)$.*

Proof:
From the positive definiteness of the matrix A it follows that $a_{ii} \in \mathbb{R}^{+}$ for $i = 1, \ldots, n$, which ensures the well-definedness of the Gauss-Seidel relaxation method.

" $\Rightarrow$ " Assume the Gauss-Seidel relaxation method is convergent.

In this case, $\omega \in (0,2)$ follows immediately from Corollary 4.24.

" $\Leftarrow$ " Assume $\omega \in (0,2)$.
Let λ be an eigenvalue of $M_{GS}(\omega)$ with eigenvector $x \in \mathbb{C}^n$. Since A is Hermitian, it follows that $L^* = R$ and thus

$$((1 - \omega)D - \omega L^*)x = \lambda(D + \omega L)x.$$

With

$$
\begin{aligned}
2\left[(1 - \omega)D - \omega L^*\right] &= (2 - \omega)D + \omega(-D - 2L^*) \\
&= (2 - \omega)D - \omega A + \omega(L - L^*)
\end{aligned}
$$

and

$$2(D + \omega L) = (2 - \omega)D + \omega(D + 2L)$$
$$= (2 - \omega)D + \omega A + \omega(L - L^*)$$

we obtain for the eigenvector $x \in \mathbb{C}^n$ the equation

$$\lambda((2 - \omega)x^* D x + \omega x^* A x + \omega x^*(L - L^*)x)$$
$$= 2\lambda x^*(D + \omega L)x$$
$$= 2x^*[(1 - \omega)D - \omega L^*]x$$
$$= (2 - \omega)x^* D x - \omega x^* A x + \omega x^*(L - L^*)x.$$

Using the imaginary unit i, we write

$$x^*(L - L^*)x = x^* L x - x^* L^* x = x^* L x - \overline{x^* L x} = i \cdot s, \ s \in \mathbb{R}.$$

Furthermore, we have

$$d := x^* D x \in \mathbb{R}^+,$$
$$a := x^* A x \in \mathbb{R}^+,$$

so that

$$\lambda((2 - \omega)d + \omega a + i\omega s) = (2 - \omega)d - \omega a + i\omega s$$

follows. Dividing by ω and substituting $\mu = \frac{2-\omega}{\omega}$ yields

$$\lambda(\mu d + a + is) = \mu d - a + is.$$

From the assumption $\omega \in (0,2)$ we obtain $\mu \in \mathbb{R}^+$, so that with $a, d \in \mathbb{R}^+$ the inequality $|\mu d + is - a| < |\mu d + is - (-a)|$ holds and therefore the estimate

$$|\lambda| = \frac{|\mu d + is - a|}{|\mu d + is + a|} < 1$$

follows. Since the eigenvalue was chosen arbitrarily from the spectrum of the iteration matrix, we obtain the necessary and sufficient condition for the convergence of the method due to

$$\rho(M_{GS}(\omega)) < 1.$$

$\square$

For determining the optimal relaxation parameter, consistently ordered matrices prove to be advantageous, and are therefore introduced with the following definition.

Definition 4.26 Let $L \in \mathbb{C}^{n \times n}$ be a strict lower left and $R \in \mathbb{C}^{n \times n}$ a strict upper right triangular matrix, then the matrix $A = D + L + R \in \mathbb{C}^{n \times n}$ with non-singular diagonal part D is called consistently ordered if the eigenvalues of

$$C(\alpha) = -(\alpha D^{-1}L + \alpha^{-1}D^{-1}R) \text{ with } \alpha \in \mathbb{C} \setminus \{0\}$$

are independent of α.

Example 4.27 Every matrix $A \in \mathbb{C}^{n \times n}$ of the form

$$A = \begin{pmatrix} I & A_{12} \\ A_{21} & I \end{pmatrix}$$

is consistently ordered. We have

$$L = \begin{pmatrix} 0 & 0 \\ A_{21} & 0 \end{pmatrix} \quad \text{and} \quad R = \begin{pmatrix} 0 & A_{12} \\ 0 & 0 \end{pmatrix},$$

so that $C(\alpha)$ has the representation

$$C(\alpha) = \begin{pmatrix} 0 & -\alpha^{-1}A_{12} \\ -\alpha A_{21} & 0 \end{pmatrix} = \begin{pmatrix} I & 0 \\ 0 & -\alpha I \end{pmatrix} \begin{pmatrix} 0 & A_{12} \\ A_{21} & 0 \end{pmatrix} \begin{pmatrix} I & 0 \\ 0 & -\alpha^{-1}I \end{pmatrix}.$$

Thus, λ is an eigenvalue of $C(\alpha)$ if and only if λ is an eigenvalue of $C(1) = A - I$, which gives the required independence of the eigenvalues from the chosen value $\alpha \in \mathbb{C} \setminus \{0\}$.

Example 4.28 Every tridiagonal matrix

$$A = \begin{pmatrix} a_1 & b_1 & & & \\ c_2 & a_2 & b_2 & & \\ & \ddots & \ddots & \ddots & \\ & & \ddots & \ddots & b_{n-1} \\ & & & c_n & a_n \end{pmatrix} \in \mathbb{C}^{n \times n}$$

with $a_i \neq 0$ for $i = 1, \ldots, n$ is consistently ordered, since

$$C(1) = -D^{-1}(L + R) = \begin{pmatrix} 0 & d_1 & & \\ \ell_2 & \ddots & \ddots & \\ & \ddots & \ddots & d_{n-1} \\ & & \ell_n & 0 \end{pmatrix}$$

with $d_i = -\dfrac{b_i}{a_i}$ for $i = 1, \ldots, n-1$ and $\ell_i = -\dfrac{c_i}{a_i}$ for $i = 2, \ldots, n$, so that with the invertible matrix

$$S(\alpha) = \begin{pmatrix} 1 & & & & \\ & \alpha & & & \\ & & \alpha^2 & & \\ & & & \ddots & \\ & & & & \alpha^{n-1} \end{pmatrix}, \quad \alpha \in \mathbb{C} \setminus \{0\}$$

the equation $S(\alpha)C(1)S^{-1}(\alpha) = C(\alpha)$ follows.

Theorem 4.29 *Let* $A \in \mathbb{C}^{n \times n}$ *be consistently ordered and* $\omega \in (0,2)$, *then* $\mu \in \mathbb{C} \setminus \{0\}$ *is an eigenvalue of* $M_{GS}(\omega)$ *if and only if*

$$\lambda = \frac{\mu + \omega - 1}{\omega \mu^{1/2}}$$

is an eigenvalue of M_J.

Proof:

Let $\mu \in \mathbb{C}\backslash\{0\}$, then

$$(I + \omega D^{-1}L)(\mu I - M_{GS}(\omega))$$

$$= \mu(I + \omega D^{-1}L) - D^{-1}(D + \omega L)\underbrace{(D + \omega L)^{-1}((1 - \omega)D - \omega R)}_{M_{GS}(\omega)=}$$

$$= (\mu - (1 - \omega))I + \omega D^{-1}(\mu L + R)$$

$$= (\mu - (1 - \omega))I + \omega \mu^{1/2} D^{-1}\left(\mu^{1/2}L + \mu^{-1/2}R\right)$$

is satisfied. With $\det(I + \omega D^{-1}L) = 1$, it follows from the above equation that $\mu \in \mathbb{C}\backslash\{0\}$ is an eigenvalue of $M_{GS}(\omega)$ if and only if

$$\det\left((\mu - (1 - \omega))I + \omega\mu^{1/2}D^{-1}\left(\mu^{1/2}L + \mu^{-1/2}R\right)\right) = 0$$

holds, that is,

$$\frac{\mu - (1 - \omega)}{\omega\mu^{1/2}}$$

is an eigenvalue of $-D^{-1}\left(\mu^{1/2}L + \mu^{-1/2}R\right)$. With the assumption that the matrix A is consistently ordered, the eigenvalues of the two matrices $-D^{-1}\left(\mu^{1/2}L + \mu^{-1/2}R\right)$ and $-D^{-1}(L + R) = M_J$ coincide, which proves the assertion of the theorem. $\qquad\square$

For consistently ordered matrices $A \in \mathbb{C}^{n \times n}$ it thus holds that

$$\rho(M_{GS}) = \rho(M_J)^2,$$

so that the Gauss-Seidel method, compared to the Jacobi method, usually requires half as many iterations to reach a given accuracy threshold. Furthermore, in the case of a consistently ordered matrix A, we always have

$$\sigma(M_J) = -\sigma(M_J),$$

so that according to Theorem 4.22, for such matrices, no acceleration of convergence can be achieved by a relaxation of the Jacobi method.

Theorem 4.30 *Let $A \in \mathbb{C}^{n \times n}$ be consistently ordered. Let the eigenvalues of M_J be real, and suppose that*

$$\rho := \rho(M_J) < 1.$$

Then the following holds:

(a) The Gauss-Seidel relaxation method converges for all $\omega \in (0,2)$.

(b) The spectral radius of the iteration matrix $M_{GS}(\omega)$ is minimal for

$$\omega_{opt} = \frac{2}{1 + \sqrt{1 - \rho^2}}$$

with

$$\rho(M_{GS}(\omega_{opt})) = \omega_{opt} - 1 = \frac{1 - \sqrt{1 - \rho^2}}{1 + \sqrt{1 - \rho^2}}$$

holding.

Proof:

Let $\lambda_1,\ldots,\lambda_n \in \mathbb{R}$ be the eigenvalues of M_J, then by Theorem 4.29, μ is an eigenvalue of $M_{GS}(\omega)$ if and only if

$$\lambda = \frac{\mu + \omega - 1}{\omega \mu^{1/2}} \in \{\lambda_1,\ldots,\lambda_n\} \tag{4.1.20}$$

holds. Since the matrix A is consistently ordered, with $\lambda \in \mathbb{R}$, $-\lambda$ is also an eigenvalue of M_J, so the sign in (4.1.20) is irrelevant. We can therefore consider the equation

$$\lambda^2 \omega^2 \mu = (\mu + \omega - 1)^2 \tag{4.1.21}$$

and, without loss of generality, assume $\lambda \geq 0$.

From $\rho(M_J) < 1$ it follows that $\lambda \in [0,1)$. Furthermore, due to Corollary 4.24, we always consider $\omega \in (0,2)$. From (4.1.21) we obtain the two eigenvalues in the form

$$\mu^{\pm} = \mu^{\pm}(\omega,\lambda) = \frac{1}{2}\lambda^2\omega^2 - (\omega - 1) \pm \lambda\omega\sqrt{\frac{1}{4}\lambda^2\omega^2 - (\omega - 1)}. \tag{4.1.22}$$

We define

$$g(\omega,\lambda) := \frac{1}{4}\lambda^2\omega^2 - (\omega - 1).$$

For a given $\lambda \in [0,1)$, the zeros of this function are

$$\omega^{\pm} = \omega^{\pm}(\lambda) = \frac{2}{1 \pm \sqrt{1 - \lambda^2}}. \tag{4.1.23}$$

With $\omega \in (0,2)$ we can neglect ω^-, and it follows that $\omega^+(\lambda) > 1$ for all $\lambda \in [0,1)$. Moreover, for all $\lambda \in [0,1)$ and $\omega \in (0,2)$, we have

$$\frac{\partial g}{\partial \omega}(\omega,\lambda) = \frac{1}{2}\lambda^2\omega - 1 < 0.$$

We obtain the following three cases:

(1) $2 > \omega > \omega^+(\lambda)$:

    The two eigenvalues $\mu^+(\omega,\lambda)$ and $\mu^-(\omega,\lambda)$ are complex, and

$$|\mu^+(\omega,\lambda)| = |\mu^-(\omega,\lambda)| = |\omega - 1| = \omega - 1.$$

(2) $\omega = \omega^+(\lambda)$:

    From (4.1.23) it follows that $\lambda^2 = \dfrac{4}{\omega} - \dfrac{4}{\omega^2}$, which yields

$$|\mu^+(\omega,\lambda)| = |\mu^-(\omega,\lambda)| = \frac{1}{2}\lambda^2\omega^2 - (\omega - 1) = 2\omega - 2 - (\omega - 1) = \omega - 1.$$

(3) $0 < \omega < \omega^+(\lambda)$:

    Equation (4.1.22) yields two real eigenvalues

$$\mu^{\pm}(\omega,\lambda) = \underbrace{\frac{1}{2}\lambda^2\omega^2 - (\omega - 1)}_{>0} \pm \underbrace{\lambda\omega\sqrt{\frac{1}{4}\lambda^2\omega^2 - (\omega - 1)}}_{\geq 0}$$

    with

$$\max\{|\mu^+(\omega,\lambda)|,|\mu^-(\omega,\lambda)|\} = \mu^+(\omega,\lambda).$$

To determine $\rho(\boldsymbol{M}_{GS}(\omega))$, in all three cases we are only interested in $\mu^+(\omega,\lambda)$. Consequently, for $\lambda \in [0,1)$, we consider

$$\mu(\omega,\lambda) = \begin{cases} \mu^+(\omega,\lambda) & \text{for } 0 < \omega < \omega^+(\lambda) \\ \omega - 1 & \text{for } \omega^+(\lambda) \le \omega < 2. \end{cases} \tag{4.1.24}$$

Thus, for $0 < \omega < \omega^+(\lambda)$ and $\lambda \in [0,1)$, we have

$$\frac{\partial \mu}{\partial \lambda}(\omega,\lambda) = \underbrace{\lambda \omega^2}_{\ge 0} + \underbrace{\omega\sqrt{\frac{1}{4}\lambda^2\omega^2 - (\omega - 1)}}_{>0} + \underbrace{\lambda \omega \frac{1}{2}\frac{\frac{1}{2}\lambda\omega^2}{\sqrt{\frac{1}{4}\lambda^2\omega^2 - (\omega - 1)}}}_{\ge 0} > 0, \tag{4.1.25}$$

and since

$$\mu(\omega,\lambda) = \left(\frac{\omega\lambda}{2} + \sqrt{\frac{1}{4}\lambda^2\omega^2 - (\omega - 1)} \right)^2$$

it follows that

$$\frac{\partial \mu}{\partial \omega}(\omega,\lambda) = 2 \underbrace{\left(\frac{\omega\lambda}{2} + \sqrt{\frac{1}{4}\lambda^2\omega^2 - (\omega - 1)} \right)}_{>0} \underbrace{\left[\frac{\lambda}{2} + \frac{1}{2}\frac{\frac{1}{2}\lambda^2\omega - 1}{\sqrt{\frac{1}{4}\lambda^2\omega^2 - (\omega - 1)}} \right]}_{q(\omega,\lambda):=}.$$

We write

$$q(\omega,\lambda) = \underbrace{\frac{1}{2\sqrt{\frac{1}{4}\lambda^2\omega^2 - (\omega - 1)}}}_{>0} \left(\underbrace{\lambda\sqrt{\frac{1}{4}\lambda^2\omega^2 - (\omega - 1)}}_{q_1(\omega,\lambda):=} + \underbrace{\frac{1}{2}\lambda^2\omega - 1}_{q_2(\omega,\lambda):=} \right).$$

For the functions q_1 and q_2, the following holds for all $\lambda \in [0,1)$ and $\omega \in (0,\omega^+(\lambda))$:

$$q_1(\omega,\lambda) \ge 0 \quad \text{and} \quad q_2(\omega,\lambda) < 0.$$

Furthermore, we have

$$[q_1(\omega,\lambda)]^2 = \frac{\omega^2\lambda^4}{4} + \lambda^2 - \omega\lambda^2 < \frac{\omega^2\lambda^4}{4} + 1 - \omega\lambda^2 = [q_2(\omega,\lambda)]^2$$

which yields the inequality

$$\frac{\partial \mu}{\partial \omega}(\omega,\lambda) < 0 \quad \text{for all} \quad \lambda \in [0,1) \quad \text{and} \quad \omega \in (0,\omega^+(\lambda)).$$

From (4.1.24) we also obtain $\mu(0,\lambda) = 1 = \mu(2,\lambda)$, so that $|\mu(\omega,\lambda)| < 1$ for all $\lambda \in [0,1)$ and $\omega \in (0,2)$, which directly yields $\rho(\boldsymbol{M}_{GS}(\omega)) < 1$. For each eigenvalue λ, $|\mu(\omega,\lambda)|$ is minimal for $\omega_{\mathrm{opt}} = \omega^+(\lambda)$.

Equation (4.1.24) hence gives

$$
\begin{aligned}
\rho(\boldsymbol{M}_{GS}(\omega_{\mathrm{opt}})) &= |\mu(\omega_{\mathrm{opt}},\rho(\boldsymbol{M}_J))| \\
&= \omega_{\mathrm{opt}}(\rho(\boldsymbol{M}_J)) - 1 \\
&\stackrel{(4.1.23)}{=} \frac{2}{1+\sqrt{1-\rho^2}} - 1 \\
&= \frac{1-\sqrt{1-\rho^2}}{1+\sqrt{1-\rho^2}}.
\end{aligned}
$$

$\square$

Remark:
Since for all $\lambda \in (0,1)$

$$
\lim_{\omega \searrow \omega_{\mathrm{opt}}} \frac{\partial \mu}{\partial \omega}(\omega,\lambda) = 1 \quad \text{and} \quad \lim_{\omega \nearrow \omega_{\mathrm{opt}}} \frac{\partial \mu}{\partial \omega}(\omega,\lambda) = -\infty \tag{4.1.26}
$$

holds, ω should, in case of doubt, be chosen rather larger than ω_{opt} than smaller than ω_{opt}.

Moreover, the optimal relaxation parameter for matrices satisfying the assumptions of Theorem 4.30 lies in the interval $[1,2)$, which is why this relaxation method is also called the SOR method (*successive overrelaxation method*). In the case of a damped Gauss-Seidel method ($\omega < 1$), one also speaks of the *successive underrelaxation method*.

Example 4.31 Let us again consider the model problem $\boldsymbol{Ax} = \boldsymbol{b}$ with

$$
\boldsymbol{A} = \begin{pmatrix} 0.7 & -0.4 \\ -0.2 & 0.5 \end{pmatrix}, \quad \boldsymbol{b} = \begin{pmatrix} 0.3 \\ 0.3 \end{pmatrix}.
$$

The matrix $\boldsymbol{A}$ is consistently ordered as a tridiagonal matrix, and the eigenvalues of

$$
\boldsymbol{M}_J = -\boldsymbol{D}^{-1}(\boldsymbol{L}+\boldsymbol{R})
$$

are, according to Example 4.17,

$$
\lambda_{1,2} = \pm\sqrt{\frac{8}{35}} \in \mathbb{R},
$$

so that $\rho(\boldsymbol{M}_J) < 1$ holds. Using these properties, Theorem 4.30 yields the convergence of the Gauss-Seidel relaxation method

$$
\boldsymbol{x}_{m+1} = \boldsymbol{M}_{GS}(\omega)\boldsymbol{x}_m + \boldsymbol{N}_{GS}(\omega)\boldsymbol{b}
$$

for all $\omega \in (0,2)$. The optimal relaxation parameter is

$$
\omega_{\mathrm{opt}} = \frac{2}{1+\sqrt{1-\frac{8}{35}}} \approx 1.0648
$$

and yields

$$\rho(\boldsymbol{M}_{GS}(\omega^*)) = \frac{1 - \sqrt{1 - \frac{8}{35}}}{1 + \sqrt{1 - \frac{8}{35}}} \approx 0.0648.$$

Figure 4.3 shows the spectral radius as a function of the relaxation parameter for $\lambda = \sqrt{\frac{8}{35}}$. Moreover, the table presents the convergence behavior of the Gauss-Seidel relaxation method with optimal relaxation parameter for the model problem using the initial vector $\boldsymbol{x}_0 = (21, -19)^T$.

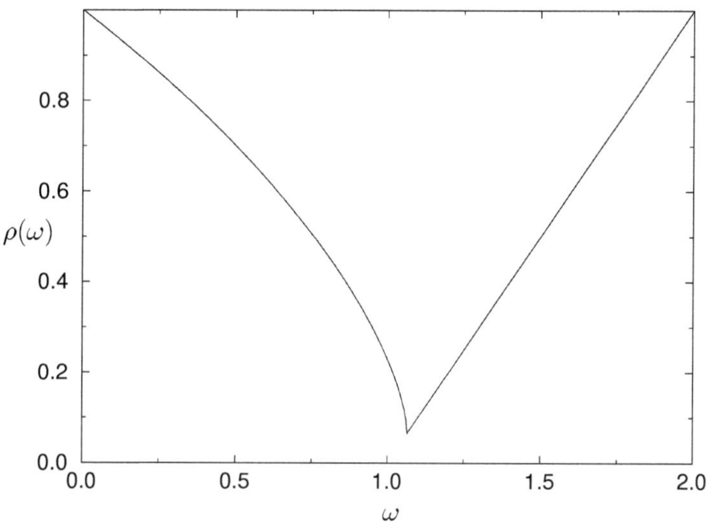

Figure 4.3   Spectral radius as a function of the relaxation parameter

SOR-Gauss-Seidel Method				
m	$x_{m,1}$	$x_{m,2}$	$\varepsilon_m := \|\boldsymbol{x}_m - \boldsymbol{A}^{-1}\boldsymbol{b}\|_\infty$	$\varepsilon_m/\varepsilon_{m-1}$
0	2.100000e+01	-1.900000e+01	2.000000e+01	
5	9.987226e-01	9.997003e-01	1.277401e-03	8.134709e-02
10	1.000000e-00	1.000000e-00	2.942099e-09	7.205638e-02
15	1.000000e-00	1.000000e-00	4.884981e-15	6.727829e-02

Figure 4.3 clearly illustrates the behavior of the spectral radius near the optimal relaxation parameter, thereby emphasizing the analytically determined behavior.

4.1.4   Richardson Method

The Richardson method is based on the basic algorithm

$$\boldsymbol{x}_{m+1} = (\boldsymbol{I} - \boldsymbol{A})\boldsymbol{x}_m + \boldsymbol{b}$$

(see Example 4.10) and a weighting of the correction vector

$$\boldsymbol{r}_m = \boldsymbol{b} - \boldsymbol{A}\boldsymbol{x}_m$$

with a number $\Theta \in \mathbb{C}$. We obtain

$$\boldsymbol{x}_{m+1} = \underbrace{(\boldsymbol{I} - \Theta\boldsymbol{A})}_{\boldsymbol{M}_R(\Theta):=}\boldsymbol{x}_m + \underbrace{\Theta\boldsymbol{I}}_{\boldsymbol{N}_R(\Theta):=}\boldsymbol{b} \tag{4.1.27}$$

or in component notation

$$x_{m+1,i} = x_{m,i} + \Theta\left(b_i - \sum_{j=1}^{n} a_{ij}x_{m,j}\right)$$

for $i = 1,\ldots,n$.

Lemma 4.32 *Let $\boldsymbol{A} \in \mathbb{C}^{n\times n}$ with $\sigma(\boldsymbol{A}) \subset \mathbb{R}$ be given, and let $\lambda_{\max} = \max\limits_{\lambda\in\sigma(\boldsymbol{A})} \lambda$ as well as $\lambda_{\min} = \min\limits_{\lambda\in\sigma(\boldsymbol{A})} \lambda$, then for all $\Theta \in \mathbb{R}$ it holds that*

$$\sigma(\boldsymbol{M}_R(\Theta)) \subset \mathbb{R}$$

and for all $\Theta \in \mathbb{C}$

$$\rho(\boldsymbol{M}_R(\Theta)) = \max\{|1 - \Theta\lambda_{\max}|, |1 - \Theta\lambda_{\min}|\}.$$

Proof:
With $\boldsymbol{M}_R(\Theta) = \boldsymbol{I} - \Theta\boldsymbol{A}$, all eigenvalues of the matrix $\boldsymbol{M}_R(\Theta)$ have the form $\mu = 1 - \Theta\lambda$ with $\lambda \in \sigma(\boldsymbol{A})$. With $\Theta \in \mathbb{R}$ and $\sigma(\boldsymbol{A}) \subset \mathbb{R}$, all eigenvalues of $\boldsymbol{M}_R(\Theta)$ are real, and it holds that $\sigma(\boldsymbol{M}_R(\Theta)) \subset \mathbb{R}$.

For a given $\Theta \in \mathbb{C}$ we consider the function defined by

$$\begin{aligned} p &: & [a,b] &\to & \mathbb{R} \\ & & \lambda &\mapsto & p(\lambda) = |1 - \Theta\lambda| \end{aligned}$$

which does not have a local maximum on $[a,b]$. Thus,

$$\max_{\lambda\in[a,b]} p(\lambda) = \max\{|1 - \Theta a|, |1 - \Theta b|\},$$

and we obtain

$$\rho(\boldsymbol{M}_R(\Theta)) = \max_{\lambda\in\sigma(\boldsymbol{A})} p(\lambda) = \max\left\{|1 - \Theta\lambda_{\min}|, |1 - \Theta\lambda_{\max}|\right\}.$$

$\square$

Analogous to the Gauss-Seidel relaxation method, we will determine the convergence region of the Richardson iteration depending on the parameter Θ as well as its optimal value Θ_{opt} with the following two theorems.

Theorem 4.33 *Let $\boldsymbol{A} \in \mathbb{C}^{n\times n}$ with $\sigma(\boldsymbol{A}) \subset \mathbb{R}^+$ and $\lambda_{\max} = \max\limits_{\lambda\in\sigma(\boldsymbol{A})} \lambda$, $\lambda_{\min} = \min\limits_{\lambda\in\sigma(\boldsymbol{A})} \lambda$, then the Richardson method (4.1.27) converges for $\Theta \in \mathbb{R}$ if and only if*

$$0 < \Theta < \frac{2}{\lambda_{\max}}$$

holds.

Proof:

" $\Rightarrow$ "   Assume the Richardson method is convergent.

In this case, by Lemma 4.32 we have

$$1 > \rho(\boldsymbol{M}_R(\Theta)) \geq |1 - \Theta\lambda_{\max}| \geq 1 - \Theta\lambda_{\max},$$

which implies $\Theta\lambda_{\max} > 0$ and thus, due to the positive real spectrum of $\boldsymbol{A}$, the inequality $\Theta > 0$ follows. Furthermore,

$$-1 < -\rho(\boldsymbol{M}_R(\Theta)) \leq -|1 - \Theta\lambda_{\max}| \leq 1 - \Theta\lambda_{\max},$$

which yields $\Theta\lambda_{\max} < 2$ and with $\lambda_{\max} > 0$ the estimate $\Theta < \frac{2}{\lambda_{\max}}$.

" $\Leftarrow$ "   Let $0 < \Theta < \frac{2}{\lambda_{\max}}$.

Using $\sigma(\boldsymbol{A}) \subset \mathbb{R}^+$ we obtain

$$-1 < 1 - \Theta\lambda_{\max} \leq 1 - \Theta\lambda_{\min} < 1,$$

so that by Lemma 4.32

$$\rho(\boldsymbol{M}_R(\Theta)) = \max\left\{|1 - \Theta\lambda_{\min}|, |1 - \Theta\lambda_{\max}|\right\} < 1$$

follows. $\square$

Theorem 4.34 *Let* $\boldsymbol{A} \in \mathbb{C}^{n \times n}$ *with* $\sigma(\boldsymbol{A}) \subset \mathbb{R}^+$ *, and let* $\lambda_{\min} = \min\limits_{\lambda \in \sigma(\boldsymbol{A})} \lambda$ *and* $\lambda_{\max} = \max\limits_{\lambda \in \sigma(\boldsymbol{A})} \lambda$ *, then the spectral radius of* $\boldsymbol{M}_R(\Theta)$ *is minimal for*

$$\Theta_{opt} = \frac{2}{\lambda_{\min} + \lambda_{\max}},$$

and it holds that

$$\rho(\boldsymbol{M}_R(\Theta_{opt})) = \frac{\lambda_{\max} - \lambda_{\min}}{\lambda_{\max} + \lambda_{\min}}.$$

Proof:

For $\Theta \in \mathbb{C}$ and $\lambda \in \mathbb{R}^+$ it follows that

$$|1 - \Theta\lambda|^2 = (1 - \lambda\mathrm{Re}(\Theta))^2 + (\lambda\mathrm{Im}(\Theta))^2 \geq (1 - \lambda\mathrm{Re}(\Theta))^2,$$

which implies that $\Theta_{opt} \in \mathbb{R}$ results. Furthermore, let the functions $g_{\max}, g_{\min} : \mathbb{R} \to \mathbb{R}$ be given by $g_{\max}(\Theta) = |1 - \Theta\lambda_{\max}|$, $g_{\min}(\Theta) = |1 - \Theta\lambda_{\min}|$, then, taking into account

$$\Theta_{opt} = \arg\min_{\Theta \in \mathbb{C}} \rho(\boldsymbol{M}_R(\Theta)) = \arg\min_{\Theta \in \mathbb{R}} \rho(\boldsymbol{M}_R(\Theta)) = \arg\min_{\Theta \in \mathbb{R}} \max\left\{g_{\max}(\Theta), g_{\min}(\Theta)\right\}$$

from Figure 4.4 we can deduce the condition

$$\Theta_{opt}\lambda_{\max} - 1 = 1 - \Theta_{opt}\lambda_{\min}$$

for the optimal Θ , which yields

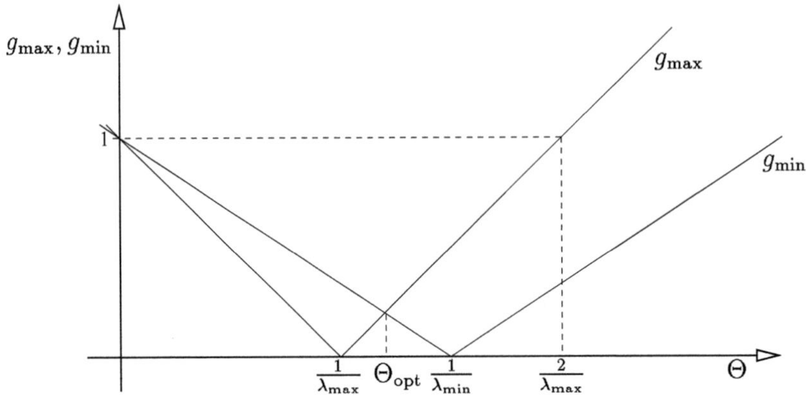

Figure 4.4   Determination of the optimal parameter for the Richardson method

$$\Theta_{\text{opt}} = \frac{2}{\lambda_{\max} + \lambda_{\min}}$$

and

$$\rho(\boldsymbol{M}_R(\Theta_{\text{opt}})) = \frac{\lambda_{\max} - \lambda_{\min}}{\lambda_{\max} + \lambda_{\min}}$$

as results.

$\square$

Example 4.35 Let us consider the already familiar model problem $\boldsymbol{Ax} = \boldsymbol{b}$ with

$$\boldsymbol{A} = \begin{pmatrix} 0.7 & -0.4 \\ -0.2 & 0.5 \end{pmatrix} \quad \text{and} \quad \boldsymbol{b} = \begin{pmatrix} 0.3 \\ 0.3 \end{pmatrix}.$$

With

$$0 = \det(\boldsymbol{A} - \lambda \boldsymbol{I}) = (0.7 - \lambda)(0.5 - \lambda) - 0.08 = (\lambda - 0.6)^2 - 0.09$$

we obtain the eigenvalues $\lambda_{1,2} = 0.6 \pm 0.3$ and thus a positive real spectrum $\sigma(\boldsymbol{A})$. Theorem 4.33 guarantees the convergence of the Richardson method for all $\Theta \in \mathbb{R}$ with $0 < \Theta < \frac{2}{0.9} = 2.\bar{2}$. With Theorem 4.34 we obtain

$$\Theta_{\text{opt}} = \frac{2}{0.9 + 0.3} = \frac{5}{3}$$

and

$$\rho(\boldsymbol{M}_R(\Theta_{\text{opt}})) = \frac{0.9 - 0.3}{0.9 + 0.3} = 0.5.$$

For the usual initial vector $\boldsymbol{x}_0 = (21, -19)^T$ we obtain the convergence history listed in the following table:

Richardson Method				
m	$x_{m,1}$	$x_{m,2}$	$\varepsilon_m := \|\boldsymbol{x}_m - \boldsymbol{A}^{-1}\boldsymbol{b}\|_\infty$	$\varepsilon_m / \varepsilon_{m-1}$
0	2.100000e+01	-1.900000e+01	2.000000e+01	
15	9.989827e-01	1.000203e+00	1.017253e-03	8.333333e-01
30	1.000000e+00	1.000000e-00	1.862645e-08	3.000000e-01
45	1.000000e-00	1.000000e+00	9.473533e-13	8.331381e-01
52	1.000000e+00	1.000000e-00	4.662937e-15	3.157895e-01

Example 4.36 The convergence history in the maximum norm presented in the above Example 4.35 shows, as already discussed in Example 4.17, a monotonic character with varying factors in the error reduction. We now want to turn to a system of equations in which the Richardson method converges for a chosen parameter $\Theta \in \mathbb{R}$, although with respect to the absolute sum norm, a temporary increase in the error can even be observed. For this, we consider the system of equations

$$\underbrace{\begin{pmatrix} 2 - \frac{9}{5}\cos\left(\frac{\pi}{8}\right) & \frac{9}{5}\sin\left(\frac{\pi}{8}\right) \\ -\frac{9}{5}\sin\left(\frac{\pi}{8}\right) & 2 - \frac{9}{5}\cos\left(\frac{\pi}{8}\right) \end{pmatrix}}_{=A} x = \underbrace{\begin{pmatrix} 0 \\ 0 \end{pmatrix}}_{=b}$$

and use $\Theta = 0.5$ as well as the initial vector $x_0 = (1,0)^T$. Thus, we obtain the iteration formula

$$x_{m+1} = \underbrace{\frac{9}{10}\begin{pmatrix} \cos\left(\frac{\pi}{8}\right) & -\sin\left(\frac{\pi}{8}\right) \\ \sin\left(\frac{\pi}{8}\right) & \cos\left(\frac{\pi}{8}\right) \end{pmatrix}}_{=M_R\left(\frac{1}{2}\right)} x_m \quad \text{for} \quad m = 0,1,\ldots \tag{4.1.28}$$

With $\rho\left(M_R\left(\frac{1}{2}\right)\right) = \frac{9}{10} < 1$, the sequence of iterates converges to the uniquely determined solution $x = (0,0)^T$. From the following table, it is clearly visible that the quotient $\varepsilon_m/\varepsilon_{m-1}$ sometimes takes a value greater than 1, and thus the error in the 1-norm increases through the corresponding iteration step.

Richardson Method				
m	$x_{m,1}$	$x_{m,2}$	$\varepsilon_m := \left\| x_m - A^{-1}b \right\|_1$	$\varepsilon_m/\varepsilon_{m-1}$
0	1.000000e+00	0.000000e+00	1.000000e+00	
1	8.314916e-01	3.444151e-01	1.175907e+00	1.175907e+00
2	5.727565e-01	5.727565e-01	1.145513e+00	9.741530e-01
3	2.789762e-01	6.735082e-01	9.524844e-01	8.314916e-01
4	-1.387779e-16	6.561000e-01	6.561000e-01	6.888302e-01
5	-2.259707e-01	5.455416e-01	7.715124e-01	1.175907e+00
10	-2.465529e-01	-2.465529e-01	4.931058e-01	9.741530e-01
20	-7.632783e-17	1.215767e-01	1.215767e-01	6.888302e-01
45	3.340047e-03	-8.063587e-03	1.140363e-02	1.175907e+00
70	-4.430581e-04	4.430581e-04	8.861162e-04	9.741530e-01
100	-1.118083e-19	2.656140e-05	2.656140e-05	6.888302e-01
150	-9.679689e-08	9.679689e-08	1.935938e-07	9.741530e-01
200	-7.055079e-10	-5.686867e-24	7.055079e-10	6.888302e-01
250	-2.571061e-12	-2.571061e-12	5.142122e-12	9.741530e-01
300	2.374471e-28	-1.873928e-14	1.873928e-14	6.888302e-01
325	-5.148198e-16	1.242885e-15	1.757705e-15	1.175907e+00

The possibility of such behavior is due to the value of the iteration matrix with respect to the corresponding induced matrix norm, since

$$\varepsilon_m \leq \left\| M_R\left(\frac{1}{2}\right) \right\|_1 \varepsilon_{m-1} \quad \text{with} \quad \left\| M_R\left(\frac{1}{2}\right) \right\|_1 = \frac{9}{10}\left(\cos\left(\frac{\pi}{8}\right) + \sin\left(\frac{\pi}{8}\right)\right) \approx 1.176 > 1.$$

On the other hand, it is obvious that multiplying an approximate solution by the iteration matrix $\boldsymbol{M}_R\left(\frac{1}{2}\right)$ results in a rotation about the origin by an angle of $\frac{\pi}{8}$ combined with a contraction by a factor of $\frac{9}{10}$, and thus the sequence of approximate solutions exhibits the spiral-shaped behavior toward the origin visible in Figure 4.5. Due to the contour line of the 1-norm at value 1, which is also drawn in the figure, one can directly see the occasional increase of the corresponding error in this norm. It should also be mentioned that the geometric mean of the quotients $\varepsilon_m/\varepsilon_{m-1}$ over 4 iterations is exactly equal to the spectral radius of the iteration matrix $\boldsymbol{M}_R\left(\frac{1}{2}\right)$. If, on the other hand, the error is measured in the Euclidean norm, a monotonic convergence behavior results. This behavior can be proven by

$$\boldsymbol{M}_R\left(\frac{1}{2}\right)^T \boldsymbol{M}_R\left(\frac{1}{2}\right) = \left(\frac{9}{10}\right)^2 \boldsymbol{I}$$

such that according to Theorem 2.35 one can write

$$\left\|\boldsymbol{M}_R\left(\frac{1}{2}\right)\right\|_2 = \sqrt{\rho\left(\boldsymbol{M}_R\left(\frac{1}{2}\right)^T \boldsymbol{M}_R\left(\frac{1}{2}\right)\right)} = \frac{9}{10}\sqrt{\rho(\boldsymbol{I})} = \frac{9}{10}.$$

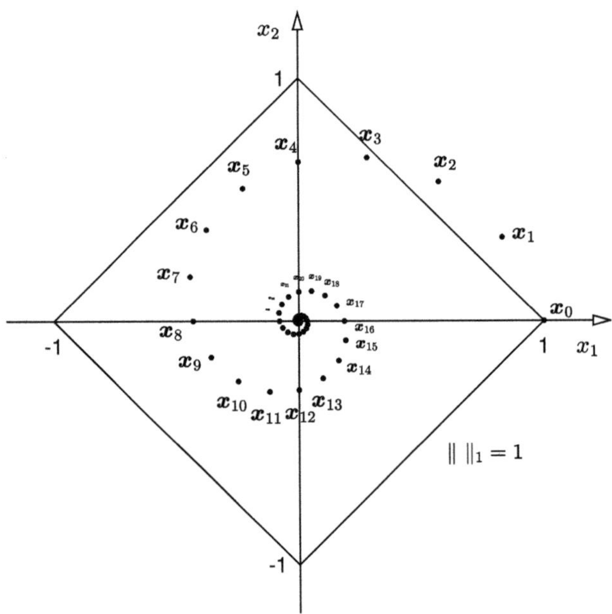

Figure 4.5   Spiral-shaped convergence behavior for the sequence of iterations according to rule (4.1.28)

4.1.5   Symmetric Splitting Methods

The matrix $\boldsymbol{B}$ present in the considered splitting methods

$$x_{m+1} = B^{-1}(B - A)x_m + B^{-1}b$$

can be understood as an easily invertible approximation of the matrix A or at least a matrix B that enables simple matrix-vector multiplications $B^{-1}x$. Due to this, B^{-1} represents a possible preconditioning matrix for further iterative methods. If an iterative method requires the symmetry and positive definiteness of the matrix A, then the preconditioned system, for example

$$\underbrace{B^{-1/2}AB^{-1/2}}_{\tilde{A}\,:=}y = B^{-1/2}b,$$

should also have a symmetric and positive definite matrix $\tilde{A}$. To achieve this goal, symmetric splitting methods are of interest.

Definition 4.37 The splitting method

$$x_{m+1} = B^{-1}(B - A)x_m + B^{-1}b$$

is called symmetric if, for every symmetric, positive definite matrix A, the matrix B is also symmetric and positive definite.

Before we proceed to the construction of symmetric splitting methods, we first want to analyze some well-known methods with respect to this property.

Theorem 4.38

(1) The Jacobi relaxation method is symmetric.

(2) The Richardson method is symmetric if and only if $\Theta \in \mathbb{R}^+$.

(3) The Gauss-Seidel relaxation method is not symmetric.

Proof:
Let $A = D + L + R$ be a positive definite and symmetric matrix, then $a_{ii} > 0$ for $i = 1,\ldots,n$, so that in addition to the Richardson method, the Jacobi and Gauss-Seidel methods are also well-defined.

(1) $B_J(\omega) = \frac{1}{\omega}D = \frac{1}{\omega}\operatorname{diag}\{a_{11},\ldots,a_{nn}\}$ is symmetric and, with $\omega,a_{ii} \in \mathbb{R}^+$ for $i = 1,\ldots,n$, also positive definite[1].

(2) With $B_R(\Theta) = \frac{1}{\Theta}I$, the Richardson method is a symmetric splitting method if and only if $\Theta \in \mathbb{R}^+$.

(3) The Gauss-Seidel relaxation method is, regardless of the specific choice of the relaxation parameter $\omega \in (0,2)$, not a symmetric splitting method, since the matrix $B_{GS}(\omega) = \frac{1}{\omega}(D + \omega L)$ is nonsymmetric for all $L \neq 0$.

$\square$

In the following, we want to present, using the example of the Gauss-Seidel relaxation method, a possibility for symmetrizing nonsymmetric splitting methods. If $A = D + L + R$ corresponds to the usual decomposition of the matrix A, then we define

---

[1] It should be noted here that for all relaxation methods, $\omega > 0$ was assumed.

$$\boldsymbol{x}_{m+1} = \underbrace{-(\boldsymbol{D}+\boldsymbol{R})^{-1}\boldsymbol{L}}_{\boldsymbol{M}_{RGS}\,:=}\boldsymbol{x}_m + \underbrace{(\boldsymbol{D}+\boldsymbol{R})^{-1}}_{\boldsymbol{N}_{RGS}\,:=}\boldsymbol{b} \ \text{for } m = 0,1,\ldots,$$

which is called the backward Gauss-Seidel method. The composition of the two Gauss-Seidel methods corresponds to performing two successive iteration schemes. By combining the two individual steps into a single iteration step, using the intermediate vector

$$\boldsymbol{x}_{m+1/2} = \boldsymbol{M}_{GS}\boldsymbol{x}_m + \boldsymbol{N}_{GS}\boldsymbol{b}$$

one can write the entire method as

$$\boldsymbol{x}_{m+1} = \boldsymbol{M}_{RGS}\boldsymbol{x}_{m+1/2} + \boldsymbol{N}_{RGS}\boldsymbol{b} = \underbrace{\boldsymbol{M}_{RGS}\boldsymbol{M}_{GS}}_{\boldsymbol{M}_{SGS}\,:=}\boldsymbol{x}_m + \underbrace{(\boldsymbol{M}_{RGS}\boldsymbol{N}_{GS} + \boldsymbol{N}_{RGS})}_{\boldsymbol{N}_{SGS}\,:=}\boldsymbol{b}.$$

For the resulting symmetric Gauss-Seidel method the following hold:

$$\begin{aligned}\boldsymbol{M}_{SGS} &= \left(-(\boldsymbol{D}+\boldsymbol{R})^{-1}\boldsymbol{L}\right)\left(-(\boldsymbol{D}+\boldsymbol{L})^{-1}\boldsymbol{R}\right)\\[2mm] &= (\boldsymbol{D}+\boldsymbol{R})^{-1}\boldsymbol{L}(\boldsymbol{D}+\boldsymbol{L})^{-1}\boldsymbol{R}\end{aligned}$$

and

$$\begin{aligned}\boldsymbol{N}_{SGS} &= \left(-(\boldsymbol{D}+\boldsymbol{R})^{-1}\boldsymbol{L}\right)(\boldsymbol{D}+\boldsymbol{L})^{-1} + (\boldsymbol{D}+\boldsymbol{R})^{-1}\\[2mm] &= (\boldsymbol{D}+\boldsymbol{R})^{-1}\left(\boldsymbol{I} - \boldsymbol{L}(\boldsymbol{D}+\boldsymbol{L})^{-1}\right)\\[2mm] &= (\boldsymbol{D}+\boldsymbol{R})^{-1}\boldsymbol{D}(\boldsymbol{D}+\boldsymbol{L})^{-1}.\end{aligned}$$

Analogously, we obtain the symmetric Gauss-Seidel relaxation method (symmetric SOR method = SSOR method) as a combination of the relaxed individual methods in the form

$$\boldsymbol{x}_{m+1} = \boldsymbol{M}_{SGS}(\omega)\boldsymbol{x}_m + \boldsymbol{N}_{SGS}(\omega)\boldsymbol{b}$$

with

$$\boldsymbol{M}_{SGS}(\omega) = (\boldsymbol{D}+\omega\boldsymbol{R})^{-1}\left((1-\omega)\boldsymbol{D}-\omega\boldsymbol{L}\right)(\boldsymbol{D}+\omega\boldsymbol{L})^{-1}\left((1-\omega)\boldsymbol{D}-\omega\boldsymbol{R}\right)$$

and

$$\boldsymbol{N}_{SGS}(\omega) = \omega(2-\omega)\,(\boldsymbol{D}+\omega\boldsymbol{R})^{-1}\boldsymbol{D}(\boldsymbol{D}+\omega\boldsymbol{L})^{-1}. \tag{4.1.29}$$

For every positive definite and Hermitian matrix $\boldsymbol{A} \in \mathbb{C}^{n\times n}$ there exists a unitary matrix $\boldsymbol{Q} \in \mathbb{C}^{n\times n}$ such that

$$\boldsymbol{D} = \boldsymbol{Q}^{*}\boldsymbol{A}\boldsymbol{Q} = \text{diag}\,\{d_{11},\ldots,d_{nn}\} \ \text{with } d_{ii} > 0 \ \text{for } i = 1,\ldots,n.$$

Then, in the following, let

$$\boldsymbol{A}^{1/\alpha} := \boldsymbol{Q}^{*}\boldsymbol{D}^{1/\alpha}\boldsymbol{Q} = \boldsymbol{Q}^{*}\text{diag}\left\{d_{11}^{1/\alpha},\ldots,d_{nn}^{1/\alpha}\right\}\boldsymbol{Q}.$$

In the case of a positive definite, symmetric matrix $\boldsymbol{A} \in \mathbb{R}^{n\times n}$, $\boldsymbol{Q} \in \mathbb{R}^{n\times n}$ is orthogonal.

Theorem 4.39 *The symmetric Gauss-Seidel relaxation method is, for $\omega \in (0,2)$, a consistent symmetric splitting method.*

Proof:
With
$$\boldsymbol{M}_{GS}(\omega) = \boldsymbol{I} - \boldsymbol{N}_{GS}(\omega)\boldsymbol{A} \text{ and } \boldsymbol{M}_{RGS}(\omega) = \boldsymbol{I} - \boldsymbol{N}_{RGS}(\omega)\boldsymbol{A}$$
we obtain
$$\boldsymbol{I} - \boldsymbol{N}_{SGS}(\omega)\boldsymbol{A}$$

$$= \boldsymbol{I} - \left(\boldsymbol{M}_{RGS}(\omega)\boldsymbol{N}_{GS}(\omega) + \boldsymbol{N}_{RGS}(\omega)\right)\boldsymbol{A}$$

$$= \underbrace{\boldsymbol{I} - \boldsymbol{N}_{RGS}(\omega)\boldsymbol{A}}_{=\boldsymbol{M}_{RGS}(\omega)} - \boldsymbol{M}_{RGS}(\omega)\,\underbrace{\boldsymbol{N}_{GS}(\omega)\boldsymbol{A}}_{=\boldsymbol{I}-\boldsymbol{M}_{GS}(\omega)}$$

$$= \boldsymbol{M}_{RGS}(\omega)\boldsymbol{M}_{GS}(\omega)$$

$$= \boldsymbol{M}_{SGS}(\omega), \tag{4.1.30}$$

which shows that the SSOR represents a consistent splitting method.

Let $\boldsymbol{A}$ be positive definite and symmetric, then $a_{ii} > 0$ ($i = 1,\ldots,n$) and $\boldsymbol{L} = \boldsymbol{R}^T$.
Let $\boldsymbol{D}^{1/2} = \mathrm{diag}\left\{\sqrt{a_{11}},\ldots,\sqrt{a_{nn}}\right\}$, then, with $\boldsymbol{B}_{SGS}(\omega) = \boldsymbol{N}_{SGS}^{-1}(\omega)$, we obtain the equation

$$\boldsymbol{B}_{SGS}(\omega) = \frac{1}{\omega(2-\omega)}(\boldsymbol{D} + \omega\boldsymbol{L})\boldsymbol{D}^{-1}(\boldsymbol{D} + \omega\boldsymbol{R})$$

$$= \frac{1}{\sqrt{\omega(2-\omega)}}\left(\boldsymbol{D} + \omega\boldsymbol{R}^T\right)\boldsymbol{D}^{-1/2}\underbrace{\frac{1}{\sqrt{\omega(2-\omega)}}\boldsymbol{D}^{-1/2}(\boldsymbol{D} + \omega\boldsymbol{R})}_{\boldsymbol{C}:=}$$

$$= \boldsymbol{C}^T\boldsymbol{C}.$$

The assumptions $\omega \in (0,2)$ and $a_{ii} > 0$ ($i = 1,\ldots,n$) ensure the invertibility of the matrix $\boldsymbol{C}$, which, in addition to symmetry, also yields the positive definiteness of the matrix $\boldsymbol{B}_{SGS} = \boldsymbol{C}^T\boldsymbol{C}$. $\qquad\square$

Theorem and Definition 4.40 *Let $\boldsymbol{A} \in \mathbb{C}^{n\times n}$ be a positive definite and Hermitian matrix, then*

$$\|\cdot\|_{\boldsymbol{A}} : \mathbb{C}^n \;\rightarrow\; \mathbb{R}$$

$$\boldsymbol{x} \;\mapsto\; \|\boldsymbol{x}\|_{\boldsymbol{A}} := \left\|\boldsymbol{A}^{1/2}\boldsymbol{x}\right\|_2$$

defines a norm. This norm is called the energy norm (with respect to $\boldsymbol{A}$). The corresponding matrix norm is given by

$$\|\cdot\|_{\boldsymbol{A}} : \mathbb{C}^{n\times n} \;\rightarrow\; \mathbb{R}$$

$$\boldsymbol{B} \;\mapsto\; \|\boldsymbol{B}\|_{\boldsymbol{A}} = \left\|\boldsymbol{A}^{1/2}\boldsymbol{B}\boldsymbol{A}^{-1/2}\right\|_2.$$

Proof:

The proof of the first statement is based on straightforward verification of the vector norm axioms and is left as an exercise. For the matrix norm, we obtain

$$
\|\boldsymbol{B}\|_{\boldsymbol{A}} \;=\; \sup_{\|\boldsymbol{x}\|_{\boldsymbol{A}}=1} \|\boldsymbol{B}\boldsymbol{x}\|_{\boldsymbol{A}} = \sup_{\|\boldsymbol{A}^{1/2}\boldsymbol{x}\|_2=1} \|\boldsymbol{A}^{1/2}\boldsymbol{B}\boldsymbol{x}\|_2
$$

$$
\;=\; \sup_{\|\boldsymbol{y}\|_2=1} \|\boldsymbol{A}^{1/2}\boldsymbol{B}\boldsymbol{A}^{-1/2}\boldsymbol{y}\|_2 = \left\|\boldsymbol{A}^{1/2}\boldsymbol{B}\boldsymbol{A}^{-1/2}\right\|_2.
$$

$\square$

Theorem 4.41 *Let $\boldsymbol{A} \in \mathbb{C}^{n\times n}$ be Hermitian and positive definite, then the symmetric Gauss-Seidel relaxation method converges for all $\omega \in (0,2)$, and it holds that*

$$
\rho\left(\boldsymbol{M}_{SGS}(\omega)\right) = \|\boldsymbol{M}_{SGS}(\omega)\|_{\boldsymbol{A}} = \|\boldsymbol{M}_{GS}(\omega)\|_{\boldsymbol{A}}^2
$$

as well as

$$
\sigma\left(\boldsymbol{M}_{SGS}(\omega)\right) \subset \left[0, \rho\left(\boldsymbol{M}_{SGS}(\omega)\right)\right].
$$

Proof:

Since $\boldsymbol{A}^{1/2} \in \mathbb{C}^{n\times n}$ is non-singular, $\boldsymbol{M}_{SGS}(\omega)$ is similar to

$$
\boldsymbol{A}^{1/2}\boldsymbol{M}_{SGS}(\omega)\boldsymbol{A}^{-1/2} = \left(\boldsymbol{A}^{1/2}\boldsymbol{M}_{RGS}(\omega)\boldsymbol{A}^{-1/2}\right)\left(\boldsymbol{A}^{1/2}\boldsymbol{M}_{GS}(\omega)\boldsymbol{A}^{-1/2}\right). \tag{4.1.31}
$$

With

$$
\boldsymbol{M}_{RGS}(\omega) = (\boldsymbol{D}+\omega\boldsymbol{R})^{-1}\left((1-\omega)\boldsymbol{D}-\omega\boldsymbol{L}\right) = \boldsymbol{I} - \omega(\boldsymbol{D}+\omega\boldsymbol{R})^{-1}\boldsymbol{A}
$$

and

$$
\boldsymbol{M}_{GS}(\omega) = (\boldsymbol{D}+\omega\boldsymbol{L})^{-1}\left((1-\omega)\boldsymbol{D}-\omega\boldsymbol{R}\right) = \boldsymbol{I} - \omega(\boldsymbol{D}+\omega\boldsymbol{L})^{-1}\boldsymbol{A}
$$

we obtain

$$
\begin{aligned}
\boldsymbol{A}^{1/2}\boldsymbol{M}_{RGS}(\omega)\boldsymbol{A}^{-1/2} &= \boldsymbol{I} - \omega\boldsymbol{A}^{1/2}(\boldsymbol{D}+\omega\boldsymbol{R})^{-1}\boldsymbol{A}^{1/2} \\[2mm]
&= \left(\boldsymbol{I} - \omega\boldsymbol{A}^{1/2}\left((\boldsymbol{D}+\omega\boldsymbol{R})^{-1}\right)^{*}\boldsymbol{A}^{1/2}\right)^{*} \\[2mm]
&= \left(\boldsymbol{I} - \omega\boldsymbol{A}^{1/2}(\boldsymbol{D}+\omega\boldsymbol{L})^{-1}\boldsymbol{A}^{1/2}\right)^{*} \\[2mm]
&= \left(\boldsymbol{A}^{1/2}\boldsymbol{M}_{GS}(\omega)\boldsymbol{A}^{-1/2}\right)^{*}. \tag{4.1.32}
\end{aligned}
$$

From $\boldsymbol{A}^{1/2}\boldsymbol{M}_{SGS}(\omega)\boldsymbol{A}^{-1/2} = \left(\boldsymbol{A}^{1/2}\boldsymbol{M}_{GS}(\omega)\boldsymbol{A}^{-1/2}\right)^{*}\left(\boldsymbol{A}^{1/2}\boldsymbol{M}_{GS}(\omega)\boldsymbol{A}^{-1/2}\right)$ it follows that

$$
\sigma\left(\boldsymbol{M}_{SGS}(\omega)\right) = \sigma\left(\boldsymbol{A}^{1/2}\boldsymbol{M}_{SGS}(\omega)\boldsymbol{A}^{-1/2}\right) \subset \mathbb{R}_0^{+}.
$$

Thus, it holds that

$$\sigma\left(M_{SGS}(\omega)\right) \subset \left[0, \rho\left(M_{SGS}(\omega)\right)\right]. \tag{4.1.33}$$

From (4.1.31) and (4.1.32) it is immediately apparent that the matrix $A^{1/2}M_{SGS}(\omega)A^{-1/2}$ is Hermitian and thus

$$
\begin{aligned}
\rho(M_{SGS}(\omega)) \quad &= \quad \rho\left(A^{1/2}M_{SGS}(\omega)A^{-1/2}\right) \\[2mm]
&\stackrel{\substack{\text{Theorem 2.36}}}{=} \quad \left\|A^{1/2}M_{SGS}(\omega)A^{-1/2}\right\|_2 \;=\; \left\|M_{SGS}(\omega)\right\|_A \\[2mm]
&\stackrel{\substack{(4.1.31),(4.1.32)}}{=} \quad \left\|\left(A^{1/2}M_{GS}(\omega)A^{-1/2}\right)^*\left(A^{1/2}M_{GS}(\omega)A^{-1/2}\right)\right\|_2 \\[2mm]
&= \quad \left\|A^{1/2}M_{GS}(\omega)A^{-1/2}\right\|_2^2 = \left\|M_{GS}(\omega)\right\|_A^2
\end{aligned}
$$

follows.

For Hermitian matrices, we introduce the following order relation: Let the matrices $F, G \in \mathbb{C}^{n\times n}$ be Hermitian, then we write $G > F$ if $G - F$ is positive definite. Let $F, G \in \mathbb{C}^{n\times n}$ be Hermitian with $G > F$ and F given as positive definite, then let $\lambda \in \mathbb{R}^+$ be an eigenvalue of F with $\lambda = \rho(F)$ and corresponding eigenvector $x \in \mathbb{C}^n$ with $\|x\|_2 = 1$. Thus, we obtain

$$
\begin{aligned}
0 \;&<\; \rho(F) = (Fx, x)_2 \;\stackrel{(G>F)}{<}\; (Gx, x)_2 \leq \|Gx\|_2\|x\|_2 \\[2mm]
&\leq\; \|G\|_2\|x\|_2^2 = \|G\|_2 = \rho(G). \tag{4.1.34}
\end{aligned}
$$

By assumption, A is positive definite and Hermitian, which for $\omega \in (0,2)$ yields the equation

$$
\begin{aligned}
N_{GS}^{-1}(\omega) + \underbrace{N_{GS}^{-*}(\omega)}_{:=\left(N_{GS}^{-1}(\omega)\right)^*} \;&=\; \frac{1}{\omega}(D + \omega L) + \frac{1}{\omega}(D + \omega R) \\[4mm]
&=\; L + R + D + \left(\frac{2}{\omega} - 1\right)D \\[4mm]
&=\; A + \left(\frac{2}{\omega} - 1\right)D > A. \tag{4.1.35}
\end{aligned}
$$

Let us consider the matrix

$$
\begin{aligned}
\widehat{M}_{GS}(\omega) \;&:=\; A^{1/2}M_{GS}(\omega)A^{-1/2} \\[2mm]
&=\; A^{1/2}\left(I - N_{GS}(\omega)A\right)A^{-1/2} \\[2mm]
&=\; I - A^{1/2}N_{GS}(\omega)A^{1/2},
\end{aligned}
$$

then it follows that

$$\widehat{\boldsymbol{M}}^{*}_{GS}(\omega)\widehat{\boldsymbol{M}}_{GS}(\omega)$$

$$= \left(\boldsymbol{I} - \boldsymbol{A}^{1/2}\boldsymbol{N}^{*}_{GS}(\omega)\boldsymbol{A}^{1/2}\right)\left(\boldsymbol{I} - \boldsymbol{A}^{1/2}\boldsymbol{N}_{GS}(\omega)\boldsymbol{A}^{1/2}\right)$$

$$= \boldsymbol{I} - \boldsymbol{A}^{1/2}\left(\boldsymbol{N}^{*}_{GS}(\omega) + \boldsymbol{N}_{GS}(\omega)\right)\boldsymbol{A}^{1/2}$$

$$+ \boldsymbol{A}^{1/2}\boldsymbol{N}^{*}_{GS}(\omega)\boldsymbol{A}\boldsymbol{N}_{GS}(\omega)\boldsymbol{A}^{1/2}$$

$$= \boldsymbol{I} - \boldsymbol{A}^{1/2}\boldsymbol{N}^{*}_{GS}(\omega)\left(\boldsymbol{N}^{-1}_{GS}(\omega) + \boldsymbol{N}^{-*}_{GS}(\omega)\right)\boldsymbol{N}_{GS}(\omega)\boldsymbol{A}^{1/2}$$

$$+ \boldsymbol{A}^{1/2}\boldsymbol{N}^{*}_{GS}(\omega)\boldsymbol{A}\boldsymbol{N}_{GS}(\omega)\boldsymbol{A}^{1/2}$$

$$= \boldsymbol{I} - \boldsymbol{A}^{1/2}\boldsymbol{N}^{*}_{GS}(\omega)\left(\boldsymbol{N}^{-1}_{GS}(\omega) + \boldsymbol{N}^{-*}_{GS}(\omega) - \boldsymbol{A}\right)\boldsymbol{N}_{GS}(\omega)\boldsymbol{A}^{1/2}$$

$$\overset{(4.1.35)}{<} \boldsymbol{I}. \tag{4.1.36}$$

With (4.1.34) we obtain $\rho\left(\boldsymbol{M}_{SGS}(\omega)\right) = \rho\left(\widehat{\boldsymbol{M}}^{*}_{GS}(\omega)\widehat{\boldsymbol{M}}_{GS}(\omega)\right) < \rho(\boldsymbol{I}) = 1.$ $\square$

The above proof also contains the statement that for a positive definite matrix $\boldsymbol{A}$, the Gauss-Seidel relaxation method for $\omega \in (0,2)$ converges strictly monotonically in the energy norm.

4.2 Multigrid Methods

Multigrid methods are frequently used to solve linear systems of equations that arise, for example, from the discretization of elliptic differential equations (Poisson equation). In addition, they are also suitable for accelerating convergence within explicit numerical methods for solving hyperbolic (Euler equations) and hyperbolic-parabolic (Navier-Stokes equations) differential equations [11, 4, 31]. Although in the case of these partial differential equations, a linear system is usually not considered, the basic approach remains identical.

In this section, we will introduce the basic idea of multigrid methods and study their effect in solving a specific model problem. Statements about convergence can be found, for example, in [38, 36].

We simplify the Poisson equation presented in Example 1.1 to the one-dimensional Dirichlet boundary value problem.

Given: $\Omega = (0,1)$ and $f \in C(\Omega; \mathbb{R})$

Find: $u \in C^{2}(\Omega; \mathbb{R}) \cap C(\overline{\Omega}; \mathbb{R})$ such that

$$\begin{aligned} -u''(x) &= f(x) &\text{for } x \in \Omega, \\ u(x) &= 0 &\text{for } x \in \partial\Omega = \{0,1\}. \end{aligned} \tag{4.2.1}$$

We define the sequence of step sizes $\{h_\ell\}_{\ell=0}^{\infty}$ with $h_0 = \frac{1}{2}$ and $h_\ell = h_0/2^\ell = 2^{-(\ell+1)}$, as well as the sequence of grids

$$\Omega_\ell := \Omega_{h_\ell} = \{jh_\ell \mid j = 1,\ldots,2^{\ell+1} - 1\} \text{ for } \ell = 0,1,\ldots ,$$

where ℓ is referred to as the level number or level index.

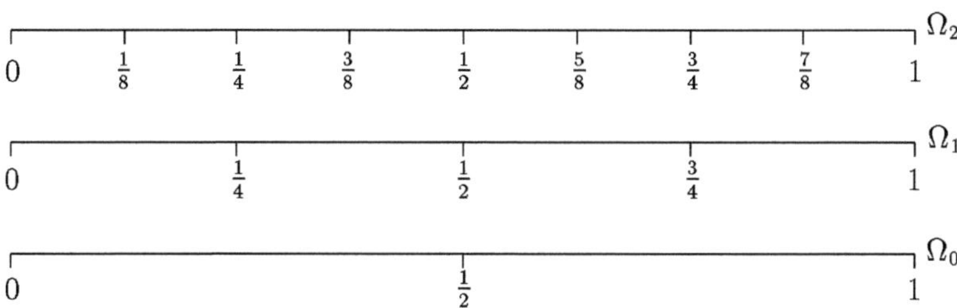

Figure 4.6   Grid hierarchy

If we again approximate the derivative of the sought function u using a central difference quotient, we obtain, with

$$u_j^\ell := u\,(jh_\ell) \text{ for } j = 1,\ldots,N_\ell := 2^{\ell+1} - 1$$

and

$$f_j^\ell := f\,(jh_\ell) \text{ for } j = 1,\ldots,N_\ell$$

taking into account the boundary conditions, the linear system of equations at level ℓ in the form

$$A_\ell u^\ell = f^\ell$$

with

$$u^\ell = \left(u_1^\ell,\ldots,u_{N_\ell}^\ell\right)^T , \quad f^\ell = \left(f_1^\ell,\ldots,f_{N_\ell}^\ell\right)^T$$

and

$$A_\ell = \frac{1}{h_\ell^2}\begin{pmatrix} 2 & -1 & & & \\ -1 & 2 & -1 & & \\ & \ddots & \ddots & \ddots & \\ & & \ddots & \ddots & -1 \\ & & & -1 & 2 \end{pmatrix} \in \mathbb{R}^{N_\ell \times N_\ell}. \tag{4.2.2}$$

The matrix A_ℓ is irreducible and diagonally dominant, so that both the Jacobi and the Gauss-Seidel method compute, for all initial values $u_0 \in \mathbb{R}^{N_\ell}$, a sequence of approximate solutions converging to the solution of the system.

Remark:

This model problem is very well suited for describing the multigrid algorithm. However, it should be mentioned at this point that such systems of equations should be solved directly in practice. The reason for this lies in the special structure of the matrix A_ℓ. First, $\det A_\ell[k] \neq 0$ for $k = 1,\ldots,n$, which, according to Theorem 3.7, means that an LR factorization of the matrix A_ℓ can be determined without pivoting. If one takes into account the tridiagonal structure of the matrix during the decomposition, the system of equations can be solved using Gaussian elimination in such a way that the computational effort is proportional to the number of unknowns. An explicit derivation of this method is described in [65].

We consider the Jacobi relaxation method

$$\boldsymbol{u}^\ell_{m+1} = \boldsymbol{u}^\ell_m + \tilde{\omega}\boldsymbol{D}^{-1}_\ell\left(\boldsymbol{f}^\ell - \boldsymbol{A}_\ell\boldsymbol{u}^\ell_m\right) \text{ for } m = 0,1,\ldots,$$

which, due to the special form of the diagonal matrix D_ℓ with $\omega = \frac{1}{2}\tilde{\omega}$, takes the form

$$\begin{aligned}
\boldsymbol{u}^\ell_{m+1} &= \boldsymbol{u}^\ell_m + \omega h^2_\ell\left(\boldsymbol{f}^\ell - \boldsymbol{A}_\ell\boldsymbol{u}^\ell_m\right) \\
&= \underbrace{\left(\boldsymbol{I} - \omega h^2_\ell\boldsymbol{A}_\ell\right)}_{\boldsymbol{M}_\ell(\omega)\,:=}\boldsymbol{u}^\ell_m + \underbrace{\omega h^2_\ell\boldsymbol{I}}_{\boldsymbol{N}_\ell(\omega)\,:=}\boldsymbol{f}^\ell
\end{aligned} \tag{4.2.3}$$

and corresponds to the Richardson method with $\Theta = \omega h^2_\ell$. We will first investigate the convergence properties of the method.

Since the eigenfunctions of the homogeneous boundary value problem (4.2.1) are given by

$$u(x) = c\,\sin(j\pi x) \text{ with } j \in \mathbb{N} \text{ and } c \in \mathbb{R},$$

we can already anticipate that the eigenvectors $\boldsymbol{e}^{\ell,j}$ of the matrix $\boldsymbol{A}_\ell$ are represented by

$$\boldsymbol{e}^{\ell,j} = \sqrt{2h_\ell}\begin{pmatrix} \sin(j\pi h_\ell) \\ \vdots \\ \sin(j\pi N_\ell h_\ell) \end{pmatrix} \text{ for } j = 1,\ldots,N_\ell \tag{4.2.4}$$

which represent their discrete formulation. Following the highly recommended book by Briggs [15], we will refer to the vectors $\boldsymbol{e}^{\ell,j}$ as *Fourier modes* or simply *modes* for the respective wavenumber $j = 1,\ldots,N_\ell$ with respect to the grid Ω_ℓ. The property of the modes as eigenvectors of the matrix $\boldsymbol{A}_\ell$, already anticipated above, will now be mathematically demonstrated, including the specification of the eigenvalues.

Lemma 4.42 *The Fourier modes $\boldsymbol{e}^{\ell,j}$, $j = 1,\ldots,N_\ell$ given by (4.2.4) are eigenvectors of the matrix $\boldsymbol{A}_\ell$ with the corresponding eigenvalues $\lambda^{\ell,j} = 4h^{-2}_\ell\sin^2\left(\frac{j\pi h_\ell}{2}\right)$.*

Proof:
Starting from the trigonometric identities

$$\sin(x \pm y) = \sin x \cos y \pm \cos x \sin y \tag{4.2.5}$$
$$\cos(x \pm y) = \cos x \cos y \mp \sin x \sin y$$

we obtain by simple combination

$$\sin(x - y) + \sin(x + y) = 2\sin x \cos y \quad \text{and} \quad \cos(2x) = \cos^2 x - \sin^2 x \tag{4.2.6}$$

respectively

$$1 - \cos(2x) = \sin^2 x + \cos^2 x - (\cos^2 x - \sin^2 x) = 2\sin^2 x.$$

Taking into account that

$$\sin(j\pi \cdot 0 \cdot h_\ell) = \sin(j\pi \underbrace{(N_\ell + 1)h_\ell}_{=1}) = 0$$

for $k = 1, \ldots, N_\ell$, we obtain the representation

$$
\begin{aligned}
(\boldsymbol{A}_\ell e^{\ell,j})_k &= \frac{\sqrt{2h_\ell}}{h_\ell^2} \big[-\sin(j\pi(k-1)h_\ell) + 2\sin(j\pi k h_\ell) - \sin(j\pi(k+1)h_\ell) \big] \\
&= \frac{\sqrt{2h_\ell}}{h_\ell^2} \big[2\sin(j\pi k h_\ell) - 2\sin(j\pi k h_\ell)\cos(j\pi h_\ell) \big] \\
&= \frac{2}{h_\ell^2} \big[1 - \cos(j\pi h_\ell) \big] e_k^{\ell,j} = \frac{4}{h_\ell^2} \sin^2\left(\frac{j\pi h_\ell}{2}\right) e_k^{\ell,j}.
\end{aligned}
$$

$$\square$$

With $\boldsymbol{M}_\ell(\omega) = \boldsymbol{I} - \omega h_\ell^2 \boldsymbol{A}_\ell$ the eigenvectors of $\boldsymbol{A}_\ell$ and $\boldsymbol{M}_\ell(\omega)$ coincide, and the eigenvalues of the iteration matrix are

$$\lambda^{\ell,j}(\omega) = 1 - 4\omega \sin^2\left(\frac{j\pi h_\ell}{2}\right) \quad \text{for } j = 1, \ldots, N_\ell. \tag{4.2.7}$$

Lemma 4.43 *The Fourier modes $\{e^{\ell,1}, \ldots, e^{\ell,N_\ell}\}$ given by (4.2.4) form an orthonormal basis of $\mathbb{R}^{N_\ell}$.*

Proof:
Let i be the imaginary unit and $z = e^{i\frac{2\pi j}{N_\ell+1}} \in \mathbb{C}$ with $j \in \mathbb{Z}$. For $\frac{j}{N_\ell+1} \in \mathbb{Z}$ it follows directly that $z = 1$ and thus

$$\sum_{k=1}^{N_\ell+1} z^k = N_\ell + 1. \tag{4.2.8}$$

If $\frac{j}{N_\ell+1} \notin \mathbb{Z}$, then $z \neq 1 = z^{N_\ell+1}$ and consequently

$$\sum_{k=1}^{N_\ell+1} z^k = z\frac{z^{N_\ell+1} - 1}{z - 1} = 0. \tag{4.2.9}$$

In summary, from equations (4.2.8) and (4.2.9) we obtain for the real part of the considered sums the representation

$$\sum_{k=1}^{N_\ell+1} \cos\left(j\frac{2\pi k}{N_\ell + 1}\right) = \begin{cases} 0, & \text{if } \frac{j}{N_\ell+1} \notin \mathbb{Z}, \\ N_\ell + 1 & \text{otherwise.} \end{cases}$$

The orthogonality of the vectors is obtained for $j,m \in \{1, \ldots, N_\ell\}$ by using

$$\cos((j-m)\pi) - \cos((j+m)\pi) = 0 \tag{4.2.10}$$

via

$$(e^{\ell,j}, e^{\ell,m})_2 \quad = \quad \frac{2}{N_\ell + 1} \sum_{k=1}^{N_\ell} \sin\left(j\frac{\pi k}{N_\ell + 1} \right) \sin\left(m\frac{\pi k}{N_\ell + 1} \right)$$

$$= \quad \frac{1}{N_\ell + 1} \sum_{k=1}^{N_\ell} \left\{ \cos\left((j-m)\frac{\pi k}{N_\ell + 1} \right) - \cos\left((j+m)\frac{\pi k}{N_\ell + 1} \right) \right\}$$

$$\stackrel{(4.2.10)}{=} \quad \frac{1}{N_\ell + 1} \sum_{k=1}^{N_\ell+1} \left\{ \cos\left((j-m)\frac{\pi k}{N_\ell + 1} \right) - \cos\left((j+m)\frac{\pi k}{N_\ell + 1} \right) \right\}$$

$$= \quad \begin{cases} 0 & \text{for } j \neq m, \\ 1 & \text{for } j = m. \end{cases}$$

The basis property follows directly from the orthonormality of the vectors. $\qquad\square$

Since the matrix A is, as a tridiagonal matrix, consistently ordered according to Example 4.28, there is a distribution of the eigenvalues of the iteration matrix M_J of the Jacobi method that is symmetric about zero. Taking this property into account, a relaxation of the method according to Theorem 4.22 cannot achieve a reduction of the spectral radius $\rho(M_J)$. The advantage of relaxation lies rather in the damping properties of the method with respect to different error frequencies, which we will examine in the following.

With the previous lemma, the error between the initial vector u_0^ℓ and the exact solution $u^{\ell,*} = A_\ell^{-1} f^\ell$ can be written as

$$u_0^\ell - u^{\ell,*} = \sum_{j=1}^{N_\ell} \alpha_j e^{\ell,j}, \quad \alpha_j \in \mathbb{R}$$

and we obtain

$$u_1^\ell - u^{\ell,*} \quad = \quad M_\ell(\omega) u_0^\ell + N_\ell(\omega) f^\ell - \left(M_\ell(\omega) u^{\ell,*} + N_\ell(\omega) f^\ell \right)$$

$$= \quad M_\ell(\omega) \left(u_0^\ell - u^{\ell,*} \right) = \sum_{j=1}^{N_\ell} \alpha_j \lambda^{\ell,j}(\omega) e^{\ell,j}$$

and accordingly

$$u_m^\ell - u^{\ell,*} = \sum_{j=1}^{N_\ell} \alpha_j \left[\lambda^{\ell,j}(\omega) \right]^m e^{\ell,j} \quad \text{for } m = 0,1,\dots .$$

This representation of the iteration-dependent error behavior clearly shows the connection between the eigenvalues and the convergence behavior of the Jacobi method. Since each eigenvector $e^{\ell,j}$ represents a discrete sample of the sine waves $\sin(j\pi x)$ for $x = h_\ell, 2h_\ell, \dots, N_\ell h_\ell$, we will refer to $1 \leq j < \frac{N_\ell+1}{2}$ as *long-wavelength* or *smooth* modes and to $\frac{N_\ell+1}{2} \leq j \leq N_\ell$ as *short-wavelength* or *oscillatory* modes.

For the eigenvalues $\lambda^{\ell,j}(\omega)$, in the case of the classical Jacobi method ($\tilde{\omega} = 1$, i.e., $\omega = \tilde{\omega}/2 = 1/2$), we obtain the graphical distribution shown in Figure 4.7, where the discrete values for $\ell = 3$ are marked by $\circ$.

The symmetry within the eigenvalue distribution, which is visually illustrated in Figure 4.7 for the case $\ell = 3$, will now be proven in general.

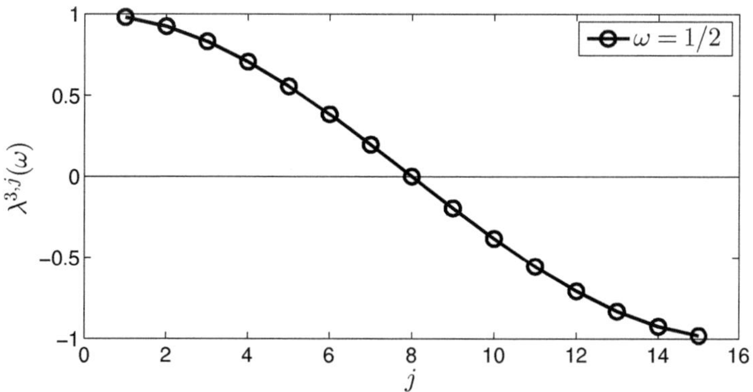

Figure 4.7   Eigenvalue distribution of the iteration matrix of the classical Jacobi method

Lemma 4.44 *The eigenvalues of the iteration matrix of the classical Jacobi method for the matrix $\boldsymbol{A}_\ell$ according to (4.2.2) satisfy the symmetry property*

$$\lambda^{\ell,j+k} = -\lambda^{\ell,j-k}$$

for $k = 0,\ldots,2^\ell - 1$ and $j = \frac{N_\ell+1}{2}$.

From the lemma, we see that $\lambda^{\ell,\frac{N_\ell+1}{2}} = 0$ holds, and the remaining eigenvalues always appear in pairs with positive and negative signs.

Proof:
With $j = \frac{N_\ell+1}{2}$ we obtain $jh_\ell = \frac{1}{2}$ and thus

$$\sin\left(\frac{j\pi h_\ell}{2}\right) = \sin\left(\frac{\pi}{4}\right) = \frac{\sqrt{2}}{2} = \cos\left(\frac{\pi}{4}\right) = \cos\left(\frac{j\pi h_\ell}{2}\right).$$

This property, in combination with the trigonometric identity (4.2.5), yields

$$\sin\left(\frac{(j\pm k)\pi h_\ell}{2}\right) = \sin\left(\frac{j\pi h_\ell}{2}\right)\cos\left(\frac{k\pi h_\ell}{2}\right) \pm \cos\left(\frac{j\pi h_\ell}{2}\right)\sin\left(\frac{k\pi h_\ell}{2}\right)$$
$$= \frac{\sqrt{2}}{2}\left(\cos\left(\frac{k\pi h_\ell}{2}\right) \pm \sin\left(\frac{k\pi h_\ell}{2}\right)\right).$$

Simple squaring then gives

$$\sin^2\left(\frac{(j\pm k)\pi h_\ell}{2}\right) = \frac{1}{2}\left(1 \pm 2\cos\left(\frac{k\pi h_\ell}{2}\right)\sin\left(\frac{k\pi h_\ell}{2}\right)\right)$$

respectively

$$1 - 2\sin^2\left(\frac{(j\pm k)\pi h_\ell}{2}\right) = \mp 2\cos\left(\frac{k\pi h_\ell}{2}\right)\sin\left(\frac{k\pi h_\ell}{2}\right).$$

On this basis, we can now examine the eigenvalues of the iteration matrix of the Jacobi method in more detail. With (4.2.7) and $\omega = \frac{1}{2}$, the claimed relationship follows as

$$\lambda^{\ell,j+k} = 1 - 2\sin^2\left(\frac{(j+k)\pi h_\ell}{2}\right) = -2\cos\left(\frac{k\pi h_\ell}{2}\right)\sin\left(\frac{k\pi h_\ell}{2}\right)$$

$$= -\left(1 - 2\sin^2\left(\frac{(j-k)\pi h_\ell}{2}\right)\right) = -\lambda^{\ell,j-k}.$$

$\square$

In summary, with the above considerations, we can state three essential results:

(1) The Jacobi method yields a rapid decrease of error components located in the mid-frequency range (j close to $\frac{N_\ell+1}{2}$), while the high- and low-frequency components (j close to 1 or N_ℓ) are damped much less effectively.

(2) The finer the discretization is chosen, i.e., the higher the level ℓ, the larger the spectral radius of the iteration matrix $\rho\left(M_\ell\left(\frac{1}{2}\right)\right)$ of the Jacobi method becomes.

(3) It always holds that

$$\lambda^{\ell,1}\left(\frac{1}{2}\right) = -\lambda^{\ell,N_\ell}\left(\frac{1}{2}\right) = \rho\left(M_\ell\left(\frac{1}{2}\right)\right),$$

so that the spectral radius of the iteration matrix cannot be reduced by a simple relaxation of the Jacobi method.

If we consider the Taylor expansion of the spectral radius with respect to the step size, we obtain

$$\rho\left(M_\ell(\omega)\right) = 1 - 4\omega\left(\sum_{n=0}^{\infty}(-1)^n\frac{\left(\pi\frac{h_\ell}{2}\right)^{2n+1}}{(2n+1)!}\right)^2 = 1 - 4\omega\frac{\pi^2 h_\ell^2}{4} + O\left(h_\ell^4\right),$$

so that the spectral radius converges quadratically to 1 as the step size decreases. Thus, refining the grid for improving the accuracy of the numerical solution not only increases the computational effort per iteration, but also drastically reduces the convergence rate.

Although the choice $\omega = \frac{1}{2}$ yields the optimal convergence rate for the described method, it also entails the problem that both long-wavelength and short-wavelength error components are damped very poorly.

However, for the multigrid method, it will turn out to be fundamental to damp the high-frequency error components on the finest grid. We therefore vary the relaxation parameter $\tilde{\omega} = 2/3, 1/2, 1/4$, i.e., $\omega = \tilde{\omega}/2$ takes the values $1/3, 1/4, 1/8$. The eigenvalues $\lambda^{\ell,j}(\omega)$ thus obtained for $j = 1,\ldots,N_\ell$ and $\ell = 3$ are illustrated in Figure 4.8.

By means of some numerical experiments, we want to illustrate the damping properties expected from the eigenvalue distribution. For this purpose, we apply the Jacobi method for the four relaxation parameters $\omega = 1/2$, i.e., the classical Jacobi method, as well as $\omega = 1/3, 1/4, 1/8$ to the individual modes $e^{\ell,j}$, $j = 1,\ldots,N_\ell$, where we again choose $\ell = 3$.

Figure 4.9 shows the number of iterations m that are necessary for each wave number j to satisfy the condition

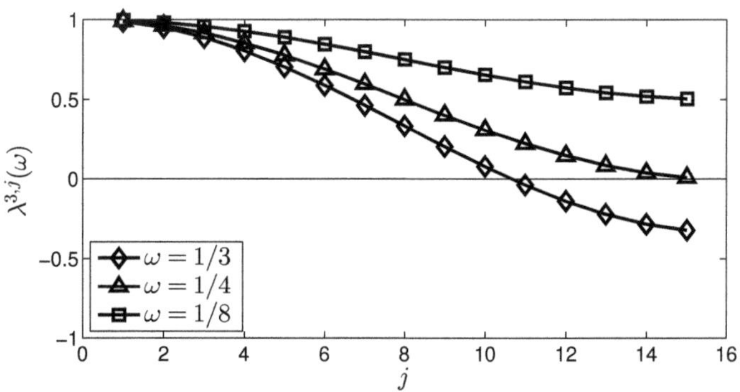

Figure 4.8 Eigenvalue distribution of the iteration matrix of the damped Jacobi method for varying relaxation parameters

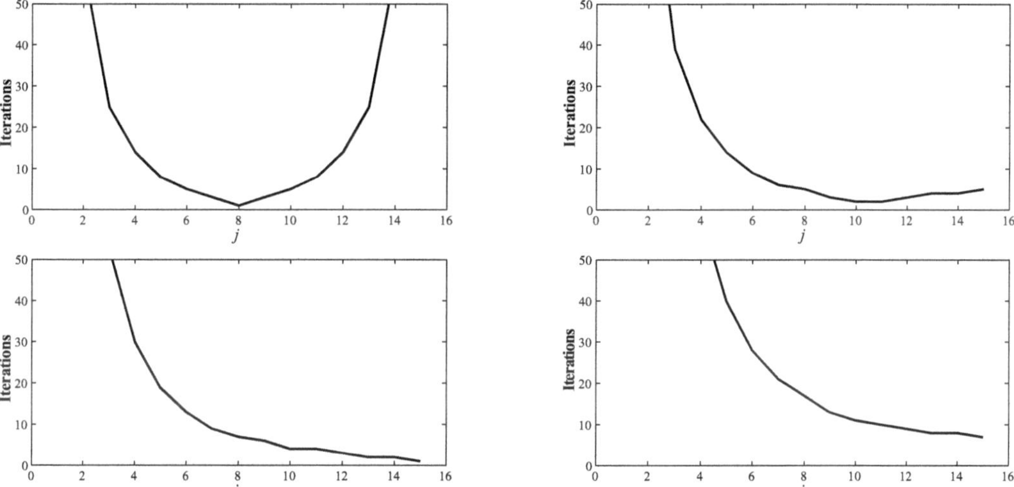

Figure 4.9 Damping behavior of the differently weighted Jacobi methods, $\omega = \frac{1}{2}$ (top left), $\omega = \frac{1}{3}$ (top right), $\omega = \frac{1}{4}$ (bottom left), $\omega = \frac{1}{8}$ (bottom right), with respect to the wave numbers j of the Fourier modes $e^{3,j}$

$$\|\boldsymbol{M}_J(\omega)^m \boldsymbol{e}^{\ell,j}\|_2 \leq 10^{-2} \underbrace{\|\boldsymbol{e}^{\ell,j}\|_2}_{=1} = 10^{-2}$$

For the classical Jacobi method ($\omega = \frac{1}{2}$), we observe the already discussed effect of weak damping for high and low wave numbers, while modes with medium wave numbers are damped quickly. By varying the relaxation parameter, this behavior can be shifted with respect to the wave number. The best damping property for oscillatory modes is exhibited by the damped Jacobi method with $\omega = \frac{1}{4}$. Using the parameter value $\omega = \frac{1}{3}$ yields the fastest decrease for this method at wave number $j = 11$, and when applying $\omega = \frac{1}{8}$, the number of iterations always decreases as the wave number increases, although the absolute number of required iterations is always higher than for the damped Jacobi method with $\omega = \frac{1}{4}$. These properties could already be expected from the eigenvalue distribution according to Figure 4.8, and we will therefore restrict ourselves in the following to the parameter value $\omega = \frac{1}{4}$ and will always refer to the corresponding method as the damped Jacobi method.

In general, the present error does not, of course, have the form of a single Fourier mode, but is given by a linear combination of all modes. As an example, we therefore consider in Figure 4.10 the development of the error for the damped Jacobi method for an error e_0^3 given at level $\ell = 3$ $(N_\ell = N_3 = 15)$ of the form

$$e_0^3 := u_0^3 - u^{3,*} = (0.75, 0.2, 0.6, 0.45, 0.9, 0.6, 0.8, 0.85, 0.55, 0.7, 0.9, 0.5, 0.6, 0.3, 0.2)^T.$$

It can be seen for the damped Jacobi method that after just two iterations, the error already exhibits a long-wavelength character.

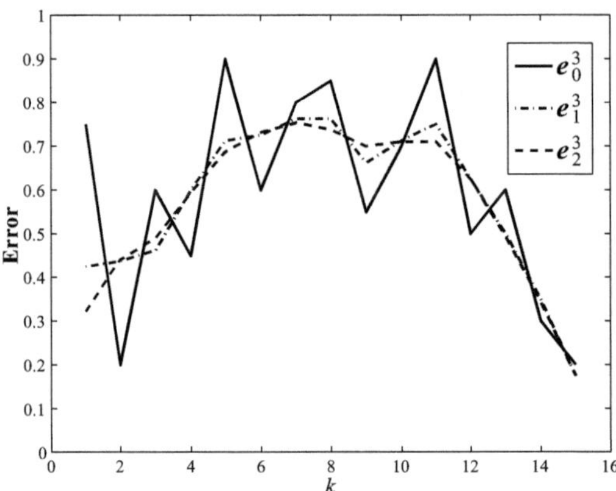

Figure 4.10   Evolution of the error for the damped Jacobi method with $\omega = 1/4$

Ideally, therefore, the Jacobi relaxation method should be coupled with an iterative scheme that has the complementary property of rapidly damping long-wavelength error terms. However, none of the methods presented so far in Section 4.1 possesses such a property. In the multigrid method, we exploit the smoothness of the error by approximating it on a coarser grid and then mapping it, for example by means of linear interpolation, onto the fine grid. For this, we need a mapping from the fine grid Ω_ℓ to the coarser grid $\Omega_{\ell-1}$ as well as a mapping from $\Omega_{\ell-1}$ to Ω_ℓ. In general, arbitrary mappings between the individual grids can be introduced. However, we want to transfer the linearity of the Jacobi method to the multigrid method and will therefore restrict ourselves to linear mappings, which are known to be representable by matrices. As a result, this also yields a very simple and efficient implementation.

Definition 4.45 A mapping

$$F : \mathbb{R}^{N_\ell} \to \mathbb{R}^{N_{\ell-1}}$$

is called a restriction from Ω_ℓ to $\Omega_{\ell-1}$ if it is linear and surjective.

The requirement of surjectivity is based on the consideration that every possible error profile on the coarser grid $\Omega_{\ell-1}$ can be represented under the mapping by at least one element of the preimage on the finer grid Ω_ℓ.

For the special nesting of the grid sequence, for example, the trivial restriction, also called injection, according to Figure 4.11 can be used, which is given by

$$
\boldsymbol{u}^{\ell-1} = \begin{pmatrix} u_1^{\ell-1} \\ \vdots \\ u_{N_{\ell-1}}^{\ell-1} \end{pmatrix} = \boldsymbol{R}_\ell^{\ell-1} \boldsymbol{u}^\ell = \begin{pmatrix} u_2^\ell \\ u_4^\ell \\ \vdots \\ u_{N_\ell-1}^\ell \end{pmatrix}
$$

and is represented by the matrix

$$
\boldsymbol{R}_\ell^{\ell-1} = \begin{pmatrix} 0 & 1 & 0 & & & & & & \\ & & 0 & 1 & 0 & & & & \\ & & & \ddots & \ddots & \ddots & & & \\ & & & & \ddots & \ddots & \ddots & & \\ & & & & & \ddots & \ddots & \ddots & \\ & & & & & & 0 & 1 & 0 \end{pmatrix} \in \mathbb{R}^{N_{\ell-1} \times N_\ell}. \qquad (4.2.11)
$$

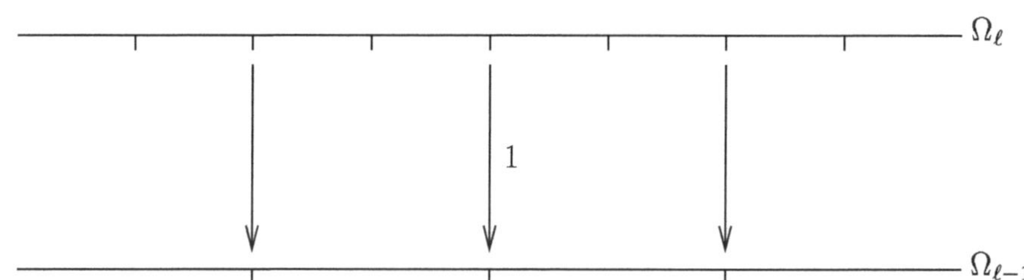

Figure 4.11   Trivial restriction or injection

This simple restriction often proves to be disadvantageous, since the values at the grid points $\Omega_\ell \setminus \Omega_{\ell-1}$ and thus their information are lost. Therefore, the linear restriction according to Figure 4.12 with the corresponding matrix representation

$$
\boldsymbol{R}_\ell^{\ell-1} = \frac{1}{4} \begin{pmatrix} 1 & 2 & 1 & & & & & & \\ & & 1 & 2 & 1 & & & & \\ & & & \ddots & \ddots & \ddots & & & \\ & & & & \ddots & \ddots & \ddots & & \\ & & & & & \ddots & \ddots & \ddots & \\ & & & & & & 1 & 2 & 1 \end{pmatrix} \qquad (4.2.12)
$$

is often used.

Lemma 4.46 *The injection and the linear restriction are restrictions in the sense of Definition 4.45.*

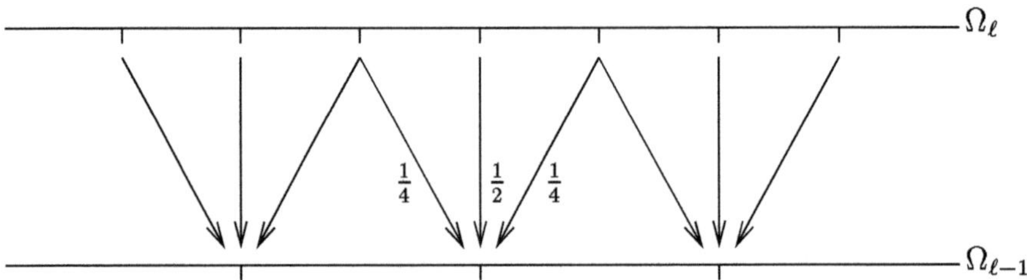

Figure 4.12   Linear restriction

Proof:
The linearity of both mappings is already established by the two given matrices. Furthermore, the columns with even numbering in both matrices are obviously linearly independent, so that $\mathrm{rang}(\boldsymbol{R}_\ell^{\ell-1}) = N_{\ell-1} = \dim \mathbb{R}^{N_{\ell-1}}$ follows, and thus the mapping is surjective.   □

Let us now turn to the mapping from $\Omega_{\ell-1}$ to Ω_ℓ.

Definition 4.47 A mapping

$$\boldsymbol{G} : \mathbb{R}^{N_\ell} \to \mathbb{R}^{N_{\ell-1}}$$

is called a prolongation from $\Omega_{\ell-1}$ to Ω_ℓ if it is linear and injective.

We want to ensure that no information from the coarse grid is lost when transferring to the finer grid. That is, the image space of a prolongation should always have the maximal dimension $N_{\ell-1}$, which leads to the requirement of injectivity.

As a mapping, for example, a linear interpolation can be used to define the values at the intermediate points. In our model case, the graphical representation is given in Figure 4.13, and from the algebraic point of view by the matrix

$$\boldsymbol{P}_{\ell-1}^\ell = \frac{1}{2}\begin{pmatrix} 1 & & & \\ 2 & & & \\ 1 & 1 & & \\ & 2 & & \\ & 1 & & \\ & & \ddots & \\ & & & 1 \\ & & & 2 \\ & & & 1 \end{pmatrix} \in \mathbb{R}^{N_\ell \times N_{\ell-1}}. \tag{4.2.13}$$

Lemma 4.48 *The linear prolongation is a prolongation according to Definition 4.47.*

Proof:
Linearity follows from the definition $\boldsymbol{G}(\boldsymbol{u}) = \boldsymbol{P}_{\ell-1}^\ell \boldsymbol{u}$. Moreover, since the rows of the matrix with even row numbers are already linearly independent, we have $\mathrm{kern}(\boldsymbol{P}_{\ell-1}^\ell) = \{\boldsymbol{0}\}$, and injectivity is thus established.   □

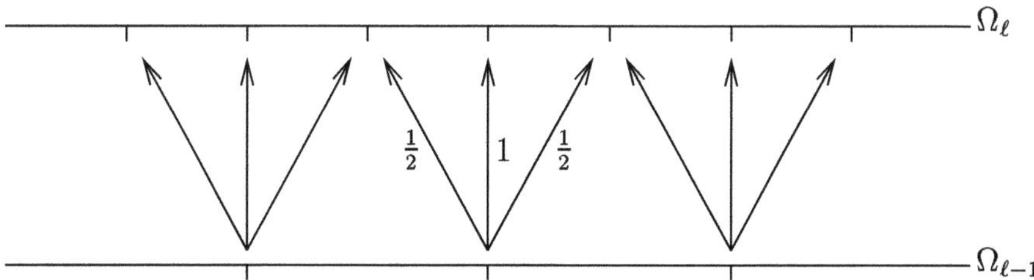

Figure 4.13   Linear prolongation

Since the error vectors can always be written as a linear combination of the Fourier modes, understanding the effect of the restriction and prolongation operators used on these modes will prove essential for the overall understanding of the multigrid method. If we consider the results for all modes on the grid Ω_3 presented in Figures 4.14 and 4.15, we notice that, in the case of injection, the modes with wavenumbers j and $\bar{j} = N_\ell + 1 - j$ for $j = 1, \ldots, N_{\ell-1}$ yield, apart from the sign, apparently identical images. In contrast, for linear reconstruction, the image vectors for small wavenumbers j show good agreement with the results of injection, whereas for larger wavenumbers, no visual correlation can be observed. The mode with the central wavenumber $j = \frac{N_\ell + 1}{2} = N_{\ell-1} + 1$ is clearly in the kernel of both mappings. We aim to gain a deeper insight into this observation, which is initially based only on visual inspection, through a mathematical analysis of the effect of both restriction operators on the Fourier modes.

Theorem 4.49 *For the images of the modes $e^{\ell,j}$, $j = 1, \ldots, N_\ell$ on Ω_ℓ, the following holds when applying the injection given by (4.2.11):*

$$R_\ell^{\ell-1} e^{\ell,j} = \frac{1}{\sqrt{2}} e^{\ell-1,j} \quad \text{for } j \in \{1, \ldots, N_{\ell-1}\}, \tag{4.2.14}$$

$$R_\ell^{\ell-1} e^{\ell,j} = 0 \quad \text{for } j = N_{\ell-1} + 1, \tag{4.2.15}$$

$$R_\ell^{\ell-1} e^{\ell,j} = -\frac{1}{\sqrt{2}} e^{\ell-1,\bar{j}} \quad \text{for } j = N_\ell + 1 - \bar{j} \text{ with } \bar{j} \in \{1, \ldots, N_{\ell-1}\}. \tag{4.2.16}$$

Proof:
First, we obtain, independently of the wavenumber $j = 1, \ldots, N_\ell$, the representation

$$(R_\ell^{\ell-1} e^{\ell,j})_k = \sqrt{2h_\ell} \sin(j\pi 2kh_\ell) = \frac{1}{\sqrt{2}} \sqrt{2h_{\ell-1}} \sin(j\pi kh_{\ell-1}).$$

For $j = 1, \ldots, N_{\ell-1}$, it follows that

$$R_\ell^{\ell-1} e^{\ell,j} = \frac{1}{\sqrt{2}} e^{\ell-1,j}.$$

In the special case $j = N_{\ell-1} + 1$, we have, since $h_{\ell-1} = \frac{1}{N_{\ell-1}+1}$ for $k = 1, \ldots, N_{\ell-1}$,

$$(R_\ell^{\ell-1} e^{\ell,j})_k = \frac{1}{\sqrt{2}} \sqrt{2h_{\ell-1}} \sin(\pi k) = 0.$$

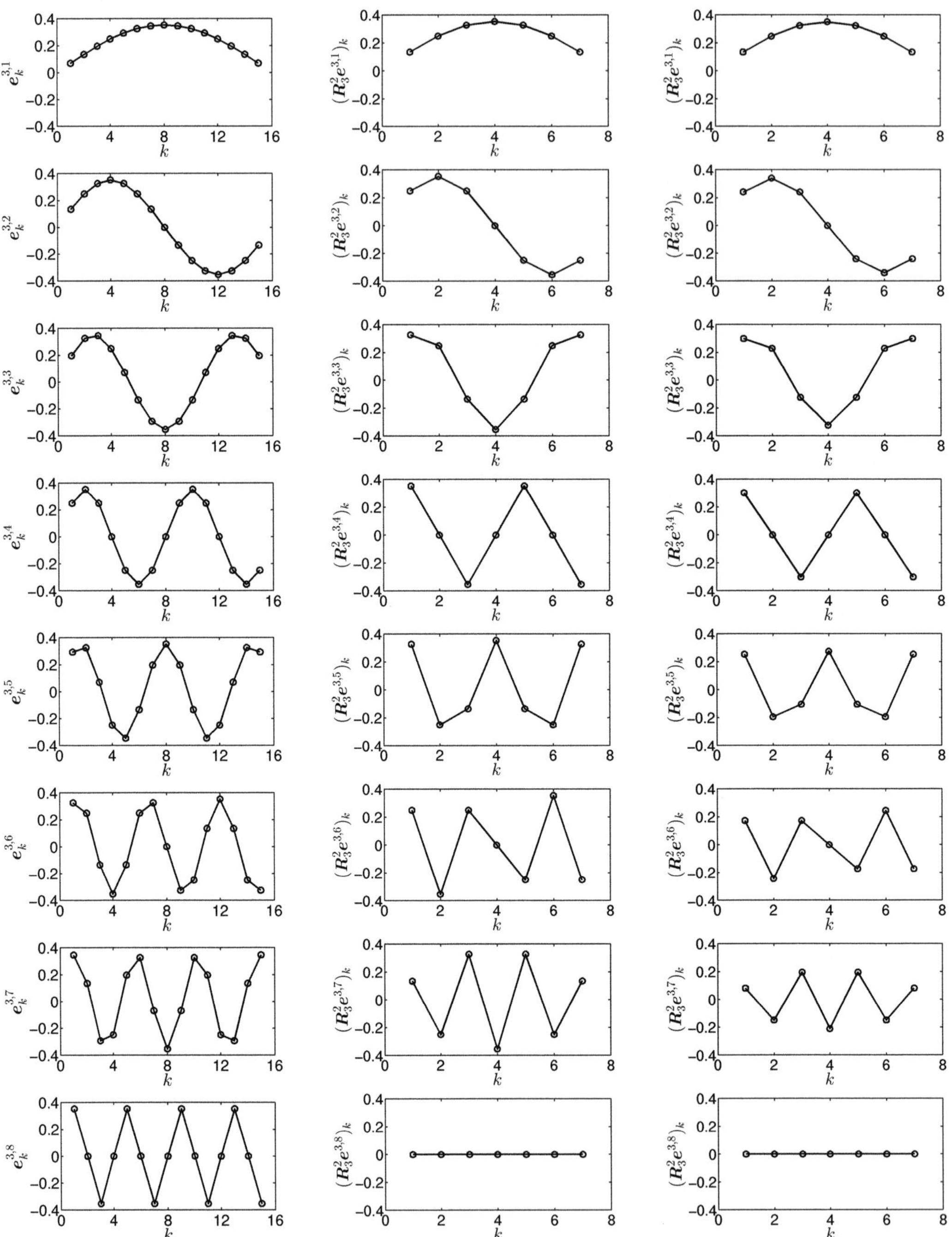

Figure 4.14 Representation of the Fourier modes $e^{3,j}$, $j = 1, \ldots, 8$ (first column) as well as the corresponding images under injection (second column) and linear restriction (third column). The continuation for $j = 9, \ldots, 15$ can be found in Figure 4.15.

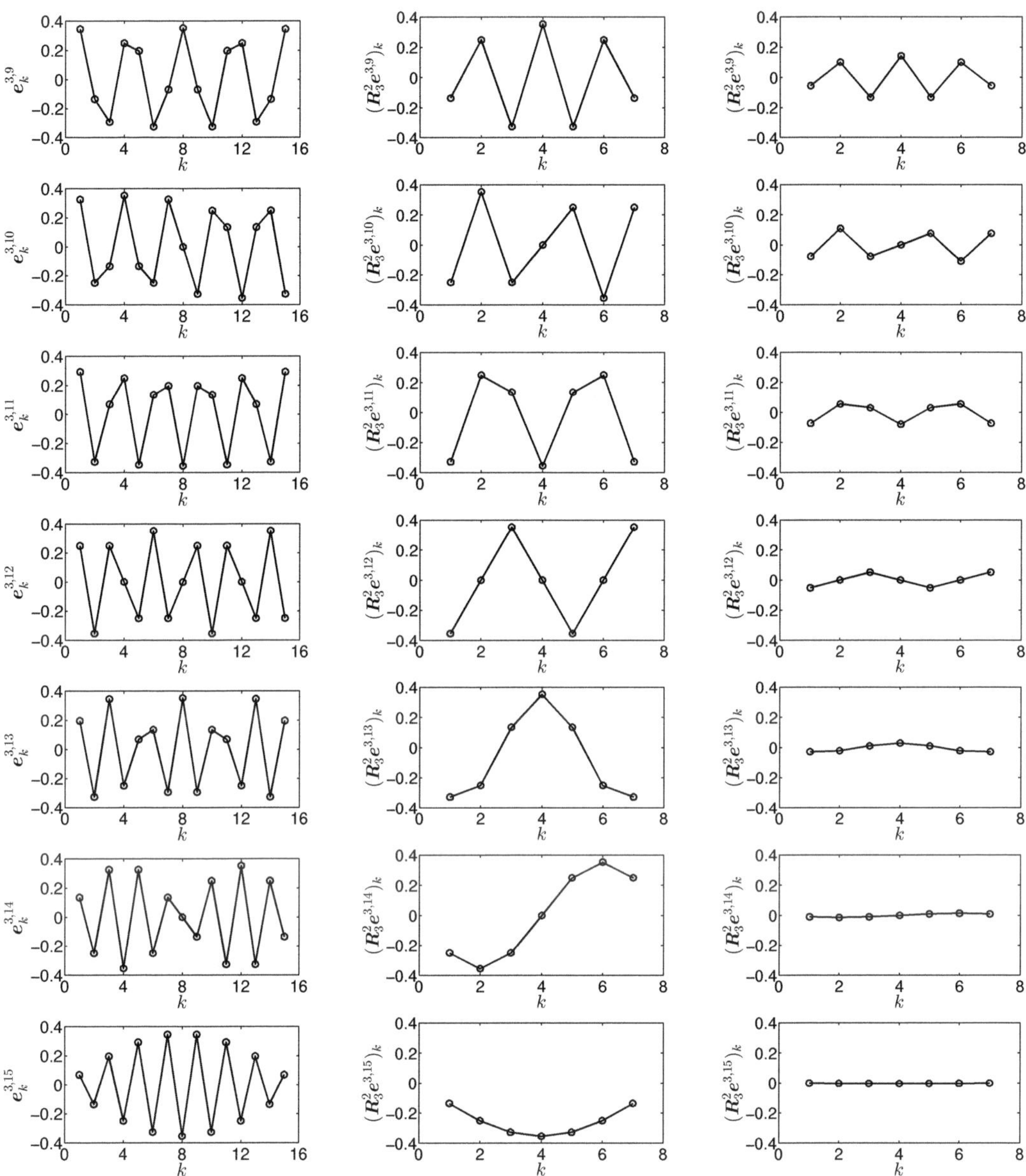

Figure 4.15 Representation of the Fourier modes $e^{3,j}$, $j = 9, \ldots, 15$ (first column) as well as the corresponding images under injection (second column) and linear restriction (third column)

Finally, for $j = N_\ell + 1 - \bar{j}$ with $\bar{j} \in \{1, \ldots, N_{\ell-1}\}$, taking into account that $(N_\ell + 1)h_{\ell-1} = 2$, we write

$$(\boldsymbol{R}_\ell^{\ell-1} \boldsymbol{e}^{\ell,j})_k = \frac{1}{\sqrt{2}} \sqrt{2h_{\ell-1}} \sin((N_\ell + 1 - \bar{j})\pi k h_{\ell-1}) = \frac{1}{\sqrt{2}} \sqrt{2h_{\ell-1}} \sin(2\pi k - \bar{j}\pi k h_{\ell-1})$$

$$= -\frac{1}{\sqrt{2}} \sqrt{2h_{\ell-1}} \sin(\bar{j}\pi k h_{\ell-1}) = -\frac{1}{\sqrt{2}} (\boldsymbol{e}^{\ell-1,\bar{j}})_k,$$

which covers all cases and completes the proof. $\square$

Except for the scaling factors, we will derive in the following theorem a statement analogous to the injection for the effect of the linear restriction.

Theorem 4.50 *For the images of the modes $e^{\ell,j}$, $j = 1,\ldots,N_\ell$ on Ω_ℓ, the following holds when applying the linear restriction given by (4.2.12):*

$$R_\ell^{\ell-1} e^{\ell,j} = \frac{c_j}{\sqrt{2}}\, e^{\ell-1,j} \quad \textit{for } j \in \{1,\ldots,N_{\ell-1}\},$$

$$R_\ell^{\ell-1} e^{\ell,j} = \qquad 0 \qquad \textit{for } j = N_{\ell-1}+1,$$

$$R_\ell^{\ell-1} e^{\ell,j} = -\frac{s_{\bar{j}}}{\sqrt{2}}\, e^{\ell-1,\bar{j}} \quad \textit{for } j = N_\ell+1-\bar{j} \textit{ with } \bar{j} \in \{1,\ldots,N_{\ell-1}\},$$

where $c_j = \cos^2\left(j\pi\frac{h_\ell}{2}\right)$ and $s_{\bar{j}} = \sin^2\left(\bar{j}\pi\frac{h_\ell}{2}\right).$

Proof:
Using the trigonometric relations (4.2.6) as well as the step size change $h_{\ell-1} = 2h_\ell$, we first write, again in general, for the k-th component of the image vector

$$
\begin{aligned}
(R_\ell^{\ell-1} e^{\ell,j})_k &= \frac{\sqrt{2h_\ell}}{4}\Big[\sin(j\pi(2k-1)h_\ell) + 2\sin(j\pi 2kh_\ell) + \sin(j\pi(2k+1)h_\ell)\Big] \\
&= \frac{\sqrt{2h_\ell}}{4}\Big[2\sin(j\pi 2kh_\ell)\cos(j\pi h_\ell) + 2\sin(j\pi 2kh_\ell)\Big] \\
&= \frac{\sqrt{2h_\ell}}{2}\Big[\cos(j\pi h_\ell) + 1\Big]\sin(j\pi 2kh_\ell) \\
&= \frac{\sqrt{2h_\ell}}{2}\left[\cos^2\left(j\pi\frac{h_\ell}{2}\right) - \sin^2\left(j\pi\frac{h_\ell}{2}\right) + 1\right]\sin(j\pi kh_{\ell-1}) \\
&= \sqrt{2h_\ell}\cos^2\left(j\pi\frac{h_\ell}{2}\right)\sin(j\pi kh_{\ell-1}) \\
&= \frac{1}{\sqrt{2}}\cos^2\left(j\pi\frac{h_\ell}{2}\right)\sqrt{2h_{\ell-1}}\sin(j\pi kh_{\ell-1})
\end{aligned}
$$

for $k = 1,\ldots,N_{\ell-1}$. According to the approach used in the proof of the mapping property of the injection, we distinguish three cases. For $j = 1,\ldots,N_{\ell-1}$, we directly obtain

$$R_\ell^{\ell-1} e^{\ell,j} = \frac{c_j}{\sqrt{2}}\, e^{\ell-1,j}.$$

For $j = N_{\ell-1}+1$, we use $jh_{\ell-1} = (N_{\ell-1}+1)h_{\ell-1} = 1$ and hence obtain, due to $\sin(j\pi kh_{\ell-1}) = \sin(\pi k) = 0$, the equation

$$(R_\ell^{\ell-1} e^{\ell,j})_k = 0 \text{ for } k = 1,\ldots,N_{\ell-1}.$$

Finally, with $j = N_\ell+1-\bar{j}$ for $\bar{j} \in \{1,\ldots,N_{\ell-1}\}$, we can write

$$
\begin{aligned}
\cos^2\left(j\pi\frac{h_\ell}{2}\right) &= \cos^2\left((N_\ell+1)\pi\frac{h_\ell}{2} - \bar{j}\pi\frac{h_\ell}{2}\right) = \cos^2\left(\frac{\pi}{2} - \bar{j}\pi\frac{h_\ell}{2}\right) \\
&= \sin^2\left(-\bar{j}\pi\frac{h_\ell}{2}\right) = \sin^2\left(\bar{j}\pi\frac{h_\ell}{2}\right),
\end{aligned}
$$

and consequently, for $k = 1, \ldots, N_{\ell-1}$, we obtain, analogous to the injection,

$$
\begin{aligned}
(\boldsymbol{R}_\ell^{\ell-1} e^{\ell,j})_k &= \frac{1}{\sqrt{2}} \sin^2\left(\bar{j}\pi \frac{h_\ell}{2}\right) \sqrt{2 h_{\ell-1}} \sin(2\pi k - \bar{j}\pi k h_{\ell-1}) \\
&= -\underbrace{\frac{1}{\sqrt{2}} \sin^2\left(\bar{j}\pi \frac{h_\ell}{2}\right)}_{=\, s_{\bar{j}}} (e^{\ell-1,\bar{j}})_k.
\end{aligned}
$$

$\square$

Through the above theorems on the action of the injection and the linear restriction with respect to the Fourier modes, we obtain a much more detailed understanding of the results in Figures 4.14 and 4.15. For the injection, equations (4.2.14) and (4.2.16) confirm the property already observed in the initial plots, namely that the images of smooth and oscillatory modes differ only in sign, and furthermore, (4.2.15) confirms that, as expected, the middle mode lies in the kernel of the trivial restriction.

With regard to the classification of the modes in terms of their wavenumber, a very interesting picture emerges. First, the long-wavelength modes ($1 \leq j < (N_\ell + 1)/2$) retain their wavenumber j under the injection. However, since the image vector now refers to the grid $\Omega_{\ell-1}$, there is a shift with respect to the relative frequency of the modes. Only the modes $e^{\ell,j}$ with $1 \leq j < (N_\ell + 1)/4 = (N_{\ell-1} + 1)/2$ yield, with $\boldsymbol{R}_\ell^{\ell-1} e^{\ell,j}$, a smooth mode with respect to $\Omega_{\ell-1}$, while the modes $e^{\ell,j}$ with $(N_\ell + 1)/4 \leq j < (N_\ell + 1)/2$, which are long-wavelength in the context of Ω_ℓ, always result in an oscillatory mode as the image under the injection. Analogously, there is also a frequency shift for the oscillatory Fourier modes. The images of the short-wavelength modes $e^{\ell,j}$ with $(N_\ell+1)/2 < j \leq 3(N_\ell+1)/4$ exhibit, on $\Omega_{\ell-1}$, an oscillatory behavior corresponding to their preimages. In contrast, the images of the high-frequency terms $e^{\ell,j}$ with $3(N_\ell+1)/4 < j \leq N_\ell$ on $\Omega_{\ell-1}$ display a long-wavelength character. Although the first impression from the results of the linear restriction suggests a behavior different from that of the injection, at least for high wave numbers, Theorem 4.50 shows us that the properties of the images with respect to the present wave numbers are identical. However, compared to the trivial restriction, there is an additional factor c_j or $s_{\bar{j}}$ that controls the amplitude in the image space. These quantities also explain the similarity of the images under injection and linear restriction for small wave numbers j, since in these cases $\cos^2\left(j\pi \frac{h_\ell}{2}\right)$ is a value close to one. As the wave number approaches $(N_\ell + 1)/2$ from below, the corresponding factor $\cos^2\left(j\pi \frac{h_\ell}{2}\right)$ becomes smaller and smaller. This also explains why in Figure 4.14 the modes under both restrictions for wave numbers j close to $(N_\ell + 1)/2$ have the same shape but different heights. For example, for $j = 7 < (N_3 + 1)/2 = 8$, the factor is $c_7 = \cos^2\left(7\pi/32\right) \approx 0.6$. If the wave number is in the range $(N_\ell + 1)/2 < j < N_\ell$, the prefactor $\sin^2\left(\bar{j}\pi \frac{h_\ell}{2}\right) = \sin^2\left((N_\ell + 1 - j)\pi \frac{h_\ell}{2}\right)$ applies, which is strictly monotonically decreasing as j increases and takes values between 0.5 and 0.01.

The modes $e^{\ell,j}$ and $e^{\ell,N_\ell+1-j}$, $j = 1, \ldots, N_{\ell-1}$ are referred to as complementary modes due to the properties proven in Theorems 4.49 and 4.50. For pairs of complementary modes, it holds for both the injection and the linear restriction that with $\boldsymbol{R}_\ell^{\ell-1}$ we also have a surjective mapping in the form

$$
\boldsymbol{R}_\ell^{\ell-1} : \operatorname{span}\{e^{\ell,j}, e^{\ell,N_\ell+1-j}\} \to \operatorname{span}\{e^{\ell-1,j}\} \quad \text{for all } j = 1, \ldots, N_{\ell-1}
$$

since complementary modes always have the same modes in the image space.

Before we deal with the implementation of an iteration scheme spanning two grids, it is important to work out some properties of the linear prolongation. Since we perform a linear interpolation between the grid points, it is natural to assume that primarily smooth modes of the fine grid Ω_ℓ can be well approximated by the vectors of the image space. Thus, we expect small deviations on the fine grid when reproducing long-wavelength errors, while strongly oscillating error patterns are approximated less accurately.

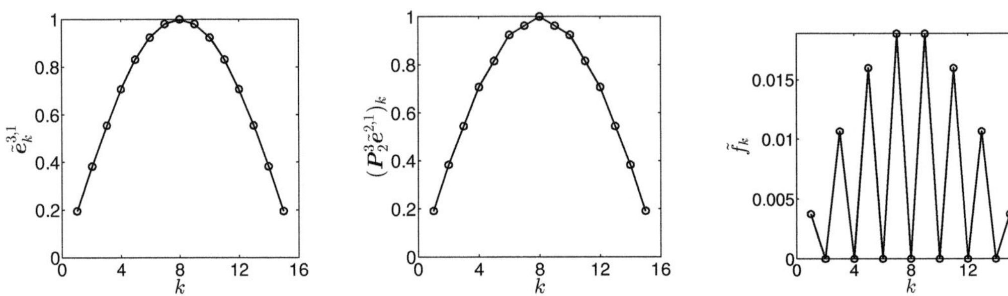

Figure 4.16   Representation of the scaled mode $\widetilde{e}^{3,1}$ (left), the scaled mode $\widetilde{e}^{2,1}$ under linear prolongation (center), and the difference $\widetilde{f} = \widetilde{e}^{3,1} - P_2^3\widetilde{e}^{2,1}$ (right)

Figure 4.16 gives an impression of the effect of linear prolongation on a long-wavelength mode of the coarse grid. We consider the scaled modes

$$\widetilde{e}^{\ell,1} = \frac{1}{\sqrt{2h_\ell}}e^{\ell,1}, \ \ \ell = 2,3$$

and compare the mode $\widetilde{e}^{3,1}$ and the linear interpolation of the smooth mode $\widetilde{e}^{2,1}$, that is,

$$\widetilde{f} = \widetilde{e}^{3,1} - P_2^3\widetilde{e}^{2,1}.$$

As the difference of these two quantities shown on the right in Figure 4.16 demonstrates, there is agreement at the nodes of the coarse grid Ω_2, while only small deviations in magnitude are visible at the intermediate points, noting that the vertical axes have different scales. An important point is also that the difference exhibits a strongly oscillatory character. Thus, the image of the smooth mode $\widetilde{e}^{2,1}$ under linear prolongation seems to contain, in addition to the smooth mode $\widetilde{e}^{3,1}$, also components of an oscillatory mode. The following theorem will show that the image space of the linear prolongation, with the exception of $e^{\ell,\frac{N_\ell+1}{2}}$, in fact contains components of all modes of the fine grid.

Theorem 4.51 *For the images of the modes* $e^{\ell-1,j}$, $j = 1,\ldots,N_{\ell-1}$ *on* $\Omega_{\ell-1}$, *the following holds when applying the linear prolongation given by* (4.2.13):

$$P_{\ell-1}^\ell e^{\ell-1,j} = \sqrt{2}\left(c_j e^{\ell,j} - s_j e^{\ell,N_\ell+1-j}\right) \tag{4.2.17}$$

with $c_j = \cos^2\left(\frac{j\pi h_\ell}{2}\right)$ *and* $s_j = \sin^2\left(\frac{j\pi h_\ell}{2}\right)$.

Proof:
Already due to the structure of the matrix corresponding to linear prolongation (4.2.13), it seems reasonable, in addition to a separate consideration of the odd and even components, to also take a closer look at the first and last vector entries. However, we first formulate, for arbitrary $k \in \mathbb{N}$, the following two useful relations:

$$\sin((N_\ell + 1 - j)\pi k h_\ell) = \sin((N_\ell + 1)\pi k h_\ell - j\pi k h_\ell)$$
$$= \sin(\pi k - j\pi k h_\ell) = (-1)^{k+1} \sin(j\pi k h_\ell) \qquad (4.2.18)$$

and

$$\cos((N_\ell + 1 - j)\pi k h_\ell) = \cos(\pi k - j\pi k h_\ell) = (-1)^k \cos(j\pi k h_\ell). \qquad (4.2.19)$$

From equation (4.2.18), we can directly read off the important relationship between the components of complementary modes of the form

$$\left(e^{\ell,j}\right)_k = (-1)^{k+1} \left(e^{\ell,N_\ell+1-j}\right)_k \qquad (4.2.20)$$

for $k = 1,\ldots,N_\ell$.

Let us now turn to the first component. Here, we obtain

$$\left(\boldsymbol{P}^\ell_{\ell-1} e^{\ell-1,j}\right)_1 = \tfrac{1}{2}\sqrt{2h_{\ell-1}}\sin(j\pi h_{\ell-1}) = \sqrt{2h_{\ell-1}}\tfrac{1}{2}\sin(j\pi 2 h_\ell)$$
$$\overset{(4.2.5)}{=} \sqrt{2h_{\ell-1}}\sin(j\pi h_\ell)\cos(j\pi h_\ell)$$
$$\overset{(4.2.6)}{=} \sqrt{2}\left(\cos^2\left(\tfrac{j\pi h_\ell}{2}\right) - \sin^2\left(\tfrac{j\pi h_\ell}{2}\right)\right)\sqrt{2h_\ell}\sin(j\pi h_\ell)$$
$$= \sqrt{2}\left(c_j\left(e^{\ell,j}\right)_1 - s_j\left(e^{\ell,j}\right)_1\right)$$
$$\overset{(4.2.20)}{=} \sqrt{2}\left(c_j\left(e^{\ell,j}\right)_1 - s_j\left(e^{\ell,N_\ell+1-j}\right)_1\right).$$

For even k we write

$$\left(\boldsymbol{P}^\ell_{\ell-1} e^{\ell-1,j}\right)_k = \tfrac{1}{2}2\sqrt{2h_{\ell-1}}\sin\left(j\pi\tfrac{k}{2}h_{\ell-1}\right) = \sqrt{2}\sqrt{2h_\ell}\sin(j\pi k h_\ell)$$
$$= \sqrt{2}\left(e^{\ell,j}\right)_k = \sqrt{2}(c_j + s_j)\left(e^{\ell,j}\right)_k$$
$$\overset{(4.2.20)}{=} \sqrt{2}\left(c_j\left(e^{\ell,j}\right)_k - s_j\left(e^{\ell,N_\ell+1-j}\right)_k\right).$$

As a third case, let us consider the odd $k \in \{3,\ldots,N_\ell-2\}$. With $k = 2m+1$ we obtain

$$\left(\boldsymbol{P}^\ell_{\ell-1} e^{\ell-1,j}\right)_k = \tfrac{1}{2}\sqrt{2h_{\ell-1}}\left(\sin\left(j\pi m h_{\ell-1}\right) + \sin\left(j\pi(m+1)h_{\ell-1}\right)\right)$$
$$= \sqrt{2}\sqrt{2h_\ell}\tfrac{1}{2}\left(\sin\left(j\pi 2m h_\ell\right) + \sin\left(j\pi 2(m+1)h_\ell\right)\right)$$
$$\overset{(4.2.6)}{=} \sqrt{2}\sqrt{2h_\ell}\cos(j\pi h_\ell)\sin\left(j\pi(2m+1)h_\ell\right)$$
$$\overset{(4.2.6)}{=} \sqrt{2}\sqrt{2h_\ell}(c_j - s_j)\sin\left(j\pi k h_\ell\right)$$
$$\overset{(4.2.20)}{=} \sqrt{2}\left(c_j\left(e^{\ell,j}\right)_k - s_j\left(e^{\ell,N_\ell+1-j}\right)_k\right).$$

If we consider equations (4.2.18) and (4.2.19) for $j = N_\ell$, then

$$\cos(k\pi N_\ell h_\ell)\sin(k\pi h_\ell) = (-1)^k \cos(k\pi h_\ell) \cdot (-1)^{k+1}\sin(k\pi N_\ell h_\ell)$$
$$= -\cos(k\pi h_\ell)\sin(k\pi N_\ell h_\ell) \tag{4.2.21}$$

can be written for arbitrary $k \in \mathbb{N}$. Furthermore, we use the relation

$$N_{\ell-1}h_{\ell-1} = (N_\ell - 1)h_\ell,$$

thus, for the last component, based on the above relations with $k = j$, one can write the remaining statement

$$
\begin{aligned}
\left(\boldsymbol{P}^\ell_{\ell-1}e^{\ell-1,j}\right)_{N_\ell}
&= \tfrac{1}{2}\sqrt{2h_{\ell-1}}\sin\left(j\pi N_{\ell-1}h_{\ell-1}\right) = \tfrac{1}{2}\sqrt{2h_{\ell-1}}\sin\left(j\pi(N_\ell - 1)h_\ell\right) \\
&= \sqrt{2h_{\ell-1}}\tfrac{1}{2}\sin\left(j\pi N_\ell h_\ell - j\pi h_\ell\right) \\
&\overset{(4.2.5)}{=} \sqrt{2h_{\ell-1}}\tfrac{1}{2}\left(\sin(j\pi N_\ell h_\ell)\cos(j\pi h_\ell) - \cos(j\pi N_\ell h_\ell)\sin(j\pi h_\ell)\right) \\
&\overset{(4.2.21)}{=} \sqrt{2}\sqrt{2h_\ell}\cos(j\pi h_\ell)\sin(j\pi N_\ell h_\ell) \\
&= \sqrt{2}\cos(j\pi h_\ell)\left(e^{\ell,j}\right)_{N_\ell} \overset{(4.2.6)}{=} \sqrt{2}(c_j - s_j)\left(e^{\ell,j}\right)_{N_\ell} \\
&\overset{(4.2.20)}{=} \sqrt{2}\left(c_j\left(e^{\ell,j}\right)_{N_\ell} - s_j\left(e^{\ell,N_\ell+1-j}\right)_{N_\ell}\right).
\end{aligned}
$$

□

To make the described effects of prolongation and restriction a bit clearer, let us consider the composition of the injection with the linear prolongation applied to the first, smooth mode $e^{\ell,1}$ on the grid Ω_ℓ. For the difference

$$\boldsymbol{f} := e^{\ell,1} - \boldsymbol{P}^\ell_{\ell-1}\boldsymbol{R}^{\ell-1}_\ell e^{\ell,1} \in \mathbb{R}^{N_\ell}$$

we expect that the vector $\boldsymbol{f} = (f_1,\dots,f_{N_\ell})^T$ has vanishing even components, since at those positions both mappings do not change the values. In addition, nonzero values should appear at the components with odd indices, since these values are omitted during injection and then linearly reconstructed, which leads to an error as shown in Figure 4.16, because the underlying mode is a sampling of a nonlinear sine wave. Furthermore, the absolute values of the error vector must be relatively large where the curvature of the sine wave underlying the mode $e^{\ell,1}$ attains its maximum, since here the linear reconstruction produces a larger deviation than in regions with low curvature. Thus, for f_k with odd k near $(N_\ell + 1)/2$, we expect the largest entries. Taking into account Theorems 4.49 and 4.51, we write

$$\boldsymbol{P}^\ell_{\ell-1}\boldsymbol{R}^{\ell-1}_\ell e^{\ell,1} = \frac{1}{\sqrt{2}}\boldsymbol{P}^\ell_{\ell-1}e^{\ell-1,1} = c_1 e^{\ell,1} - s_1 e^{\ell,N_\ell}$$

with

$$c_1 = \cos^2\left(\frac{\pi h_\ell}{2}\right) \quad \text{and} \quad s_1 = \sin^2\left(\frac{\pi h_\ell}{2}\right).$$

and, using (4.2.18), we obtain for the k-th component of the above image vector

$$
\begin{aligned}
\left(\boldsymbol{P}^\ell_{\ell-1}\boldsymbol{R}^{\ell-1}_\ell e^{\ell,1}\right)_k &= \sqrt{2h_\ell}\left[c_1\sin(\pi k h_\ell) - s_1\sin(N_\ell\pi k h_\ell)\right] \\
&= \sqrt{2h_\ell}\sin(\pi k h_\ell)\left(\cos^2\left(\frac{\pi h_\ell}{2}\right) - (-1)^{k+1}\sin^2\left(\frac{\pi h_\ell}{2}\right)\right).
\end{aligned}
$$

Consequently, taking (4.2.6) into account, for $k = 1, \ldots, N_\ell$ we have

$$f_k = \left(e^{\ell,1}\right)_k - \left(\boldsymbol{P}^\ell_{\ell-1}\boldsymbol{R}^{\ell-1}_\ell e^{\ell,1}\right)_k = \begin{cases} 0, & \text{if } k \text{ is even,} \\ \sqrt{2h_\ell}\,\sin(\pi k h_\ell)\cos(\pi h_\ell), & \text{if } k \text{ is odd.} \end{cases}$$

Thus, we see that all the above conjectures about the entries of the difference vector $\boldsymbol{f}$ are confirmed. From Theorem 4.51 we can deduce that, due to the present prolongation, even in the case of an error representation on the coarse grid $\Omega_{\ell-1}$ consisting exclusively of long-wavelength modes $e^{\ell-1,j}$, oscillatory modes are always generated on the fine grid, since the image of a smooth mode can be written as a linear combination of complementary modes; see also Figure 4.16. For the mode $e^{\ell-1,1}$ considered above as an example, this means that under linear prolongation, the oscillatory mode $e^{\ell,15}$ appears in the image. Based on our intuitive understanding of the interpolation of long-wavelength modes, we already expect that the proportion of the oscillatory mode in the image space should be small. A Taylor expansion of the sine and cosine functions about zero yields

$$\sin x = x + \mathcal{O}(x^3) \quad \text{and} \quad \cos x = 1 + \mathcal{O}(x^2)$$

and thus

$$\sin^2 x = x^2 + \mathcal{O}(x^4) = \mathcal{O}(x^2) \quad \text{and} \quad \cos^2 x = 1 + \mathcal{O}(x^2) \tag{4.2.22}$$

for $x \to 0$. Consequently, from (4.2.17) we obtain the representation

$$\boldsymbol{P}^\ell_{\ell-1}e^{\ell-1,j} = \left(\sqrt{2} + \mathcal{O}(j^2 h_\ell^2)\right) e^{\ell,j} - \mathcal{O}(j^2 h_\ell^2)e^{\ell,N_\ell+1-j}$$

for $jh_\ell \to 0$. Thus, linear interpolation does generate oscillatory modes, but in the case of smooth initial modes, these have a significantly smaller amplitude. The image under linear prolongation is therefore essentially determined by modes whose wavenumber is identical to that of the original mode. Nevertheless, the representation (4.2.17) shows that post-smoothing on the fine grid can be useful.

4.2.1 Two-Grid Method

With the smoother, restriction, and prolongation, we are now able to derive a first iterative method that operates over two grid levels.

After m steps of the iterative method (4.2.3) with $\omega = 1/4$, we obtain a largely smooth error

$$e^\ell_m = u^\ell_m - u^{\ell,*}. \tag{4.2.23}$$

This vector can therefore be well approximated on the next coarser grid $\Omega_{\ell-1}$ with significantly less computational effort. With the defect

$$d^\ell_m := \boldsymbol{A}_\ell u^\ell_m - \boldsymbol{f}^\ell \tag{4.2.24}$$

we obtain

$$\boldsymbol{A}_\ell e^\ell_m \overset{(4.2.23)}{=} \boldsymbol{A}_\ell\left(u^\ell_m - u^{\ell,*}\right) = \boldsymbol{A}_\ell u^\ell_m - \boldsymbol{A}_\ell u^{\ell,*} = d^\ell_m. \tag{4.2.25}$$

We now determine an approximation to e^ℓ_m by considering the equation (4.2.25) on the grid $\Omega_{\ell-1}$, solving it exactly there, and then prolongating the computed vector back to the original grid Ω_ℓ.

We therefore consider the equation

$$A_{\ell-1}e^{\ell-1} = d^{\ell-1} \tag{4.2.26}$$

with the restricted defect

$$d^{\ell-1} = R_\ell^{\ell-1}d_m^\ell. \tag{4.2.27}$$

If (4.2.26) is solved exactly, then

$$P_{\ell-1}^\ell e^{\ell-1} = P_{\ell-1}^\ell A_{\ell-1}^{-1} d^{\ell-1} \tag{4.2.28}$$

provides an approximation to the desired error vector e_m^ℓ.

The correction of the current approximate solution u_m^ℓ, determined using the coarse grid, can thus be summarized using (4.2.23) to (4.2.28) as

$$u_m^{\ell,\text{new}} = u_m^\ell - P_{\ell-1}^\ell A_{\ell-1}^{-1} R_\ell^{\ell-1} \left(A_\ell u_m^\ell - f^\ell \right).$$

Definition 4.52 Let u_m^ℓ be an approximate solution of the equation $A_\ell u^\ell = f^\ell$, then the method

$$u_m^{\ell,\text{new}} = \phi_\ell^{CGC} \left(u_m^\ell, f^\ell \right)$$

with

$$\phi_\ell^{CGC} \left(u_m^\ell, f^\ell \right) = u_m^\ell - P_{\ell-1}^\ell A_{\ell-1}^{-1} R_\ell^{\ell-1} \left(A_\ell u_m^\ell - f^\ell \right) \tag{4.2.29}$$

is called the coarse grid correction method.

Lemma 4.53 *The coarse grid correction method ϕ_ℓ^{CGC} constitutes a linear consistent iterative method with*

$$M_\ell^{CGC} = I - P_{\ell-1}^\ell A_{\ell-1}^{-1} R_\ell^{\ell-1} A_\ell \tag{4.2.30}$$

and

$$N_\ell^{CGC} = P_{\ell-1}^\ell A_{\ell-1}^{-1} R_\ell^{\ell-1}. \tag{4.2.31}$$

Proof:
With (4.2.29) we obtain

$$\phi_\ell^{CGC} (u, f) = M_\ell^{CGC} u + N_\ell^{CGC} f.$$

Here, $M_\ell^{CGC}, N_\ell^{CGC} \in \mathbb{R}^{N_\ell \times N_\ell}$ with $M_\ell^{CGC} = I - N_\ell^{CGC} A_\ell$, from which the claim follows by Theorem 4.4. $\square$

The coarse grid correction method can thus be used as a standalone iterative method. However, it turns out to be unsuitable for this purpose, as the following lemma will show.

Lemma 4.54 *The coarse grid correction method ϕ_ℓ^{CGC} is not convergent.*

Proof:
Since $N_\ell > N_{\ell-1}$, the kernel of $R_\ell^{\ell-1}$, in short $\text{kern}(R_\ell^{\ell-1})$, is nontrivial. Let $0 \neq v \in \text{kern}(R_\ell^{\ell-1})$, then with $w := A_\ell^{-1} v \neq 0$, we obtain the equation

$$M_\ell^{CGC} w = w - P_{\ell-1}^\ell A_{\ell-1}^{-1} R_\ell^{\ell-1} \underbrace{A_\ell w}_{=\,v} = w,$$

which yields $\rho\left(M_\ell^{CGC}\right) \geq 1$. $\qquad\square$

The above result is not surprising due to the grid transfer and the associated temporary reduction in dimension, and moreover, it is not problematic with regard to an effective overall method. After all, we are only looking for an algorithm that damps the long-wavelength modes of the error vector sufficiently quickly. As we will show below, this property is indeed present. In order to analyze the error development, it is helpful to relate the error vector

$$e_m^{\ell,\text{neu}} = u_m^{\ell,\text{neu}} - u^{\ell,*}$$

to the error e_m^ℓ directly before the correction step. For this, we write

$$\begin{aligned}
e_m^{\ell,\text{neu}} &= u_m^{\ell,\text{neu}} - u^{\ell,*} = \phi_\ell^{CGC}\left(u_m^\ell, f^\ell\right) - u^{\ell,*}\\
&= u_m^\ell - u^{\ell,*} - P_{\ell-1}^\ell A_{\ell-1}^{-1} R_\ell^{\ell-1}\left(A_\ell u_m^\ell - f^\ell\right)\\
&= e_m^\ell - P_{\ell-1}^\ell A_{\ell-1}^{-1} R_\ell^{\ell-1} A_\ell e_m^\ell\\
&= \Psi_\ell^{CGC}\left(e_m^\ell\right)
\end{aligned}$$

with

$$\begin{aligned}
\Psi_\ell^{CGC}: \quad \mathbb{R}^{N_\ell} &\rightarrow \mathbb{R}^{N_\ell}\\
e &\mapsto \Psi_\ell^{CGC}(e) = (I - P_{\ell-1}^\ell A_{\ell-1}^{-1} R_\ell^{\ell-1} A_\ell)e.
\end{aligned}$$

Due to the linearity of the mapping Ψ_ℓ^{CGC}, we obtain, using the basis property of the Fourier modes $e^{\ell,j}$ and $e_m^\ell = \sum_{j=1}^{N_\ell} \alpha_j e^{\ell,j}$, the representation

$$e_m^{\ell,\text{neu}} = \Psi_\ell^{CGC}\left(\sum_{j=1}^{N_\ell} \alpha_j e^{\ell,j}\right) = \sum_{j=1}^{N_\ell} \alpha_j \Psi_\ell^{CGC}\left(e^{\ell,j}\right).$$

Consequently, the effect of the mapping Ψ_ℓ^{CGC} on the modes, and in particular on the long-wavelength ones, is of central importance. First, we want to gain an initial insight into the coarse grid correction and therefore consider the numerical results shown in Figure 4.17. Selected modes of various wavenumbers and their images under the coarse grid correction using linear restriction and prolongation are displayed. The numerical results are promising, as the coarse grid correction indeed appears to have a strongly damping effect on the smooth modes, since the graphs for $\Psi_3^{CGC}\left(e^{3,1}\right)$ and $\Psi_3^{CGC}\left(e^{3,4}\right)$ shown on the right in Figure 4.17 exhibit a significantly reduced amplitude compared to the preimages $e^{3,1}$ and $e^{3,4}$ visualized on the left. Oscillatory modes, on the other hand, do not appear to be damped visually. This fundamental property of the coarse grid correction, which is important for the further design of an iteration spanning two grids, will be mathematically substantiated by the following theorem.

Theorem 4.55 *Let the coarse grid correction be given using the linear restriction* (4.2.12) *and the linear prolongation* (4.2.13). *Then, for the Fourier modes* $e^{\ell,j}$, $j = 1,\ldots,N_\ell$ *on* Ω_ℓ *under the mapping* Ψ_ℓ^{CGC}, *the following representation holds:*

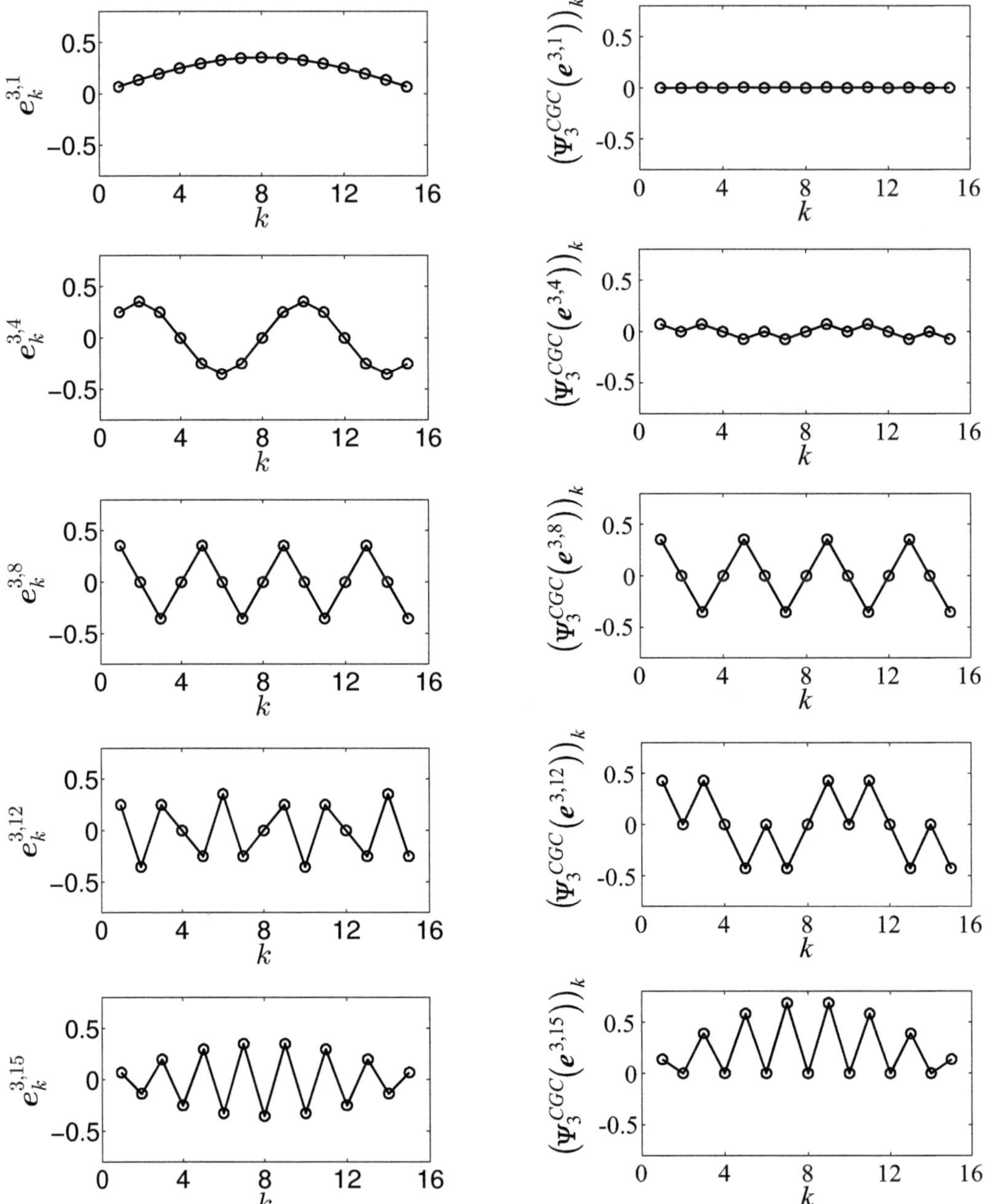

Figure 4.17   Representation of the Fourier modes $e^{3,\ell}$, $\ell = 1,4,8,12,15$ (left) as well as the images of the corresponding coarse grid correction according to Ψ_ℓ^{CGC} (right)

$$\Psi_\ell^{CGC}\left(e^{\ell,j}\right) = \quad s_j\, e^{\ell,j} + s_j\, e^{\ell,\bar{j}} \quad for \ \ j \in \{1,\ldots,N_{\ell-1}\} \ \ and \ \ \bar{j} = N_\ell + 1 - j,$$

$$\Psi_\ell^{CGC}\left(e^{\ell,j}\right) = \qquad\qquad e^{\ell,j} \qquad\qquad for \ \ j = N_{\ell-1} + 1,$$

$$\Psi_\ell^{CGC}\left(e^{\ell,j}\right) = \quad c_{\bar{j}}\, e^{\ell,j} + c_{\bar{j}}\, e^{\ell,\bar{j}} \quad for \ \ j = N_\ell + 1 - \bar{j} \ \ with \ \ \bar{j} \in \{1,\ldots,N_{\ell-1}\}$$

using $c_{\bar{\jmath}} = \cos^2\left(\bar{\jmath}\pi\frac{h_\ell}{2}\right)$ *and* $s_j = \sin^2\left(j\pi\frac{h_\ell}{2}\right)$.

Proof:
Since the Fourier modes $e^{\ell,j}$ for each grid Ω_ℓ are, according to Lemma 4.42, always eigenvectors corresponding to the eigenvalues $\mu^{\ell,j} = 4h_\ell^{-2}\sin^2\left(\frac{j\pi h_\ell}{2}\right)$ of the matrix $\boldsymbol{A}_\ell$, we obtain in addition to

$$\boldsymbol{A}_\ell e^{\ell,j} = \mu^{\ell,j} e^{\ell,j} \ \text{ for } \ j = 1,\ldots,N_\ell$$

also

$$\boldsymbol{A}_{\ell-1}^{-1} e^{\ell-1,j} = \frac{1}{\mu^{\ell-1,j}} e^{\ell-1,j} \ \text{ for } \ j = 1,\ldots,N_{\ell-1}.$$

Looking at the matrix representation of the mapping $\boldsymbol{\Psi}_\ell^{CGC}$, we expect a quotient $\mu^{\ell,j}/\mu^{\ell-1,j}$, which we therefore want to examine more closely in advance. Due to the grid spacing relation $h_{\ell-1} = 2h_\ell$, it follows that

$$\sin\left(j\pi\frac{h_{\ell-1}}{2}\right) = \sin\left(j\pi h_\ell\right) \overset{(4.2.5)}{=} 2\sin\left(j\pi\frac{h_\ell}{2}\right)\cos\left(j\pi\frac{h_\ell}{2}\right) \tag{4.2.32}$$

and thus

$$\frac{\mu^{\ell,j}}{\mu^{\ell-1,j}} = \frac{4h_\ell^{-2}\sin^2\left(\frac{j\pi h_\ell}{2}\right)}{4h_{\ell-1}^{-2}\sin^2\left(\frac{j\pi h_{\ell-1}}{2}\right)} = \frac{1}{\cos^2\left(j\pi\frac{h_\ell}{2}\right)} = \frac{1}{c_j}$$

for $j = 1,\ldots,N_{\ell-1}$.

At this point, we can use the knowledge about the effect of the linear restriction and the linear prolongation on the Fourier modes, which we formulated in Theorems 4.50 and 4.51. For $j = 1,\ldots,N_{\ell-1}$, we obtain

$$\boldsymbol{P}_{\ell-1}^\ell \boldsymbol{A}_{\ell-1}^{-1} \boldsymbol{R}_\ell^{\ell-1} \boldsymbol{A}_\ell e^{\ell,j} = \mu^{\ell,j} \boldsymbol{P}_{\ell-1}^\ell \boldsymbol{A}_{\ell-1}^{-1} \boldsymbol{R}_\ell^{\ell-1} e^{\ell,j}$$

$$= \mu^{\ell,j}\frac{c_j}{\sqrt{2}} \boldsymbol{P}_{\ell-1}^\ell \boldsymbol{A}_{\ell-1}^{-1} e^{\ell-1,j} = \frac{\mu^{\ell,j}}{\mu^{\ell-1,j}}\frac{c_j}{\sqrt{2}} \boldsymbol{P}_{\ell-1}^\ell e^{\ell-1,j}$$

$$= \underbrace{\frac{\mu^{\ell,j}}{\mu^{\ell-1,j}}\frac{c_j}{\sqrt{2}}}_{=1} \sqrt{2}\left(c_j e^{\ell,j} - s_j e^{\ell,N_\ell+1-j}\right),$$

which yields

$$\boldsymbol{\Psi}_\ell^{CGC}\left(e^{\ell,j}\right) = e^{\ell,j} - \left(c_j e^{\ell,j} - s_j e^{\ell,N_\ell+1-j}\right) = s_j e^{\ell,j} + s_j e^{\ell,N_\ell+1-j} = s_j e^{\ell,j} + s_j e^{\ell,\bar{\jmath}}$$

follows. The special case $j = N_{\ell-1}+1$ results directly from

$$\boldsymbol{R}_\ell^{\ell-1} \boldsymbol{A}_\ell e^{\ell,N_{\ell-1}+1} = \mu^{\ell,j} \boldsymbol{R}_\ell^{\ell-1} e^{\ell,N_{\ell-1}+1} \overset{\text{Theorem 4.50}}{=} \boldsymbol{0}.$$

For the complementary modes $e^{\ell,j}$, $j = \frac{N_\ell+1}{2}+1,\ldots,N_\ell$ or as described in the theorem $j = N_\ell+1-\bar{\jmath}$ with $\bar{\jmath}\in\{1,\ldots,N_{\ell-1}\}$, we first consider the relationships

$$\sin\left(j\pi\frac{h_\ell}{2}\right) = \sin\left(\frac{\pi}{2} - \bar{\jmath}\pi\frac{h_\ell}{2}\right) = \cos\left(\bar{\jmath}\pi\frac{h_\ell}{2}\right)$$

as well as

$$\frac{\mu^{\ell,j}}{\mu^{\ell-1,\bar{\jmath}}} = \frac{4h_\ell^{-2}\sin^2\left(\frac{j\pi h_\ell}{2}\right)}{4h_{\ell-1}^{-2}\sin^2\left(\frac{\bar{\jmath}\pi h_{\ell-1}}{2}\right)} \overset{(4.2.32)}{=} \frac{\sin^2\left(\frac{j\pi h_\ell}{2}\right)}{\sin^2\left(\frac{\bar{\jmath}\pi h_\ell}{2}\right)\cos^2\left(\frac{\bar{\jmath}\pi h_\ell}{2}\right)} = \frac{1}{\sin^2\left(\frac{\bar{\jmath}\pi h_\ell}{2}\right)} = \frac{1}{s_{\bar{\jmath}}}.$$

Analogous to the approach above, it follows that

$$\boldsymbol{P}_{\ell-1}^\ell \boldsymbol{A}_{\ell-1}^{-1} \boldsymbol{R}_\ell^{\ell-1} \boldsymbol{A}_\ell \boldsymbol{e}^{\ell,j}$$

$$= \mu^{\ell,j}\boldsymbol{P}_{\ell-1}^\ell \boldsymbol{A}_{\ell-1}^{-1} \boldsymbol{R}_\ell^{\ell-1}\boldsymbol{e}^{\ell,j} = -\mu^{\ell,j}\frac{s_{\bar{\jmath}}}{\sqrt{2}}\boldsymbol{P}_{\ell-1}^\ell \boldsymbol{A}_{\ell-1}^{-1}\boldsymbol{e}^{\ell-1,\bar{\jmath}}$$

$$= -\frac{\mu^{\ell,j}}{\mu^{\ell-1,\bar{\jmath}}}\frac{s_{\bar{\jmath}}}{\sqrt{2}}\boldsymbol{P}_{\ell-1}^\ell\boldsymbol{e}^{\ell-1,\bar{\jmath}} = -\underbrace{\frac{\mu^{\ell,j}}{\mu^{\ell-1,\bar{\jmath}}}\frac{s_{\bar{\jmath}}}{\sqrt{2}}\sqrt{2}}_{=1}\left(c_{\bar{\jmath}}\boldsymbol{e}^{\ell,\bar{\jmath}} - s_{\bar{\jmath}}\boldsymbol{e}^{\ell,N_\ell+1-\bar{\jmath}}\right)$$

$$= s_{\bar{\jmath}}\boldsymbol{e}^{\ell,j} - c_{\bar{\jmath}}\boldsymbol{e}^{\ell,\bar{\jmath}},$$

so that accordingly

$$\Psi_\ell^{CGC}\left(\boldsymbol{e}^{\ell,j}\right) = \boldsymbol{e}^{\ell,j} - \left(s_{\bar{\jmath}}\boldsymbol{e}^{\ell,j} - c_{\bar{\jmath}}\boldsymbol{e}^{\ell,\bar{\jmath}}\right) = c_{\bar{\jmath}}\boldsymbol{e}^{\ell,j} + c_{\bar{\jmath}}\boldsymbol{e}^{\ell,\bar{\jmath}}$$

holds. $\qquad\square$

To relate the analytical statements given in Theorem 4.55 to the numerical results, we use, as already on page 139, the order properties of the trigonometric functions. For smooth modes $\boldsymbol{e}^{\ell,j}$ with $1 \leq j \leq N_{\ell-1}$, we have

$$s_j = \sin^2\left(\frac{j\pi h_\ell}{2}\right) = \mathcal{O}\left((jh_\ell)^2\right),$$

which means that for small wave numbers j, the result is a strongly damped sum of the incoming mode and its complementary mode. In the range of oscillatory input data $\boldsymbol{e}^{\ell,j}$ with $N_{\ell-1} < j \leq N_\ell$, on the other hand, increasing j leads to a decreasing index $\bar{\jmath} = N_\ell + 1 - j$ and thus, due to

$$c_{\bar{\jmath}} = 1 + \mathcal{O}\left((\bar{\jmath}h_\ell)^2\right)$$

a coefficient approaching one from below, so that the result is a largely undamped summation of the two complementary modes and, consequently, no damping is to be expected. The analytical results thus match exactly with the numerical results presented in Figure 4.17.

To define and analyze a two-grid method, we first need a fundamental definition of the composition of iterative methods.

Definition 4.56 Let $\phi, \psi : \mathbb{C}^n \times \mathbb{C}^n \to \mathbb{C}^n$ be two iterative methods, then

$$\phi \circ \psi : \ \mathbb{C}^n \times \mathbb{C}^n \to \mathbb{C}^n$$

with

$$\boldsymbol{x}_{m+1} = (\phi \circ \psi)\,(\boldsymbol{x}_m, \boldsymbol{b}) := \phi(\psi(\boldsymbol{x}_m, \boldsymbol{b}), \boldsymbol{b})$$

is called a product iteration.

Lemma 4.57 *Let ϕ,ψ be two linear iterative methods with iteration matrices $\boldsymbol{M}_\phi$ and $\boldsymbol{M}_\psi$, then the following holds:*

    (a) If ϕ and ψ are consistent, then the product iteration $\phi \circ \psi$ is also consistent.

    (b) The iteration matrix of the product iteration $\phi \circ \psi$ has the form

$$\boldsymbol{M}_{\phi \circ \psi} = \boldsymbol{M}_\phi \boldsymbol{M}_\psi.$$

    (c) The two product iterations $\phi \circ \psi$ and $\psi \circ \phi$ have the same convergence properties in the sense that

$$\rho(\boldsymbol{M}_{\phi \circ \psi}) = \rho(\boldsymbol{M}_{\psi \circ \phi}).$$

Proof:

for (a): Let $\hat{\boldsymbol{x}} = \boldsymbol{A}^{-1}\boldsymbol{b}$, then consistency follows from

$$(\phi \circ \psi)(\hat{\boldsymbol{x}},\boldsymbol{b}) = \phi\,(\psi(\hat{\boldsymbol{x}},\boldsymbol{b}),\boldsymbol{b}) = \phi(\hat{\boldsymbol{x}},\boldsymbol{b}) = \hat{\boldsymbol{x}}.$$

for (b): With $\phi(\boldsymbol{x}_m,\boldsymbol{b}) = \boldsymbol{M}_\phi \boldsymbol{x}_m + \boldsymbol{N}_\phi \boldsymbol{b}$ and $\psi(\boldsymbol{x}_m,\boldsymbol{b}) = \boldsymbol{M}_\psi \boldsymbol{x}_m + \boldsymbol{N}_\psi \boldsymbol{b}$ the claim follows according to

$$(\phi \circ \psi)(\boldsymbol{x}_m,\boldsymbol{b}) = \boldsymbol{M}_\phi\,(\boldsymbol{M}_\psi \boldsymbol{x}_m + \boldsymbol{N}_\psi \boldsymbol{b}) + \boldsymbol{N}_\phi \boldsymbol{b} = \underbrace{\boldsymbol{M}_\phi \boldsymbol{M}_\psi}_{=\boldsymbol{M}_{\phi \circ \psi}} \boldsymbol{x}_m + \underbrace{(\boldsymbol{M}_\phi \boldsymbol{N}_\psi + \boldsymbol{N}_\phi)}_{=\boldsymbol{N}_{\phi \circ \psi}}\boldsymbol{b}.$$

for (c): Let $\boldsymbol{x}$ be an eigenvector of the matrix $\boldsymbol{M}_\phi \boldsymbol{M}_\psi$ for the eigenvalue $\lambda \neq 0$, then from $\boldsymbol{M}_\phi \boldsymbol{M}_\psi \boldsymbol{x} = \lambda \boldsymbol{x} \neq \boldsymbol{0}$ it follows that $\boldsymbol{M}_\psi \boldsymbol{x} \neq \boldsymbol{0}$, and thus from the equation $\boldsymbol{M}_\psi \boldsymbol{M}_\phi \boldsymbol{M}_\psi \boldsymbol{x} = \lambda \boldsymbol{M}_\psi \boldsymbol{x}$ we always have $\rho\,(\boldsymbol{M}_\phi \boldsymbol{M}_\psi) \leq \rho\,(\boldsymbol{M}_\psi \boldsymbol{M}_\phi)$. Analogously, we obtain $\rho\,(\boldsymbol{M}_\psi \boldsymbol{M}_\phi) \leq \rho\,(\boldsymbol{M}_\phi \boldsymbol{M}_\psi)$. Thus, the spectral radii of the iteration matrices and therefore the convergence behavior of the product iterations $\phi \circ \psi$ and $\psi \circ \phi$ coincide.

$\square$

Linear iterative methods such as the damped Jacobi method have a smoothing effect on the error progression and are therefore referred to as *smoothers* in the following. Let $\nu \in \mathbb{N}$ be the number of iteration steps on the fine grid Ω_ℓ and ϕ_ℓ the smoothing method, then we obtain the two-grid method as a product iteration in the form

$$\phi_\ell^{TGM} = \phi_\ell^{CGC} \circ \phi_\ell^\nu. \tag{4.2.33}$$

If R denotes the restriction, P the prolongation, and E the exact solution of the system of equations, then (4.2.33) can be visualized with the smoother G according to the form shown in Figure 4.18.

Let $\nu_1,\nu_2 \in \mathbb{N}$ with $\nu = \nu_1 + \nu_2$, then according to Lemma 4.57 with

$$\phi_\ell^{TGM(\nu_1,\nu_2)} := \phi_\ell^{\nu_2} \circ \phi_\ell^{CGC} \circ \phi_\ell^{\nu_1} \tag{4.2.34}$$

we have a method that exhibits the same convergence properties as (4.2.33). This is referred to as ν_1 pre- and ν_2 postsmoothing steps. We thus obtain the graphical representation presented in Figure 4.19.

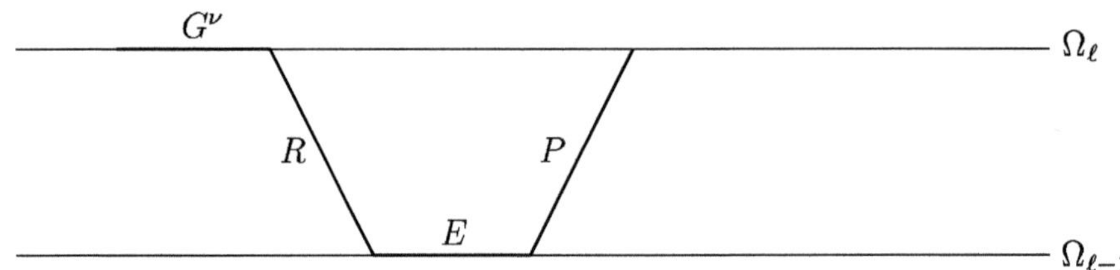

Figure 4.18 Two-grid method without postsmoothing

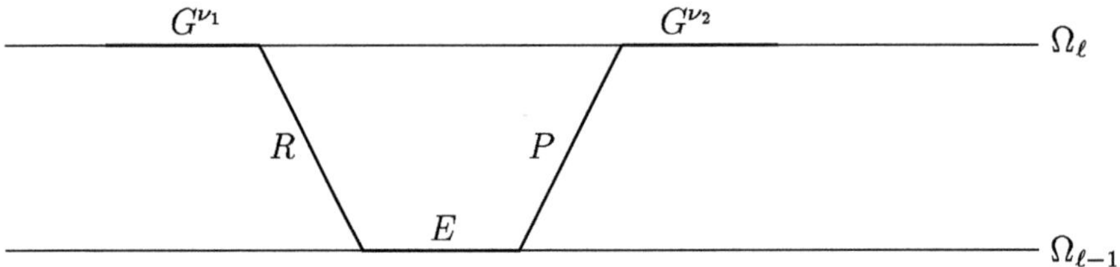

Figure 4.19 Two-grid method with postsmoothing

Theorem 4.58 *Let ϕ_ℓ be a consistent iterative method with iteration matrix M_ℓ, then the two-grid iterative method $\phi_\ell^{TGM(\nu_1,\nu_2)}$ is consistent with the iteration matrix*

$$M_\ell^{TGM(\nu_1,\nu_2)} = M_\ell^{\nu_2}\left(I - P_{\ell-1}^\ell A_{\ell-1}^{-1} R_\ell^{\ell-1} A_\ell\right) M_\ell^{\nu_1}. \qquad (4.2.35)$$

Proof:
According to Lemma 4.53, the coarse grid correction ϕ_ℓ^{CGC} is consistent with iteration matrix $M_\ell^{CGC} = I - P_{\ell-1}^\ell A_{\ell-1}^{-1} R_\ell^{\ell-1} A_\ell$. Lemma 4.57 (a) thus yields the claimed consistency, and with Lemma 4.57 (b) it follows that

$$M_\ell^{TGM(\nu_1,\nu_2)} = M_{\phi_\ell^{\nu_2}\circ\phi_\ell^{CGC}\circ\phi_\ell^{\nu_1}} = M_\ell^{\nu_2}\left(I - P_{\ell-1}^\ell A_{\ell-1}^{-1} R_\ell^{\ell-1} A_\ell\right) M_\ell^{\nu_1}.$$

$$\square$$

Of course, different smoothing algorithms can be used for the pre- and post-smoothing steps.

The two-grid method can be written in the form of a diagram as follows:

Algorithm - Two-Grid Method —

<table>
<tr><td colspan="2">For $i = 1,\dots,\nu_1$</td></tr>
<tr><td></td><td>$\boldsymbol{u}^\ell := \phi_\ell\left(\boldsymbol{u}^\ell, \boldsymbol{f}^\ell\right)$</td></tr>
<tr><td colspan="2">$\boldsymbol{d}^{\ell-1} := \boldsymbol{R}_\ell^{\ell-1}\left(\boldsymbol{A}_\ell \boldsymbol{u}^\ell - \boldsymbol{f}^\ell\right)$</td></tr>
<tr><td colspan="2">$\boldsymbol{e}^{\ell-1} := \boldsymbol{A}_{\ell-1}^{-1}\boldsymbol{d}^{\ell-1}$</td></tr>
<tr><td colspan="2">$\boldsymbol{u}^\ell := \boldsymbol{u}^\ell - \boldsymbol{P}_{\ell-1}^\ell \boldsymbol{e}^{\ell-1}$</td></tr>
<tr><td colspan="2">For $i = 1,\dots,\nu_2$</td></tr>
<tr><td></td><td>$\boldsymbol{u}^\ell := \phi_\ell\left(\boldsymbol{u}^\ell, \boldsymbol{f}^\ell\right)$</td></tr>
</table>

Let us again consider, as an example, the error

$$e_0^3 := u_0^3 - u^{3,*} = (0.75,\, 0.2,\, 0.6,\, 0.45,\, 0.9,\, 0.6,\, 0.8,\, 0.85,\, 0.55,\, 0.7,\, 0.9,\, 0.5,\, 0.6,\, 0.3,\, 0.2)^T,$$

so, with two iterations of the damped Jacobi method on Ω_3 and a subsequent coarse grid correction using linear restriction and linear prolongation, we obtain the representation shown in Figure 4.20.

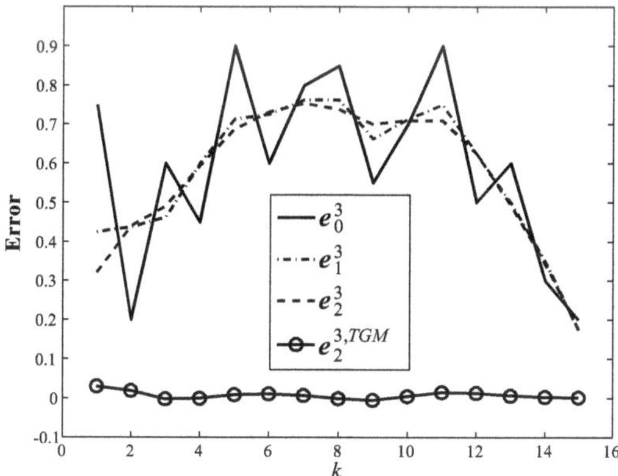

Figure 4.20 Evolution of the error in the two-grid method on Ω_3 with two smoothing steps and a subsequent coarse grid correction

In addition to the smoothing property of the damped Jacobi method already shown in Figure 4.10, the immense error reduction of the subsequent coarse grid correction is evident. Due to the value of the spectral radius of the iteration matrix of the damped Jacobi method being close to one, it is already clear here that continuing the damping steps without using the coarse grid correction would result in a significantly higher computational effort compared to the two-grid method in order to reduce the error to a prescribed maximum value.

Finally, we now want to analytically investigate the properties of the two-grid method. Through the presented composition consisting of pre- and post-smoothing steps as well as an appended coarse grid correction, a damping of all present Fourier modes should be achieved.

For the two-grid method, we have

$$\boldsymbol{u}_{m+1}^{\ell} = \phi_{\ell}^{TGM(\nu_1,\nu_2)}\left(\boldsymbol{u}_m^{\ell},\boldsymbol{f}^{\ell}\right) = \boldsymbol{M}_{\ell}^{TGM(\nu_1,\nu_2)}\boldsymbol{u}_m^{\ell} + \boldsymbol{N}_{\ell}^{TGM(\nu_1,\nu_2)}\boldsymbol{f}^{\ell},$$

where this is meant to make clear that the two-grid method is a linear iterative method and can therefore be applied repeatedly in succession. Each iteration step consists of $\nu = \nu_1 + \nu_2$ iterations of a splitting method and a coarse grid correction. For the evolution of the error $\boldsymbol{e}_m^{\ell} = \boldsymbol{u}_m^{\ell} - \boldsymbol{u}^{\ell,*}$, the consistency proven in Theorem 4.58 yields

$$\begin{aligned}
\boldsymbol{e}_{m+1}^{\ell} &= \boldsymbol{u}_{m+1}^{\ell} - \boldsymbol{u}^{\ell,*} \\
&= \boldsymbol{M}_{\ell}^{TGM(\nu_1,\nu_2)}\boldsymbol{u}_m^{\ell} + \boldsymbol{N}_{\ell}^{TGM(\nu_1,\nu_2)}\boldsymbol{f}^{\ell} - \left(\boldsymbol{M}_{\ell}^{TGM(\nu_1,\nu_2)}\boldsymbol{u}^{\ell,*} + \boldsymbol{N}_{\ell}^{TGM(\nu_1,\nu_2)}\boldsymbol{f}^{\ell}\right) \\
&= \boldsymbol{M}_{\ell}^{TGM(\nu_1,\nu_2)}(\boldsymbol{u}_m^{\ell} - \boldsymbol{u}^{\ell,*}) \\
&= \boldsymbol{M}_{\ell}^{TGM(\nu_1,\nu_2)}\boldsymbol{e}_m^{\ell}.
\end{aligned}$$

According to the coarse grid correction, we define using (4.2.35)

$$\begin{aligned}
\Psi_{\ell}^{TGM(\nu_1,\nu_2)} : \quad \mathbb{R}^{N_{\ell}} &\to \mathbb{R}^{N_{\ell}} \\
\boldsymbol{e} &\mapsto \Psi_{\ell}^{TGM(\nu_1,\nu_2)}(\boldsymbol{e}) = \boldsymbol{M}_{\ell}^{\nu_2}\left(\boldsymbol{I} - \boldsymbol{P}_{\ell-1}^{\ell}\boldsymbol{A}_{\ell-1}^{-1}\boldsymbol{R}_{\ell}^{\ell-1}\boldsymbol{A}_{\ell}\right)\boldsymbol{M}_{\ell}^{\nu_1}\boldsymbol{e}
\end{aligned}$$

and again consider the effect of this mapping on the Fourier modes on the grid Ω_{ℓ}.

Theorem 4.59 *Let the two-grid method be defined using the damped Jacobi method as well as the linear restriction (4.2.12) and the linear prolongation (4.2.13). Then, for the Fourier modes $\boldsymbol{e}^{\ell,j}$, $j = 1,\ldots,N_{\ell}$ on Ω_{ℓ} under the mapping $\Psi_{\ell}^{TGM(\nu_1,\nu_2)}$, the following representation holds:*

$$\Psi_{\ell}^{TGM(\nu_1,\nu_2)}\left(\boldsymbol{e}^{\ell,j}\right) = (\lambda^{\ell,j})^{\nu_1}s_j\left((\lambda^{\ell,j})^{\nu_2}\boldsymbol{e}^{\ell,j} + (\lambda^{\ell,\bar{j}})^{\nu_2}\boldsymbol{e}^{\ell,\bar{j}}\right)$$
$$\text{for } j \in \{1,\ldots,N_{\ell-1}\} \text{ and } \bar{j} = N_{\ell} + 1 - j,$$
$$\Psi_{\ell}^{TGM(\nu_1,\nu_2)}\left(\boldsymbol{e}^{\ell,j}\right) = (\lambda^{\ell,j})^{\nu_1+\nu_2}\boldsymbol{e}^{\ell,j} \quad \text{for } j = N_{\ell-1} + 1,$$
$$\Psi_{\ell}^{TGM(\nu_1,\nu_2)}\left(\boldsymbol{e}^{\ell,j}\right) = (\lambda^{\ell,j})^{\nu_1}c_{\bar{j}}\left((\lambda^{\ell,j})^{\nu_2}\boldsymbol{e}^{\ell,j} + (\lambda^{\ell,\bar{j}})^{\nu_2}\boldsymbol{e}^{\ell,\bar{j}}\right)$$
$$\text{for } j = N_{\ell} + 1 - \bar{j} \text{ with } \bar{j} \in \{1,\ldots,N_{\ell-1}\}$$

using $c_{\bar{j}} = \cos^2\left(\bar{j}\pi\frac{h_{\ell}}{2}\right)$, $s_j = \sin^2\left(j\pi\frac{h_{\ell}}{2}\right)$, $\lambda^{\ell,j} = \lambda^{\ell,j}(1/4)$ and $\lambda^{\ell,\bar{j}} = \lambda^{\ell,\bar{j}}(1/4)$.

Proof:
As already done in the previous proofs of theorems of this type, we consider three different cases. Let us first turn to the index range $j = 1,\ldots,N_{\ell-1}$ with $\bar{j} = N_{\ell-1} + 1 - j$. Using Theorem 4.55, we obtain

$$\Psi_\ell^{TGM(\nu_1,\nu_2)}\left(e^{\ell,j}\right) = M_\ell^{\nu_2}\left(I - P_{\ell-1}^\ell A_{\ell-1}^{-1} R_\ell^{\ell-1} A_\ell\right) M_\ell^{\nu_1} e^{\ell,j}$$

$$= (\lambda^{\ell,j})^{\nu_1} M_\ell^{\nu_2}\left(I - P_{\ell-1}^\ell A_{\ell-1}^{-1} R_\ell^{\ell-1} A_\ell\right) e^{\ell,j}$$

$$= (\lambda^{\ell,j})^{\nu_1} M_\ell^{\nu_2}\left(s_j\, e^{\ell,j} + s_j\, e^{\ell,\bar{j}}\right)$$

$$= (\lambda^{\ell,j})^{\nu_1} s_j \left(\left(\lambda^{\ell,j}\right)^{\nu_2} e^{\ell,j} + \left(\lambda^{\ell,\bar{j}}\right)^{\nu_2} e^{\ell,\bar{j}}\right).$$

Accordingly, by the above theorem, for $j = N_{\ell-1} + 1$ we obtain the equation

$$\Psi_\ell^{TGM(\nu_1,\nu_2)}\left(e^{\ell,j}\right) = (\lambda^{\ell,j})^{\nu_1} M_\ell^{\nu_2}\left(I - P_{\ell-1}^\ell A_{\ell-1}^{-1} R_\ell^{\ell-1} A_\ell\right) e^{\ell,j} = (\lambda^{\ell,j})^{\nu_1} M_\ell^{\nu_2} e^{\ell,j}$$

$$= (\lambda^{\ell,j})^{\nu_1+\nu_2} e^{\ell,j}$$

as well as for $j = N_\ell + 1 - \bar{j}$ with $\bar{j} \in \{1,\dots,N_{\ell-1}\}$ the representation

$$\Psi_\ell^{TGM(\nu_1,\nu_2)}\left(e^{\ell,j}\right) = M_\ell^{\nu_2}\left(I - P_{\ell-1}^\ell A_{\ell-1}^{-1} R_\ell^{\ell-1} A_\ell\right) M_\ell^{\nu_1} e^{\ell,j}$$

$$= (\lambda^{\ell,j})^{\nu_1} M_\ell^{\nu_2}\left(I - P_{\ell-1}^\ell A_{\ell-1}^{-1} R_\ell^{\ell-1} A_\ell\right) e^{\ell,j}$$

$$= (\lambda^{\ell,j})^{\nu_1} M_\ell^{\nu_2}\left(c_{\bar{j}}\, e^{\ell,j} + c_{\bar{j}}\, e^{\ell,\bar{j}}\right)$$

$$= (\lambda^{\ell,j})^{\nu_1} c_{\bar{j}} \left(\left(\lambda^{\ell,j}\right)^{\nu_2} e^{\ell,j} + \left(\lambda^{\ell,\bar{j}}\right)^{\nu_2} e^{\ell,\bar{j}}\right),$$

which completes the proof. $\qquad\qquad\square$

The above theorem demonstrates very precisely the comprehensive effect of the two-grid method on all present Fourier modes. The long-wavelength modes $e^{\ell,j}$, $j \in \{1,\dots,N_{\ell-1}\}$, have a relatively small damping factor with $\frac{1}{2} < \lambda^{\ell,j} < 1$ due to the integrated Jacobi method. Here, the long-wavelength error component is reduced by the coarse grid correction because of the small value s_j. The oscillatory Fourier mode involved here exhibits two small factors, s_j and $\lambda^{\ell,\bar{j}}$, so that a very small amplitude results. The Fourier mode corresponding to the medium wave number $j = N_{\ell-1} + 1$ experiences, with $(\lambda^{\ell,j})^{\nu_1+\nu_2} = (1/2)^\nu$, exclusively a damping by the Jacobi method, which, however, quickly leads to error reduction for increasing $\nu = \nu_1 + \nu_2$. Furthermore, the oscillatory modes are primarily damped quickly by the Jacobi method via the small eigenvalue $\lambda^{\ell,j}$.

4.2.2   The Multigrid Algorithm

The two-grid method has proven to be efficient. However, it is impractical for large systems, since it requires the exact solution of the correction equation

$$A_{\ell-1} e^{\ell-1} = d^{\ell-1} \tag{4.2.36}$$

on $\Omega_{\ell-1}$. However, since the prolongation of the exact solution $e^{\ell-1}$ to the fine grid Ω_ℓ only provides an approximation to the desired error vector

$$e^\ell = u^\ell - u^{\ell,*}$$

an approximate solution of the correction equation is also sufficient.

Equation (4.2.36) has the same form as the original equation $\boldsymbol{A}_\ell \boldsymbol{u}^\ell = \boldsymbol{f}^\ell$. The idea is therefore to use a two-grid method on $\Omega_{\ell-1}$ and $\Omega_{\ell-2}$ to approximate $\boldsymbol{e}^{\ell-1} = \boldsymbol{A}_{\ell-1}^{-1} \boldsymbol{d}^{\ell-1}$. This yields a three-grid method, in which $\boldsymbol{A}_{\ell-2} \boldsymbol{e}^{\ell-2} = \boldsymbol{d}^{\ell-2}$ must be solved exactly. Continuing this idea successively leads to a method on $\ell+1$ grids $\Omega_\ell, \ldots, \Omega_0$ in which only $\boldsymbol{A}_0 \boldsymbol{e}^0 = \boldsymbol{d}^0$ must be solved exactly. With the grid refinement we have chosen, with $h_{\ell-1} = 2h_\ell$ and $h_0 = 1/2$, we have $\boldsymbol{A}_0 \in \mathbb{R}^{1\times 1}$. In practice, Ω_0 is usually chosen as a grid that allows an approximate solution of the system of equations with the matrix $\boldsymbol{A}_0$ in an efficient and simple way.

The multigrid algorithm can thus be formulated as a recursive procedure in the following form:

Algorithm - Multigrid Method $\phi_\ell^{\mathrm{MGM}(\nu_1,\nu_2)}\left(\boldsymbol{u}^\ell, \boldsymbol{f}^\ell\right)$ —

Y $\quad \ell = 0$	N
$\boldsymbol{u}^0 := \boldsymbol{A}_0^{-1} \boldsymbol{f}^0$	For $i = 1, \ldots, \nu_1$
	$\qquad \boldsymbol{u}^\ell := \phi_\ell\left(\boldsymbol{u}^\ell, \boldsymbol{f}^\ell\right)$
Return $\boldsymbol{u}^0$	$\boldsymbol{d}^{\ell-1} := \boldsymbol{R}_\ell^{\ell-1}\left(\boldsymbol{A}_\ell \boldsymbol{u}^\ell - \boldsymbol{f}^\ell\right)$
	$\boldsymbol{e}_0^{\ell-1} := \boldsymbol{0}$
	For $i = 1, \ldots, \gamma$
	$\qquad \boldsymbol{e}_i^{\ell-1} := \phi_{\ell-1}^{\mathrm{MGM}(\nu_1,\nu_2)}\left(\boldsymbol{e}_{i-1}^{\ell-1}, \boldsymbol{d}^{\ell-1}\right)$
	$\boldsymbol{u}^\ell := \boldsymbol{u}^\ell - \boldsymbol{P}_{\ell-1}^\ell \boldsymbol{e}_\gamma^{\ell-1}$
	For $i = 1, \ldots, \nu_2$
	$\qquad \boldsymbol{u}^\ell := \phi_\ell\left(\boldsymbol{u}^\ell, \boldsymbol{f}^\ell\right)$
	Return $\boldsymbol{u}^\ell$

In the representation chosen here, γ steps are used for the iterative solution of the coarse grid equation. In practice, $\gamma = 1$ or $\gamma = 2$ are usually suitable.

The case $\gamma = 1$ yields the so-called V-cycle, which for $\ell = 3$ can be represented graphically as in Figure 4.21, while for $\gamma = 2$ the iteration sequence is called the W-cycle. The algorithmic flow resulting from this approach for the case of four grids is shown in Figure 4.22.

To examine the effect of the multigrid method as an example with regard to its error damping properties, we consider, within the underlying Poisson equation (4.2.1), the right-hand side

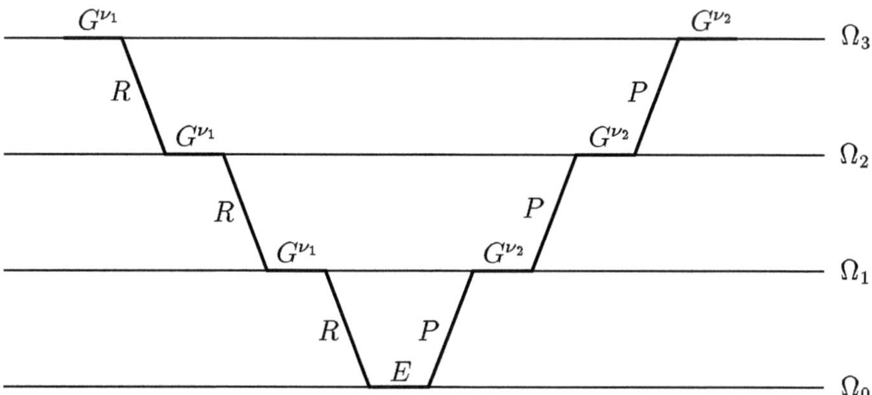

Figure 4.21   Multigrid method with V-cycle

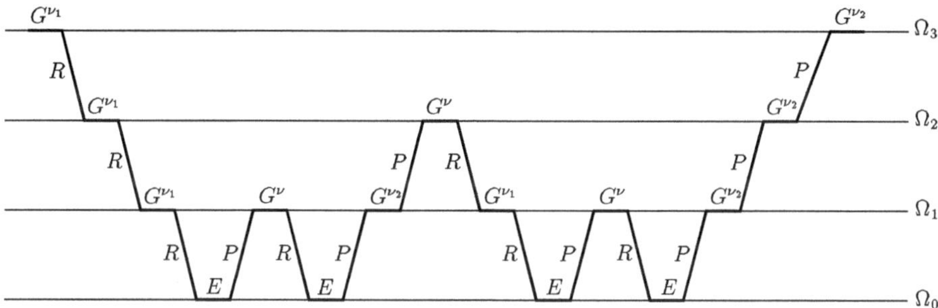

Figure 4.22   Multigrid method with W-cycle with $\nu = \nu_1 + \nu_2$

$$f(x) = \frac{\pi^2}{8}\left(9\sin\left(\frac{3\pi x}{2}\right) + 25\sin\left(\frac{5\pi x}{2}\right)\right). \tag{4.2.37}$$

A simple calculation shows that with

$$u(x) = \sin(2\pi x)\cos\left(\frac{\pi x}{2}\right)$$

we have the solution to the Dirichlet boundary value problem. We want to start on the grid Ω_3 with the error vector already used in the previous examples, see Figures 4.10 and 4.20,

$$e_0^3 := u_0^3 - u^{3,*} = (0.75,\ 0.2,\ 0.6,\ 0.45,\ 0.9,\ 0.6,\ 0.8,\ 0.85,\ 0.55,\ 0.7,\ 0.9,\ 0.5,\ 0.6,\ 0.3,\ 0.2)^T$$

so that the initial guess is given by

$$u_0^3 = \begin{pmatrix} u(x_1) \\ \vdots \\ u(x_{N_3}) \end{pmatrix} - e_0^3 \quad \text{with } x_i = ih_3.$$

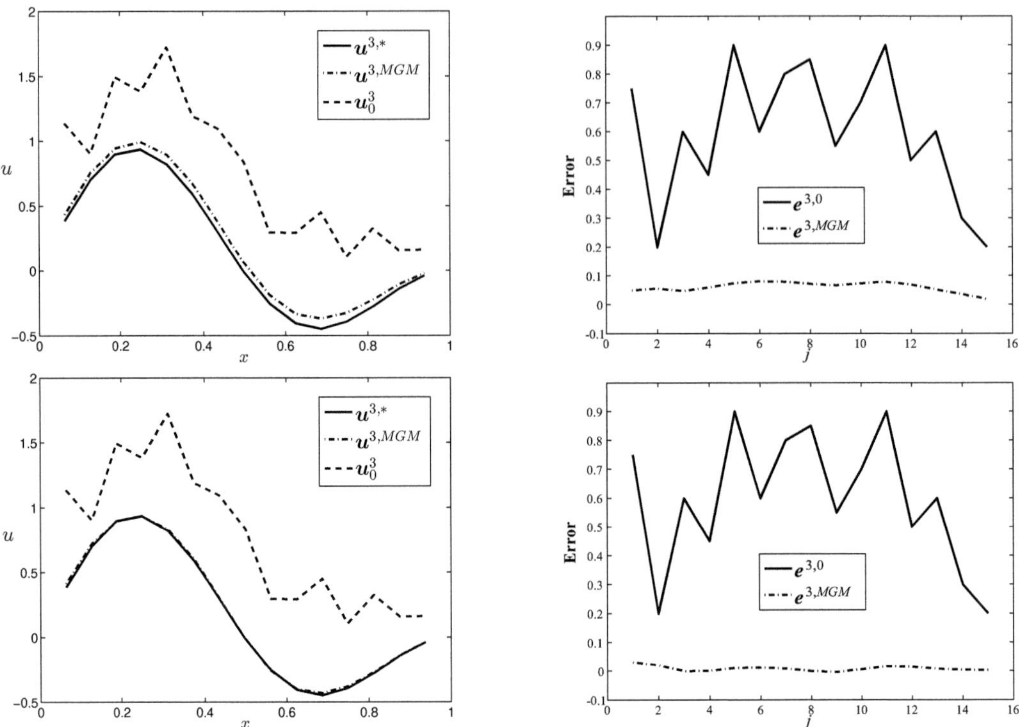

Figure 4.23   Evolution of the solution (left) and the error (right) in the multigrid method with linear restriction and prolongation using a V-cycle (top) and W-cycle (bottom)

The corresponding progression is shown in the left plots of Figure 4.23, while the error is shown in the corresponding right plot. The multigrid method with linear restriction and linear prolongation is always used, as well as $\nu_1 = 2$ pre- and $\nu_2 = 2$ post-smoothing steps. As expected, Figure 4.23 already shows a drastic error reduction after just one multigrid iteration, so that both in the case of the V-cycle shown at the top and the W-cycle shown at the bottom, a very good agreement between the solution and the numerical approximation is evident. Here, the W-cycle shows smaller absolute deviations, which, however, must also be understood in relation to the fact that one iteration step of the W-cycle requires more computational effort compared to the V-cycle.

A direct comparison of the multigrid method with the Jacobi method proves to be difficult, since one iteration step of the multigrid method includes several smoothing processes on different grid levels as well as numerous restriction and prolongation operations. Therefore, we have based the comparison on the computation time required by both algorithms to reduce a given error in the maximum norm below 10^{-6}. As a test example, we again use the one-dimensional boundary value problem (4.2.1) with the right-hand side given by (4.2.37). The multigrid method was used with $\gamma = 1$ in the form of a V-cycle, and the initialization was the same for both methods, but generated by a random number generator. This results in the comparison of relative computation times shown in the following table. Here, 100 % always indicates the faster method, and the percentage for the more time-consuming method corresponds to the ratio of the computation times of the two methods. The efficiency gain achieved by using the multigrid idea compared to the classical splitting method is impressively demonstrated.

Computation time comparison			
Grid	Number of unknowns	Multigrid method	Classical Jacobi method
Ω_2	7	100 %	117 %
Ω_4	31	100 %	838 %
Ω_6	127	100 %	9255 %
Ω_8	511	100 %	128161 %

4.2.3 The Full Multigrid Method

The idea of the full multigrid method (nested iterations, full multigrid method (FMGM)) is to use the coarser grids to improve the initial approximation u_0^ℓ on the finest grid Ω_ℓ.

The procedure can be described as follows:

(a) Solve the equation $A_0 u^0 = f^0$ exactly on Ω_0.

(b) Prolongate the result to the next finer grid and perform a few iterations with the smoothing method.

(c) Repeat (b) until an approximate solution u_0^ℓ has been obtained on Ω_ℓ.

(d) Apply the multigrid method using the initial approximation u_0^ℓ.

Graphically, for the V-cycle on three grids, we obtain the representation of the full multigrid method as shown in Figure 4.24.

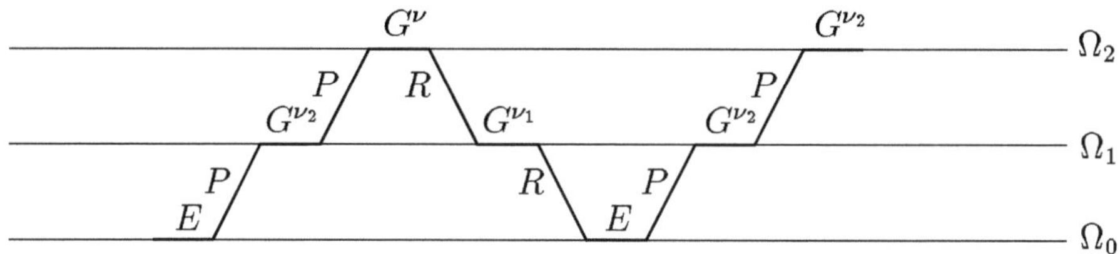

Figure 4.24   Full multigrid method with $\nu = \nu_1 + \nu_2$

Variants of the method can be obtained, for example, by

(a) using an approximate instead of an exact solution of the equation $A_0 u^0 = f^0$.

(b) using different numbers of smoothing steps.

(c) using different prolongation and restriction operators.

Finally, we analyze the full multigrid method in the context of the boundary value problem $-u''(x) = f(x)$ used on page 120 with the function f according to (4.2.37). Due to the initialization using the combined prolongation and smoothing steps, in contrast to the previous multigrid algorithm, we obtain a significantly smoother initial approximation, which can be seen as u_0^3 in the respective graphics in Figure 4.25. The corresponding error progression shown in the right-hand images is already much smoother, and we again achieve a very good reduction of the error for both the V- and W-cycle after just one multigrid iteration. As in the previously considered multigrid method, the parameters $\nu_1 = \nu_2 = 2$ were used.

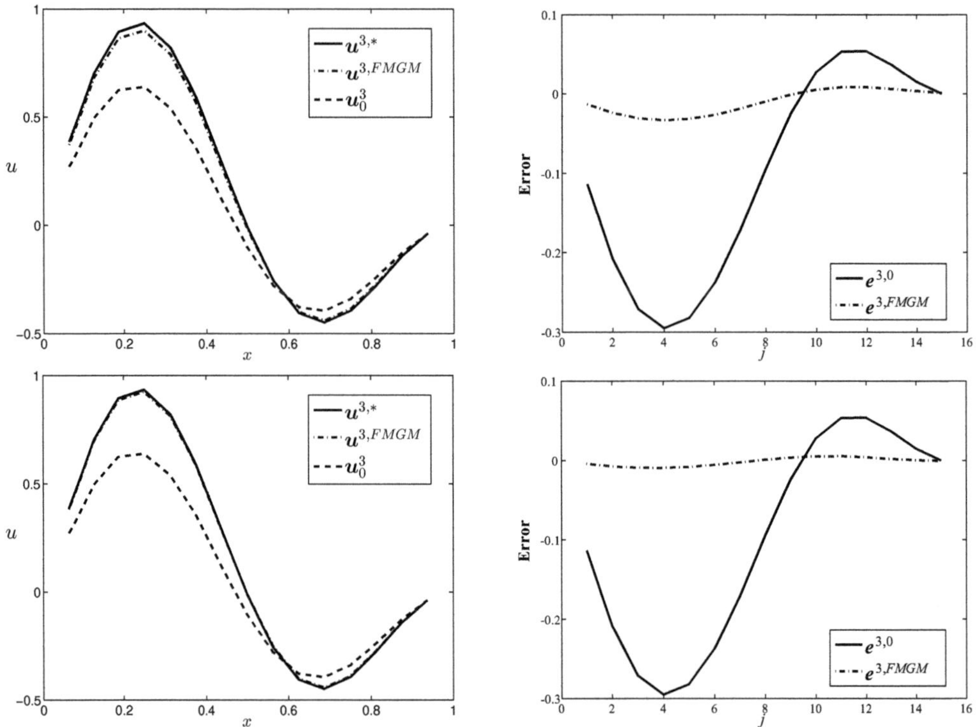

Figure 4.25   Evolution of the solution (left) and the error (right) for the full multigrid method with linear restriction and prolongation using the V-cycle (top) and W-cycle (bottom)

4.3   Projection Methods and Krylov Subspace Methods

In this section, we always consider linear systems of equations of the form

$$Ax = b \tag{4.3.1}$$

with a non-singular matrix $A \in \mathbb{R}^{n \times n}$ and a right-hand side $b \in \mathbb{R}^n$.

Definition 4.60 A projection method for solving equation (4.3.1) is a procedure for computing approximate solutions $x_m \in x_0 + K_m$ subject to the condition

$$b - Ax_m \perp L_m, \tag{4.3.2}$$

where $x_0 \in \mathbb{R}^n$ is arbitrary and K_m and L_m represent m-dimensional subspaces of $\mathbb{R}^n$. The orthogonality condition is defined here by the Euclidean inner product via

$$x \perp y \quad \Leftrightarrow \quad (x,y)_2 = 0.$$

If $K_m = L_m$, then (4.3.2) states that the residual vector $r_m = b - Ax_m$ is orthogonal to K_m. In this case, we have an orthogonal projection method, and (4.3.2) is called the Galerkin condition.

For $K_m \neq L_m$, we have a skew projection method, and (4.3.2) is referred to as the Petrov-Galerkin condition.

We want to illustrate the difference between an orthogonal and a skew projection method with a very simple graphical example. Within the system $Ax = b$, we consider the case $A = I \in \mathbb{R}^{2\times 2}$. Consequently, starting from the initial vector $x_0 \in \mathbb{R}^2$, the approximate solution $x_1 \in x_0 + K_1$ is determined by the condition

$$b - x_1 = b - Ax_1 \perp L_1.$$

In the context of an orthogonal projection method, this requirement is, due to $K_1 = L_1$, equivalent to

$$b - x_1 \perp x_0 + K_1$$

and x_1 thus represents the orthogonal projection of b onto the affine subspace $x_0 + K_1$. This situation is depicted in Figure 4.26. Even though the orthogonality of the projection here is directly correlated with the choice $A = I$, this helps to clarify the idea behind the terminology.

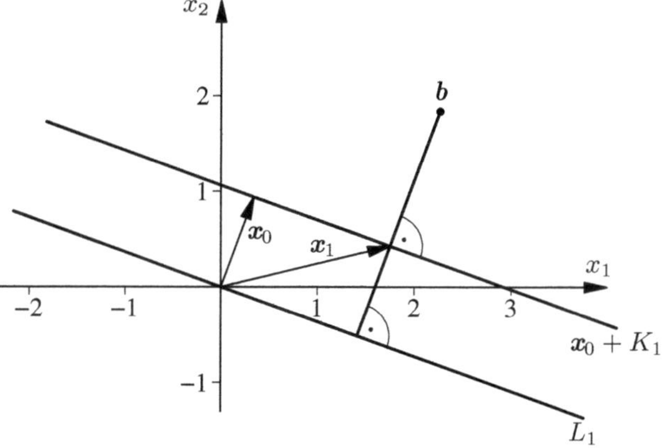

Figure 4.26　　Orthogonal projection in the case $A = I$

Furthermore, Figure 4.27 illustrates a skew projection method, where, due to $K_1 \neq L_1$, the projection of x_1 onto $x_0 + K_1$ is not orthogonal.

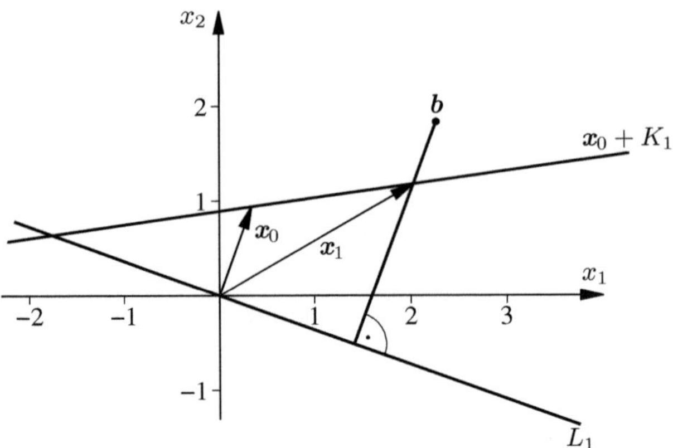

Figure 4.27   Oblique projection in the case $\boldsymbol{A} = \boldsymbol{I}$

Example 4.61 Each step of the Gauss-Seidel method can be interpreted as an orthogonal projection method with $\boldsymbol{x}_0 = (x_{m+1,1},\ldots,x_{m+1,i-1},0,x_{m,i+1},\ldots,x_{m,n})^T$ and the one-dimensional spaces $K = L = \operatorname{span}\{\boldsymbol{e}_i\}$, since

$$x_i = -\frac{1}{a_{ii}}\left(\sum_{j=1}^{i-1} a_{ij}x_{m+1,j} + \sum_{j=i+1}^{n} a_{ij}x_{m,j} - b_i\right)$$

$$\Leftrightarrow \quad b_i - (\boldsymbol{A}\boldsymbol{x})_i = 0 \quad \text{with} \quad \boldsymbol{x} \in \boldsymbol{x}_0 + K$$

$$\Leftrightarrow \quad \boldsymbol{b} - \boldsymbol{A}\boldsymbol{x} \perp L \quad \text{with} \quad \boldsymbol{x} \in \boldsymbol{x}_0 + K.$$

This property can also be demonstrated for other iterative methods of this type.

In general, however, splitting methods differ significantly from projection methods. A comparison of these two classes of methods can be found in the following table.

Splitting methods	Projection methods
Computation of approximate solutions $\boldsymbol{x}_m \in \mathbb{R}^n$	Computation of approximate solutions $\boldsymbol{x}_m \in \boldsymbol{x}_0 + K_m \subset \mathbb{R}^n$ $\dim K_m = m \leq n$
Computation rule $\boldsymbol{x}_m = \boldsymbol{M}\boldsymbol{x}_{m-1} + \boldsymbol{N}\boldsymbol{b}$	Computation rule (Orthogonality condition) $\boldsymbol{b} - \boldsymbol{A}\boldsymbol{x}_m \perp L_m \subset \mathbb{R}^n$ $\dim L_m = m \leq n$

Definition 4.62 A Krylov subspace method is a projection method for solving the equation (4.3.1), in which K_m denotes the *Krylov subspace*

$$K_m = K_m(\boldsymbol{A},\boldsymbol{r}_0) = \operatorname{span}\{\boldsymbol{r}_0,\boldsymbol{A}\boldsymbol{r}_0,\ldots,\boldsymbol{A}^{m-1}\boldsymbol{r}_0\}$$

with $r_0 = b - Ax_0$.

Krylov subspace methods are often described by a reformulation of the linear system of equations as a minimization problem. Two of the best-known representatives of this class of algorithms are the conjugate gradient method developed by Hestenes and Stiefel [39] and the GMRES method derived by Saad and Schulz [64]. Both methods determine the optimal approximation $x_m \in x_0 + K_m$ to the desired solution $A^{-1}b$ in the sense of the orthogonality condition (4.3.2), where in each iteration the dimension of the subspace is incremented by one. If one neglects the rounding errors that occur, both methods would deliver the exact solution after at most n iterations.

It should be noted here that the Krylov subspace K_m may have dimension less than, but not necessarily equal to m, which can be seen very easily, for example, in the case $A = I$. In the context of the GMRES method, we will therefore explicitly examine what consequences the case $K_j = K_{j+1}$ has for the corresponding approximate solutions x_j and x_{j+1}.

Before we deal with the derivation of the individual methods, we will first formulate a general convergence statement for Krylov subspace methods. The following lemma proves to be helpful for this purpose.

Lemma 4.63 *Let a projection method be given for solving the equation $Ax = b$ with a non-singular matrix $A \in \mathbb{R}^{n \times n}$. Let $m \in \mathbb{N}$ be fixed, and let the column vectors of the matrix $V_m \in \mathbb{R}^{n \times m}$ and $W_m \in \mathbb{R}^{n \times m}$ form a basis of the spaces K_m and L_m, respectively, such that $W_m^T A V_m \in \mathbb{R}^{m \times m}$ is non-singular, then the solution of the projection method has the representation*

$$x_m = x_0 + V_m \left(W_m^T A V_m \right)^{-1} W_m^T r_0$$

with the residual vector $r_0 = b - Ax_0$.

Proof:
Due to the basis property of the column vectors of the matrix V_m, the solution vector can be written in the form $x_m = x_0 + V_m \alpha_m$ with $\alpha_m \in \mathbb{R}^m$. From the orthogonality condition (4.3.2) we obtain

$$W_m^T \left(b - A \left(x_0 + V_m \alpha_m \right) \right) = 0,$$

which yields

$$W_m^T A V_m \alpha_m = W_m^T \left(b - A x_0 \right).$$

As a result of the assumed invertibility of the matrix $W_m^T A V_m$, it follows that

$$\alpha_m = \left(W_m^T A V_m \right)^{-1} W_m^T r_0,$$

which gives the claimed form of the iterates. $\square$

From the above lemma, we immediately obtain the representation of the associated residual vector in the form

$$r_m = b - A x_m = r_0 - A V_m \left(W_m^T A V_m \right)^{-1} W_m^T r_0. \tag{4.3.3}$$

We now come to the announced convergence theorem, which was presented in [40].

Theorem 4.64 *Let the matrix $\boldsymbol{A} \in \mathbb{R}^{n \times n}$ be invertible. Furthermore, let $\boldsymbol{v}_1, \ldots, \boldsymbol{v}_m \in \mathbb{R}^n$ and $\boldsymbol{w}_1, \ldots, \boldsymbol{w}_m \in \mathbb{R}^n$ be the basis vectors of K_m and L_m generated by an arbitrary Krylov subspace method. If $\boldsymbol{W}_m^T \boldsymbol{A} \boldsymbol{V}_m \in \mathbb{R}^{m \times m}$ composed by the matrices $\boldsymbol{V}_m = (\boldsymbol{v}_1 \ldots \boldsymbol{v}_m) \in \mathbb{R}^{n \times m}$ and $\boldsymbol{W}_m = (\boldsymbol{w}_1 \ldots \boldsymbol{w}_m) \in \mathbb{R}^{n \times m}$ is non-singular, then, with the projection*

$$\boldsymbol{P}_m = \boldsymbol{I} - \boldsymbol{A} \boldsymbol{V}_m \left(\boldsymbol{W}_m^T \boldsymbol{A} \boldsymbol{V}_m \right)^{-1} \boldsymbol{W}_m^T \tag{4.3.4}$$

the estimates for the error vector $\boldsymbol{e}_m = \boldsymbol{A}^{-1} \boldsymbol{b} - \boldsymbol{x}_m$ and the residual vector $\boldsymbol{r}_m = \boldsymbol{A} \boldsymbol{e}_m$ of the Krylov subspace method follow in the form

$$\|\boldsymbol{e}_m\| \leq \|\boldsymbol{A}^{-1} \boldsymbol{P}_m\| \min_{p \in \mathcal{P}_m^1} \|p(\boldsymbol{A}) \boldsymbol{r}_0\| \tag{4.3.5}$$

and

$$\|\boldsymbol{r}_m\| \leq \|\boldsymbol{P}_m\| \min_{p \in \mathcal{P}_m^1} \|p(\boldsymbol{A}) \boldsymbol{r}_0\|, \tag{4.3.6}$$

where $\mathcal{P}_m^1$ denotes the set of all polynomials p of degree at most m that also satisfy the side condition $p(\boldsymbol{0}) = \boldsymbol{I}$.

Proof:
From the definition of the projection $\boldsymbol{P}_m$, taking into account the invertibility of the matrix $\boldsymbol{W}_m^T \boldsymbol{A} \boldsymbol{V}_m$, we obtain the equation

$$\boldsymbol{P}_m \boldsymbol{A} \boldsymbol{V}_m = \boldsymbol{A} \boldsymbol{V}_m - \boldsymbol{A} \boldsymbol{V}_m \left(\boldsymbol{W}_m^T \boldsymbol{A} \boldsymbol{V}_m \right)^{-1} \boldsymbol{W}_m^T \boldsymbol{A} \boldsymbol{V}_m = \boldsymbol{0}. \tag{4.3.7}$$

Using equation (4.3.3), we get $\boldsymbol{r}_m = \boldsymbol{P}_m \boldsymbol{r}_0$, which, together with (4.3.7), also yields

$$\boldsymbol{r}_m = \boldsymbol{P}_m \left(\boldsymbol{r}_0 + \boldsymbol{A} \boldsymbol{V}_m \boldsymbol{\alpha} \right)$$

for any vector $\boldsymbol{\alpha} \in \mathbb{R}^m$. Since $\boldsymbol{A} \boldsymbol{V}_m \boldsymbol{\alpha} \in \boldsymbol{A} K_m$, we obtain

$$\boldsymbol{r}_m = \boldsymbol{P}_m p(\boldsymbol{A}) \boldsymbol{r}_0$$

for any polynomial $p \in \mathcal{P}_m^1$. This yields

$$\|\boldsymbol{r}_m\| = \min_{p \in \mathcal{P}_m^1} \|\boldsymbol{P}_m p(\boldsymbol{A}) \boldsymbol{r}_0\| \leq \|\boldsymbol{P}_m\| \min_{p \in \mathcal{P}_m^1} \|p(\boldsymbol{A}) \boldsymbol{r}_0\|.$$

Analogously, $\boldsymbol{e}_m = \boldsymbol{A}^{-1} \boldsymbol{r}_m$ yields the inequality (4.3.5).   □

The non-singularity of the matrix $\boldsymbol{W}_m^T \boldsymbol{A} \boldsymbol{V}_m$ required in the above theorem can be shown directly for special Krylov subspace methods. For a symmetric, positive definite matrix $\boldsymbol{A}$, according to Corollary 2.33, there exists an orthogonal matrix $\boldsymbol{U}$ with $\boldsymbol{U}^T \boldsymbol{A} \boldsymbol{U} = \boldsymbol{D} = \text{diag}\{\lambda_1, \ldots, \lambda_n\}$. From the positive definiteness it follows that $\lambda_i > 0$ for $i = 1, \ldots, n$, and we obtain $\boldsymbol{A} = \boldsymbol{U} \boldsymbol{D}^{1/2} \boldsymbol{D}^{1/2} \boldsymbol{U}^T$. If we consider an orthogonal Krylov subspace method, then, since $L_m = K_m$, the matrices $\boldsymbol{W}_m = \boldsymbol{V}_m$ can be used, which yields

$$\boldsymbol{W}_m^T \boldsymbol{A} \boldsymbol{V}_m = \left(\boldsymbol{D}^{1/2} \boldsymbol{U}^T \boldsymbol{V}_m \right)^T \left(\boldsymbol{D}^{1/2} \boldsymbol{U}^T \boldsymbol{V}_m \right) \in \mathbb{R}^{m \times m}$$

Since the columns of the matrix $\boldsymbol{V}_m$ form a basis of K_m, we have $\text{rang}(\boldsymbol{V}_m) = m$. Thus, $\boldsymbol{D}^{1/2} \boldsymbol{U} \boldsymbol{V}_m$ is injective, so that

$$\left(\boldsymbol{x}, \left(\boldsymbol{D}^{1/2}\boldsymbol{U}^{T}\boldsymbol{V}_{m} \right)^{T} \left(\boldsymbol{D}^{1/2}\boldsymbol{U}^{T}\boldsymbol{V}_{m} \right) \boldsymbol{x} \right)_{2} = \left\| \left(\boldsymbol{D}^{1/2}\boldsymbol{U}^{T}\boldsymbol{V}_{m} \right) \boldsymbol{x} \right\|_{2}^{2} \neq 0 \ \forall \boldsymbol{x} \in \mathbb{R}^{m} \setminus \{\boldsymbol{0}\},$$

which gives the invertibility of the matrix $\boldsymbol{W}_{m}^{T}\boldsymbol{A}\boldsymbol{V}_{m}$. An example of such a method is the conjugate gradient method considered below.

For invertible matrices $\boldsymbol{A} \in \mathbb{R}^{n \times n}$, we consider $\boldsymbol{L}_{m} = \boldsymbol{A}\boldsymbol{K}_{m}$, so that $\boldsymbol{W}_{m} = \boldsymbol{A}\boldsymbol{V}_{m}$ can be chosen. Analogous to the above consideration, the invertibility of the matrix $\boldsymbol{W}_{m}^{T}\boldsymbol{A}\boldsymbol{V}_{m}$ follows from

$$\boldsymbol{W}_{m}^{T}\boldsymbol{A}\boldsymbol{V}_{m} = \left(\boldsymbol{A}\boldsymbol{V}_{m} \right)^{T} \boldsymbol{A}\boldsymbol{V}_{m}.$$

These conditions are satisfied by the GMRES method derived in Section 4.3.2.4.

In general, for $m = n$, the non-singularity of the matrices $\boldsymbol{V}_{m}$ and $\boldsymbol{W}_{m}$ follows, so that $\boldsymbol{P}_{m} = \boldsymbol{0}$ holds, and from the estimates (4.3.5) and (4.3.6) we obtain in each case the statement that such Krylov subspace methods determine the exact solution after at most n steps.

The development of state-of-the-art numerical algorithms for the simulation of practically relevant problems has led to a great demand for efficient, fast, and robust iterative solvers for systems of equations, and has provided a significant impetus to the progress achieved in the field of Krylov subspace methods over the past decades. Due to the large number of these methods, in this book we will, in addition to the conjugate gradient method, essentially restrict ourselves to the very commonly used methods GMRES, CGS, BiCGSTAB, TFQMR, and QMRCGSTAB, and, as far as possible, include further algorithms in the present framework by means of remarks. The relationship between the aforementioned algorithms is illustrated schematically in Figure 4.28. For further literature on this class of methods, we refer to the books by Axelsson [7], Demmel [20], Fischer [26], Greenbaum [34], Kelley [44], Meurant [53], Saad [63], Trefethen and Bau [71], van der Vorst [74], and Weiss [76].

4.3.1 Methods for Symmetric, Positive Definite Matrices

Within this section, we assume that the considered system of equations (4.3.1) has a symmetric and positive definite matrix. To derive the methods, we consider the function

$$\begin{aligned} F : \mathbb{R}^{n} \ &\to \ \mathbb{R} \\ \boldsymbol{x} \ &\mapsto \ \frac{1}{2}(\boldsymbol{A}\boldsymbol{x},\boldsymbol{x})_{2} - (\boldsymbol{b},\boldsymbol{x})_{2} \end{aligned} \tag{4.3.8}$$

and will first study some of its fundamental properties in connection with the considered system of equations.

Lemma 4.65 *Let $\boldsymbol{A} \in \mathbb{R}^{n \times n}$ be symmetric, positive definite, and let $\boldsymbol{b} \in \mathbb{R}^{n}$ be given. Then, for the function F defined by (4.3.8), we have*

$$\hat{\boldsymbol{x}} = \arg \min_{\boldsymbol{x} \in \mathbb{R}^{n}} F(\boldsymbol{x})$$

if and only if

$$\boldsymbol{A}\hat{\boldsymbol{x}} = \boldsymbol{b}$$

holds.

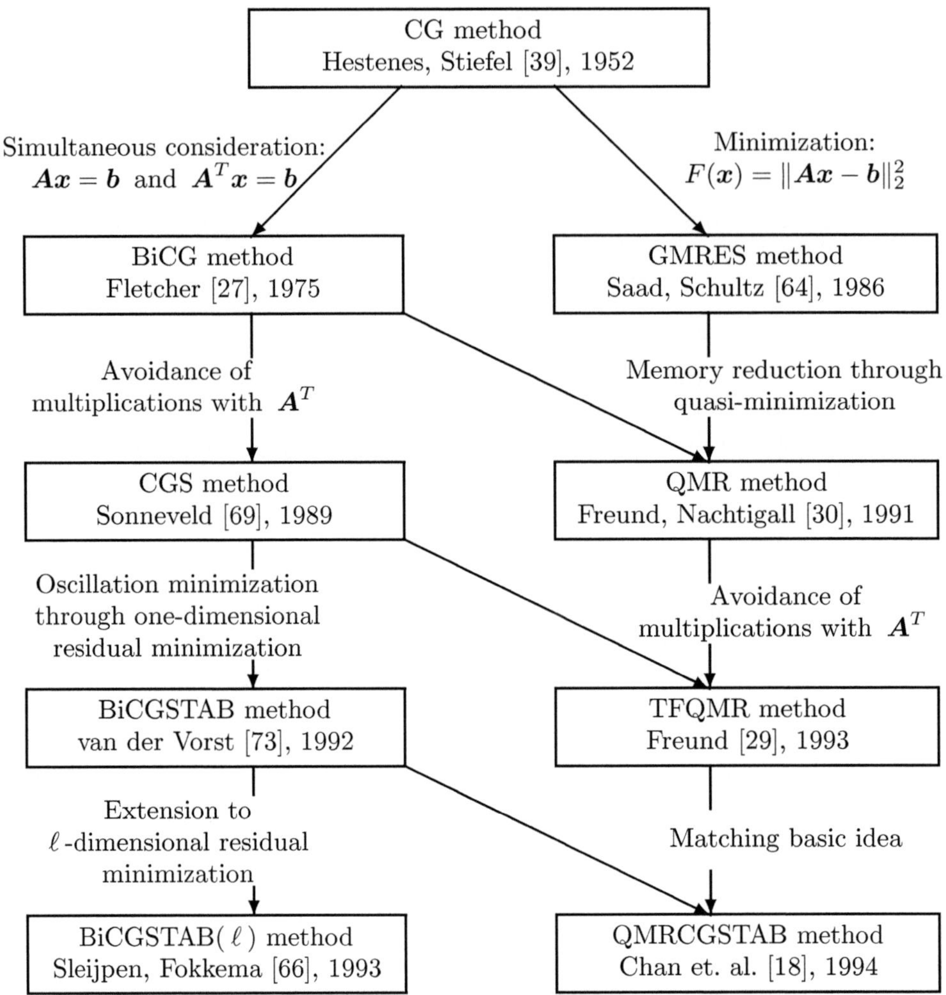

Figure 4.28   Relationships between Krylov subspace methods

Proof:
Since $\boldsymbol{A}$ is positive definite, the positive definite inverse $\boldsymbol{A}^{-1} \in \mathbb{R}^{n \times n}$ exists, and we obtain

$$\arg \min_{\boldsymbol{x} \in \mathbb{R}^n} F(\boldsymbol{x}) = \arg \min_{\boldsymbol{x} \in \mathbb{R}^n} G(\boldsymbol{x}) \tag{4.3.9}$$

with

$$
\begin{aligned}
G(\boldsymbol{x}) &= F(\boldsymbol{x}) + \frac{1}{2}\boldsymbol{b}^T \boldsymbol{A}^{-1}\boldsymbol{b} \\
&= \frac{1}{2}(\boldsymbol{A}\boldsymbol{x},\boldsymbol{x})_2 - (\boldsymbol{b},\boldsymbol{x})_2 + \frac{1}{2}\boldsymbol{b}^T \boldsymbol{A}^{-1}\boldsymbol{b} \\
&= \frac{1}{2}(\boldsymbol{A}\boldsymbol{x} - \boldsymbol{b})^T \boldsymbol{x} - \frac{1}{2}\boldsymbol{b}^T(\boldsymbol{x} - \boldsymbol{A}^{-1}\boldsymbol{b}) \\
&= \frac{1}{2}(\boldsymbol{A}\boldsymbol{x} - \boldsymbol{b})^T \boldsymbol{x} - \frac{1}{2}\left(\boldsymbol{A}^{-1}\boldsymbol{b}\right)^T (\boldsymbol{A}\boldsymbol{x} - \boldsymbol{b}) \\
&= \frac{1}{2}(\boldsymbol{A}\boldsymbol{x} - \boldsymbol{b})^T \boldsymbol{A}^{-1}(\boldsymbol{A}\boldsymbol{x} - \boldsymbol{b}).
\end{aligned}
\tag{4.3.10}
$$

Thus, $G(\boldsymbol{x}) \geq 0$ holds and

$$
G(\boldsymbol{x}) = 0 \quad \Leftrightarrow \quad \boldsymbol{x} = \boldsymbol{A}^{-1}\boldsymbol{b}.
$$

$\square$

In all methods considered in this section, we perform a successive minimization of the function F starting from the point $\boldsymbol{x} \in \mathbb{R}^n$ along special directions $\boldsymbol{p} \in \mathbb{R}^n$. We therefore define, for $\boldsymbol{x},\boldsymbol{p} \in \mathbb{R}^n$, the function

$$
\begin{aligned}
f_{\boldsymbol{x},\boldsymbol{p}} : \mathbb{R} &\to \mathbb{R} \\
\lambda &\mapsto f_{\boldsymbol{x},\boldsymbol{p}}(\lambda) := F(\boldsymbol{x} + \lambda\boldsymbol{p}).
\end{aligned}
\tag{4.3.11}
$$

Lemma and Definition 4.66 *Let the matrix $\boldsymbol{A} \in \mathbb{R}^{n\times n}$ be symmetric, positive definite, and let the vectors $\boldsymbol{x},\boldsymbol{p} \in \mathbb{R}^n$ with $\boldsymbol{p} \neq \boldsymbol{0}$ be given. Then,*

$$
\lambda_{opt} = \lambda_{opt}(\boldsymbol{x},\boldsymbol{p}) := \arg\min_{\lambda \in \mathbb{R}} f_{\boldsymbol{x},\boldsymbol{p}}(\lambda) = \frac{(\boldsymbol{r},\boldsymbol{p})_2}{(\boldsymbol{A}\boldsymbol{p},\boldsymbol{p})_2}
$$

with $\boldsymbol{r} := \boldsymbol{b} - \boldsymbol{A}\boldsymbol{x}$. The vector $\boldsymbol{r}$ is called the residual vector and its Euclidean norm $\|\boldsymbol{r}\|_2$ is called the residual.

Proof:
We have

$$
\begin{aligned}
f_{\boldsymbol{x},\boldsymbol{p}}(\lambda) &= \tfrac{1}{2}\left(\boldsymbol{A}(\boldsymbol{x} + \lambda\boldsymbol{p}),\boldsymbol{x} + \lambda\boldsymbol{p}\right)_2 - (\boldsymbol{b},\boldsymbol{x} + \lambda\boldsymbol{p})_2 \\
&= F(\boldsymbol{x}) + \lambda(\boldsymbol{A}\boldsymbol{x} - \boldsymbol{b},\boldsymbol{p})_2 + \tfrac{1}{2}\lambda^2(\boldsymbol{A}\boldsymbol{p},\boldsymbol{p})_2.
\end{aligned}
$$

Thus, it follows that

$$
f'_{\boldsymbol{x},\boldsymbol{p}}(\lambda) = (\boldsymbol{A}\boldsymbol{x} - \boldsymbol{b},\boldsymbol{p})_2 + \lambda \underbrace{(\boldsymbol{A}\boldsymbol{p},\boldsymbol{p})_2}_{>0}
$$

and

$$
f'_{\boldsymbol{x},\boldsymbol{p}}(\lambda_{\mathrm{opt}}) = (\boldsymbol{A}\boldsymbol{x} - \boldsymbol{b},\boldsymbol{p})_2 + \frac{(\boldsymbol{b} - \boldsymbol{A}\boldsymbol{x},\boldsymbol{p})_2}{(\boldsymbol{A}\boldsymbol{p},\boldsymbol{p})_2}(\boldsymbol{A}\boldsymbol{p},\boldsymbol{p})_2 = 0.
$$

With

$$
f''_{\boldsymbol{x},\boldsymbol{p}}(\lambda) = (\boldsymbol{A}\boldsymbol{p},\boldsymbol{p})_2 \overset{\substack{\boldsymbol{A}\ \text{pos.def.}\\ \boldsymbol{p}\neq 0}}{>} 0
\tag{4.3.12}
$$

one can easily see that λ_{opt} is the global minimum of $f_{\boldsymbol{x},\boldsymbol{p}}$. $\square$

Now, if we are given a sequence of search directions $\{\boldsymbol{p}_m\}_{m\in\mathbb{N}_0}$ from $\mathbb{R}^n\backslash\{\boldsymbol{0}\}$, we can construct a first algorithm:

Algorithm - Basis Solver —

<table>
<tr><td colspan="2">Choose $\boldsymbol{x}_0 \in \mathbb{R}^n$</td></tr>
<tr><td colspan="2">For $m = 0,1,\ldots$</td></tr>
<tr><td></td><td>$\boldsymbol{r}_m = \boldsymbol{b} - \boldsymbol{A}\boldsymbol{x}_m$</td></tr>
<tr><td></td><td>$\lambda_m = \dfrac{(\boldsymbol{r}_m,\boldsymbol{p}_m)_2}{(\boldsymbol{A}\boldsymbol{p}_m,\boldsymbol{p}_m)_2}$</td></tr>
<tr><td></td><td>$\boldsymbol{x}_{m+1} = \boldsymbol{x}_m + \lambda_m\boldsymbol{p}_m$</td></tr>
</table>

4.3.1.1   The Method of Steepest Descent

To complete the basic solver, we need a computational rule for determining the search directions $\boldsymbol{p}_m \in \mathbb{R}^n$. We also require, without loss of generality, that $\|\boldsymbol{p}_m\|_2 = 1$. For $\boldsymbol{x} \neq \boldsymbol{A}^{-1}\boldsymbol{b}$, we obtain a globally optimal choice by

$$\boldsymbol{p} = \frac{\hat{\boldsymbol{x}} - \boldsymbol{x}}{\|\hat{\boldsymbol{x}} - \boldsymbol{x}\|_2} \quad \text{with} \quad \hat{\boldsymbol{x}} = \boldsymbol{A}^{-1}\boldsymbol{b},$$

because, with this, by defining λ_{opt} according to Lemma 4.66, we have

$$\begin{aligned}
\tilde{\boldsymbol{x}} &= \boldsymbol{x} + \lambda_{\text{opt}}\boldsymbol{p} = \\
&= \boldsymbol{x} + \|\hat{\boldsymbol{x}} - \boldsymbol{x}\|_2 \frac{(\boldsymbol{b} - \boldsymbol{A}\boldsymbol{x},\hat{\boldsymbol{x}} - \boldsymbol{x})_2}{(\boldsymbol{b} - \boldsymbol{A}\boldsymbol{x},\hat{\boldsymbol{x}} - \boldsymbol{x})_2} \frac{\hat{\boldsymbol{x}} - \boldsymbol{x}}{\|\hat{\boldsymbol{x}} - \boldsymbol{x}\|_2} \\
&= \hat{\boldsymbol{x}}.
\end{aligned}$$

However, in this case, we already need the exact solution to define the search direction.

If we restrict ourselves to local optimality, we obtain this with the negative gradient of the function F. Here, we have

$$\nabla F(\boldsymbol{x}) = \frac{1}{2}(\boldsymbol{A} + \boldsymbol{A}^T)\boldsymbol{x} - \boldsymbol{b} \stackrel{\boldsymbol{A}\ \text{sym.}}{=} \boldsymbol{A}\boldsymbol{x} - \boldsymbol{b} = -\boldsymbol{r}.$$

Thus,

$$\boldsymbol{p} := \begin{cases} \dfrac{\boldsymbol{r}}{\|\boldsymbol{r}\|_2} & \text{for } \boldsymbol{r} \neq 0, \\[2mm] \boldsymbol{0} & \text{for } \boldsymbol{r} = 0 \end{cases} \tag{4.3.13}$$

gives the direction of steepest descent.

The function F is strictly convex because

$$\nabla^2 F(\boldsymbol{x}) = \boldsymbol{A}$$

and $\boldsymbol{A}$ is positive definite. From this, we also see that

$$\hat{\boldsymbol{x}} = \boldsymbol{A}^{-1}\boldsymbol{b}$$

because

$$\nabla F(\hat{\boldsymbol{x}}) = \boldsymbol{0}$$

is the unique and global minimum of F.

Starting from the basic solver, we obtain with (4.3.13) the method of steepest descent, which is also known as the gradient method. Due to the relationship between the residual vector $\boldsymbol{r}_m$ and the search direction $\boldsymbol{p}_m$ used here, the method can also be written as

$$\boldsymbol{x}_{m+1} = \boldsymbol{x}_m + \lambda_m \boldsymbol{p}_m = \boldsymbol{x}_m + \frac{\lambda_m}{\|\boldsymbol{r}_m\|_2}\boldsymbol{r}_m = \boldsymbol{x}_m + \tilde{\lambda}_m \boldsymbol{r}_m$$

with

$$\tilde{\lambda}_m = \frac{\lambda_m}{\|\boldsymbol{r}_m\|_2} = \frac{(\boldsymbol{r}_m,\boldsymbol{p}_m)_2}{\|\boldsymbol{r}_m\|_2(\boldsymbol{A}\boldsymbol{p}_m,\boldsymbol{p}_m)_2} \overset{\boldsymbol{p}_m=\frac{\boldsymbol{r}_m}{\|\boldsymbol{r}_m\|_2}}{=} \frac{1}{\|\boldsymbol{r}_m\|_2}\frac{\frac{1}{\|\boldsymbol{r}_m\|_2}(\boldsymbol{r}_m,\boldsymbol{r}_m)_2}{\frac{1}{\|\boldsymbol{r}_m\|_2^2}(\boldsymbol{A}\boldsymbol{r}_m,\boldsymbol{r}_m)_2}$$

$$= \frac{(\boldsymbol{r}_m,\boldsymbol{r}_m)_2}{(\boldsymbol{A}\boldsymbol{r}_m,\boldsymbol{r}_m)_2} = \frac{\|\boldsymbol{r}_m\|_2^2}{(\boldsymbol{A}\boldsymbol{r}_m,\boldsymbol{r}_m)_2}.$$

If we neglect the superscript $\sim$ used to distinguish the scalar quantities λ_m and $\tilde{\lambda}_m$, the method of steepest descent can be formulated as follows.

Algorithm - Steepest Descent Method —

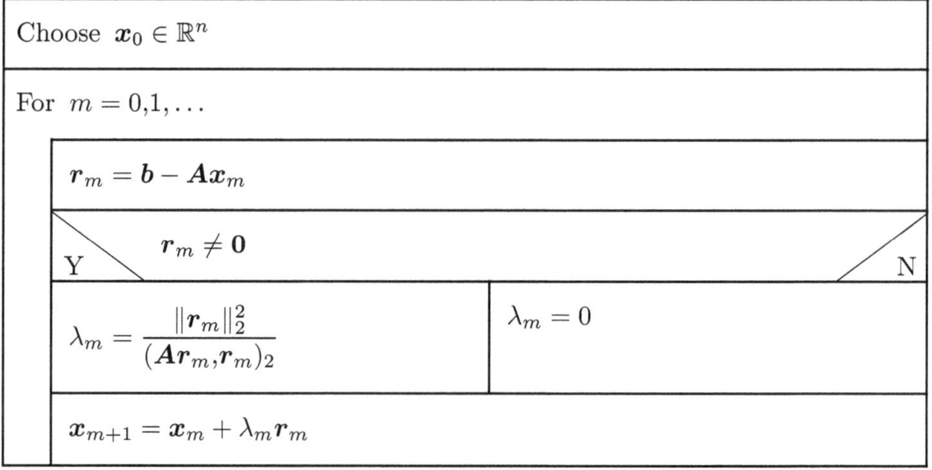

Choose $\boldsymbol{x}_0 \in \mathbb{R}^n$

For $m = 0,1,\ldots$

$\boldsymbol{r}_m = \boldsymbol{b} - \boldsymbol{A}\boldsymbol{x}_m$

Y — $\boldsymbol{r}_m \neq \boldsymbol{0}$ — N

$\lambda_m = \dfrac{\|\boldsymbol{r}_m\|_2^2}{(\boldsymbol{A}\boldsymbol{r}_m,\boldsymbol{r}_m)_2}$ | $\lambda_m = 0$

$\boldsymbol{x}_{m+1} = \boldsymbol{x}_m + \lambda_m \boldsymbol{r}_m$

Remark:
In practical applications, $\boldsymbol{r}_0$ is computed outside the loop and within the loop

$$
\begin{aligned}
r_{m+1} &= b - Ax_{m+1} = b - Ax_m - \lambda_m Ar_m \\
&= r_m - \lambda_m Ar_m
\end{aligned}
$$

is used, which avoids a matrix-vector multiplication per iteration. A MATLAB implementation is listed in Appendix A.

Theorem 4.67 *The method of steepest descent is consistent and nonlinear.*

Proof:

The iteration method $\phi : \mathbb{R}^n \times \mathbb{R}^n \to \mathbb{R}^n$ is given by

$$
\phi(x,b) = (I - \lambda(x,b)A)\,x + \lambda(x,b)b, \tag{4.3.14}
$$

where $\lambda : \mathbb{R}^n \times \mathbb{R}^n \to \mathbb{R}$ is defined as

$$
\lambda(x,b) = \begin{cases} \dfrac{\|b - Ax\|_2^2}{(A(b - Ax),b - Ax)_2} & \text{for } b - Ax \neq 0 \\[2mm] 0 & \text{otherwise.} \end{cases}
$$

Simple calculations with small examples show that λ is not constant. For $\hat{x} = A^{-1}b$, we have $\lambda(\hat{x},b) = 0$. Consequently, we obtain $\hat{x} = \phi(\hat{x},b)$ and $\hat{x}$ is a fixed point of the iteration ϕ. $\qquad\square$

Theorem 4.68 *Let A be positive definite and symmetric, then the sequence $\{x_m\}_{m\in\mathbb{N}_0}$ defined by the method of steepest descent converges for any initial vector $x_0 \in \mathbb{R}^n$ to the solution $\hat{x} = A^{-1}b$, and for the error vector $e_m = x_m - \hat{x}$ the following estimate holds:*

$$
\|e_m\|_A \leq \left(\frac{\mathrm{cond}_2(A) - 1}{\mathrm{cond}_2(A) + 1}\right)^m \|e_0\|_A. \tag{4.3.15}
$$

Proof:

If we look at the equation (4.3.14), we can see that the method can be interpreted as a special Richardson iteration with variable $\Theta = \lambda(x,b)$. Since the initial vector $x_0 \in \mathbb{R}^n$ is arbitrary, it suffices to prove the estimate (4.3.15) for $m = 1$.

Let $\lambda_{\max} = \max\limits_{\lambda\in\sigma(A)} \lambda$ and $\lambda_{\min} = \min\limits_{\lambda\in\sigma(A)} \lambda$, then for the classical Richardson method with optimal weighting parameter

$$
\Theta_{\mathrm{opt}} \overset{\text{Theorem 4.34}}{=} \frac{2}{\lambda_{\max} + \lambda_{\min}}
$$

the error vector e_1^R is given by

$$
e_1^R = M_R(\Theta_{\mathrm{opt}})e_0
$$

with

$$
M_R(\Theta_{\mathrm{opt}}) = I - \Theta_{\mathrm{opt}}A.
$$

From the symmetry of the matrix A it follows that $M_R(\Theta_{\mathrm{opt}})$ is also symmetric, and thus

$$\|\boldsymbol{M}_R(\Theta_{\mathrm{opt}})\|_2 \overset{\text{Theorem 2.36}}{=} \rho\left(\boldsymbol{M}_R(\Theta_{\mathrm{opt}})\right)$$

$$\overset{\text{Theorem 4.34}}{=} \frac{\lambda_{\max} - \lambda_{\min}}{\lambda_{\max} + \lambda_{\min}}$$

results. Furthermore, for any $p \in \mathbb{R}$ it holds that

$$\boldsymbol{M}_R(\Theta_{\mathrm{opt}})\boldsymbol{A}^p = \boldsymbol{A}^p \boldsymbol{M}_R(\Theta_{\mathrm{opt}}),$$

and with

$$\tilde{\boldsymbol{e}}_0 = \boldsymbol{A}^{1/2}\boldsymbol{e}_0 \text{ and } \tilde{\boldsymbol{e}}_1^R = \boldsymbol{A}^{1/2}\boldsymbol{e}_1^R$$

we obtain the representation

$$\tilde{\boldsymbol{e}}_1^R = \boldsymbol{A}^{1/2}\boldsymbol{M}_R(\Theta_{\mathrm{opt}})\boldsymbol{e}_0 = \boldsymbol{M}_R(\Theta_{\mathrm{opt}})\tilde{\boldsymbol{e}}_0.$$

Thus, we have

$$\|\boldsymbol{e}_1^R\|_{\boldsymbol{A}} = \|\tilde{\boldsymbol{e}}_1^R\|_2 \leq \|\boldsymbol{M}_R(\Theta_{\mathrm{opt}})\|_2\|\tilde{\boldsymbol{e}}_0\|_2 = \|\boldsymbol{M}_R(\Theta_{\mathrm{opt}})\|_2\|\boldsymbol{e}_0\|_{\boldsymbol{A}}.$$

Using equation (4.3.10) we conclude

$$G(\boldsymbol{x}) = \frac{1}{2}(\boldsymbol{A}\boldsymbol{x} - \boldsymbol{b})^T \boldsymbol{A}^{-1}(\boldsymbol{A}\boldsymbol{x} - \boldsymbol{b}) = \frac{1}{2}\|\boldsymbol{x} - \hat{\boldsymbol{x}}\|_{\boldsymbol{A}}^2$$

so that with

$$\boldsymbol{x}_1^R = \boldsymbol{x}_0 + \Theta_{\mathrm{opt}}\boldsymbol{r}_0$$

and

$$\boldsymbol{x}_1 = \boldsymbol{x}_0 + \lambda_0 \boldsymbol{r}_0$$

we obtain

$$\boldsymbol{x}_1 = \arg\min_{\boldsymbol{x} \in \boldsymbol{x}_0 + \mathrm{span}\,\{\boldsymbol{r}_0\}} F(\boldsymbol{x}) = \arg\min_{\boldsymbol{x} \in \boldsymbol{x}_0 + \mathrm{span}\,\{\boldsymbol{r}_0\}} G(\boldsymbol{x}) = \arg\min_{\boldsymbol{x} \in \boldsymbol{x}_0 + \mathrm{span}\,\{\boldsymbol{r}_0\}} \frac{1}{2}\|\boldsymbol{x} - \hat{\boldsymbol{x}}\|_{\boldsymbol{A}}^2.$$

Hence, an estimate for $\|\boldsymbol{e}_1\|_{\boldsymbol{A}}$ can be written as

$$\|\boldsymbol{e}_1\|_{\boldsymbol{A}} \leq \|\boldsymbol{e}_1^R\|_{\boldsymbol{A}} \leq \underbrace{\frac{\lambda_{\max} - \lambda_{\min}}{\lambda_{\max} + \lambda_{\min}}}_{\xi\,:=} \|\boldsymbol{e}_0\|_{\boldsymbol{A}}. \tag{4.3.16}$$

For positive definite, symmetric matrices $\boldsymbol{A} \in \mathbb{R}^{n \times n}$ it holds that $\|\boldsymbol{A}\|_2 = \rho(\boldsymbol{A}) = \lambda_{\max} > 0$ and $\|\boldsymbol{A}^{-1}\|_2 = \rho(\boldsymbol{A}^{-1}) = \lambda_{\min}^{-1} > 0$, so that on the one hand, with $|\xi| < 1$, the claimed convergence of the method is present, and on the other hand, from (4.3.16) the estimate

$$\|\boldsymbol{e}_1\|_{\boldsymbol{A}} \leq \left(\frac{\frac{\lambda_{\max}}{\lambda_{\min}} - 1}{\frac{\lambda_{\max}}{\lambda_{\min}} + 1}\right) \|\boldsymbol{e}_0\|_{\boldsymbol{A}} = \left(\frac{\|\boldsymbol{A}\|_2\|\boldsymbol{A}^{-1}\|_2 - 1}{\|\boldsymbol{A}\|_2\|\boldsymbol{A}^{-1}\|_2 + 1}\right) \|\boldsymbol{e}_0\|_{\boldsymbol{A}} = \left(\frac{\mathrm{cond}_2(\boldsymbol{A}) - 1}{\mathrm{cond}_2(\boldsymbol{A}) + 1}\right) \|\boldsymbol{e}_0\|_{\boldsymbol{A}}$$

follows. $\qquad\qquad\square$

Because $\mathrm{cond}_2(\boldsymbol{A}) \geq 1$, it is therefore always advantageous, according to the above theorem, to have the smallest possible condition number.

Example 4.69 We consider

$$Ax = b$$

with

$$A = \begin{pmatrix} 2 & 0 \\ 0 & 10 \end{pmatrix}, \quad b = \begin{pmatrix} 0 \\ 0 \end{pmatrix}, \quad x_0 = \begin{pmatrix} 4 \\ \sqrt{1.8} \end{pmatrix}.$$

For the given matrix A we have

$$\mathrm{cond}_2(A) = \frac{10}{2} = 5$$

and from Theorem 4.68 we can deduce the estimate

$$\|e_m\|_A \le \left(\frac{5-1}{5+1}\right)^m \|e_0\|_A = \left(\frac{2}{3}\right)^m \|e_0\|_A$$

respectively

$$\|e_m\|_A \le \frac{2}{3}\|e_{m-1}\|_A$$

The reduction of the error in the energy norm with respect to the matrix A must therefore be at least $\frac{2}{3}$ in each iteration step. This statement is also in good agreement with the convergence behavior presented in the following table.

Method of Steepest Descent (Gradient Method)				
m	$x_{m,1}$	$x_{m,2}$	$\varepsilon_m := \|x_m - A^{-1}b\|_A$	$\varepsilon_m/\varepsilon_{m-1}$
0	4.000000e+00	1.341641e+00	7.071068e+00	
10	3.271049e-02	1.097143e-02	5.782453e-02	6.183904e-01
40	1.788827e-08	5.999910e-09	3.162230e-08	6.183904e-01
70	9.782499e-15	3.281150e-15	1.729318e-14	6.183904e-01
72	3.740893e-15	1.254734e-15	6.613026e-15	6.183904e-01

We use the contour lines of the function F qualitatively shown in Figure 4.29 to illustrate the convergence behavior. If the considered diagonal matrix has equal and, in addition, positive diagonal entries, then the contour lines are circles, and the method yields the exact solution already in the first iteration, regardless of the chosen initial vector, since the residual vector is always pointing in the direction of the coordinate origin. This fact is also supported by the error estimate according to Theorem 4.68, because for a positive definite diagonal matrix A with identical diagonal entries, we have $\mathrm{cond}_2(A) = 1$ and consequently

$$\|e_1\|_A \le \underbrace{\left(\frac{\mathrm{cond}_2(A) - 1}{\mathrm{cond}_2(A) + 1}\right)}_{=0} \|e_0\|_A = 0$$

regardless of the chosen initial guess x_0. If the diagonal matrix has positive but very different diagonal entries, then the level curves of the function F become increasingly stretched ellipses, causing the approximate solution to alternate its sign in each iteration and possibly converge only very slowly to the minimum of the function F. The increasing reduction in the convergence rate can also be clearly explained by considering the condition number of the matrix A, which, for the underlying diagonal matrix, correlates with the stretching ratio of the ellipse. It should be noted that even for a positive definite diagonal matrix A with a very large condition number, the system $Ax = 0$ is solved exactly in one iteration step if the starting vector is chosen along one of the coordinate axes.

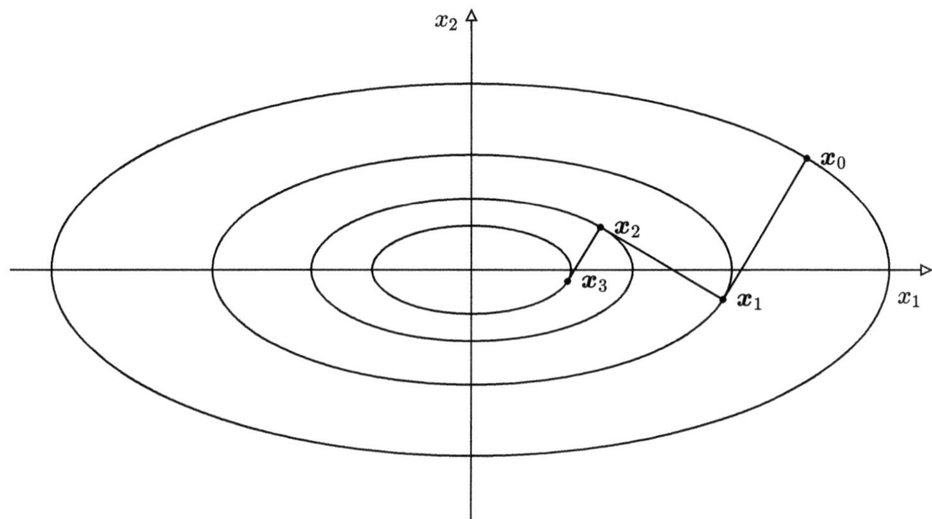

Figure 4.29   Level curves of the function F with qualitative convergence behavior

As motivation for improving the gradient method, we introduce the concepts of optimality with respect to both a direction and a subspace.

Definition 4.70 Let $F : \mathbb{R}^n \to \mathbb{R}$ be given, then $\boldsymbol{x} \in \mathbb{R}^n$ is called

(a) optimal with respect to the direction $\boldsymbol{p} \in \mathbb{R}^n$, if

$$F(\boldsymbol{x}) \le F(\boldsymbol{x} + \lambda \boldsymbol{p}) \quad \forall \, \lambda \in \mathbb{R}$$

     holds.

(b) optimal with respect to a subspace $U \subset \mathbb{R}^n$, if

$$F(\boldsymbol{x}) \le F(\boldsymbol{x} + \boldsymbol{\xi}) \quad \forall \, \boldsymbol{\xi} \in U$$

     holds.

Lemma 4.71 *Let F be given by (4.3.8), then $\boldsymbol{x} \in \mathbb{R}^n$ is optimal with respect to $U \subset \mathbb{R}^n$ if and only if*

$$\boldsymbol{r} = \boldsymbol{b} - \boldsymbol{A}\boldsymbol{x} \perp U$$

holds.

Proof:
For arbitrary $\boldsymbol{\xi} \in U \backslash \{\boldsymbol{0}\}$, we consider for $\lambda \in \mathbb{R}$

$$f_{\boldsymbol{x}, \boldsymbol{\xi}}(\lambda) = F(\boldsymbol{x} + \lambda \boldsymbol{\xi}).$$

The function $f_{x,\xi}$ is strictly convex according to (4.3.12), and from

$$f'_{x,\xi}(\lambda) = (Ax - b,\xi)_2 + \lambda(A\xi,\xi)_2$$

it follows that

$$f'_{x,\xi}(0) = 0 \quad \Leftrightarrow \quad Ax - b \perp \xi.$$

$\square$

Theorem 4.72 *The iterates* x_m, $m \in \mathbb{N}$ *of the gradient method are optimal with respect to the direction* $r_{m-1} = b - Ax_{m-1}$.

Proof:
In the case $r_{m-1} = 0$, it follows directly that $r_m \perp r_{m-1}$. If $r_{m-1} \neq 0$, we obtain with

$$\lambda_{m-1} = \frac{\|r_{m-1}\|_2^2}{(Ar_{m-1},r_{m-1})_2}$$

the orthogonality of the residual vectors by

$$\begin{aligned}
(r_m,r_{m-1})_2 &= (r_{m-1} - \lambda_{m-1}Ar_{m-1},r_{m-1})_2 \\[2mm]
&= (r_{m-1},r_{m-1})_2 - \frac{\|r_{m-1}\|_2^2}{(Ar_{m-1},r_{m-1})_2}\,(Ar_{m-1},r_{m-1})_2 \\[2mm]
&= 0.
\end{aligned}$$

Lemma 4.71 thus yields optimality. $\square$

Corollary 4.73 *The gradient method represents in each step an orthogonal projection method with* $K = L = span\,\{r_{m-1}\}$.

It would be desirable to have optimality of the iterates with respect to the entire subspace $U = span\,\{r_0,\ldots,r_{m-1}\}$, since for linearly independent residual vectors, this would ensure at the latest that

$$x_n = A^{-1}b$$

follows. However, in the method of steepest descent, the problem arises that the computed approximate solution x_m is always only optimal with respect to r_{m-1}, and the condition $r \perp p$ is not transitive, so that from $r_{m-2} \perp r_{m-1}$ and $r_{m-1} \perp r_m$ it does not necessarily follow that $r_{m-2} \perp r_m$.

4.3.1.2 The Method of Conjugate Directions

The idea of this method is to extend the optimality of the computed approximations x_m to the entire subspace $U_m = span\,\{p_0,\ldots,p_{m-1}\}$ with linearly independent search directions $p_0,\ldots,p_{m-1}$. To this end, we will formulate, with the following theorem, a condition on the search directions to be used, which guarantees the preservation of optimality with respect to U_m in the $(m+1)$-th step.

Theorem 4.74 *Let F be given as in (4.3.8) and let $\boldsymbol{x} \in \mathbb{R}^n$ be optimal with respect to the subspace $U = \operatorname{span}\{\boldsymbol{p}_0,\ldots,\boldsymbol{p}_{m-1}\} \subset \mathbb{R}^n$, then $\tilde{\boldsymbol{x}} = \boldsymbol{x} + \boldsymbol{\xi}$ is optimal with respect to U if and only if*

$$\boldsymbol{A}\boldsymbol{\xi} \perp U$$

holds.

Proof:
Let $\boldsymbol{\eta} \in U$ be arbitrary, then the claim follows directly from

$$(\boldsymbol{b} - \boldsymbol{A}\tilde{\boldsymbol{x}},\boldsymbol{\eta})_2 = \underbrace{(\boldsymbol{b} - \boldsymbol{A}\boldsymbol{x},\boldsymbol{\eta})_2}_{=\,0} - (\boldsymbol{A}\boldsymbol{\xi},\boldsymbol{\eta})_2 \,.$$

$\square$

If a search direction $\boldsymbol{p}_m$ is chosen such that

$$\boldsymbol{A}\boldsymbol{p}_m \perp U_m = \operatorname{span}\{\boldsymbol{p}_0,\ldots,\boldsymbol{p}_{m-1}\}$$

or equivalently

$$\boldsymbol{A}\boldsymbol{p}_m \perp \boldsymbol{p}_j,\ j = 0,\ldots,m-1$$

holds, then the approximate solution

$$\boldsymbol{x}_{m+1} = \boldsymbol{x}_m + \lambda_m \boldsymbol{p}_m$$

inherits, according to Theorem 4.74, the optimality of $\boldsymbol{x}_m$ with respect to U_m, independently of the choice of the scalar weighting parameter λ_m. This degree of freedom is subsequently used to extend the optimality to

$$U_{m+1} = \operatorname{span}\{\boldsymbol{p}_0,\ldots,\boldsymbol{p}_m\} \,.$$

Definition 4.75 Let $\boldsymbol{A} \in \mathbb{R}^{n \times n}$, then the vectors $\boldsymbol{p}_0,\ldots,\boldsymbol{p}_m \in \mathbb{R}^n$ are called pairwise conjugate or $\boldsymbol{A}$-orthogonal if

$$(\boldsymbol{p}_i,\boldsymbol{p}_j)_{\boldsymbol{A}} := (\boldsymbol{A}\boldsymbol{p}_i,\boldsymbol{p}_j)_2 = 0 \quad \forall\, i,j \in \{0,\ldots,m\} \text{ and } i \neq j$$

holds.

An important property for the method lies in the successive increase in dimension within the considered sequence of subvector spaces $\{U_m\}_{m=1,2,\ldots}$, which is demonstrated by the following lemma.

Lemma 4.76 *Let $\boldsymbol{A} \in \mathbb{R}^{n \times n}$ be symmetric and positive definite, and let $\boldsymbol{p}_0,\ldots,\boldsymbol{p}_{m-1} \in \mathbb{R}^n \setminus \{\boldsymbol{0}\}$ be pairwise $\boldsymbol{A}$-orthogonal, then*

$$\dim \operatorname{span}\{\boldsymbol{p}_0,\ldots,\boldsymbol{p}_{m-1}\} = m$$

for $m = 1,\ldots,n$.

Proof:

Suppose $\sum\limits_{j=0}^{m-1} \alpha_j \boldsymbol{p}_j = \boldsymbol{0}$ with $\alpha_j \in \mathbb{R}$, then for $i = 0,\ldots,m-1$ we obtain

$$0 = (\boldsymbol{0},\boldsymbol{Ap}_i)_2 = \left(\sum_{j=0}^{m-1} \alpha_j \boldsymbol{p}_j, \boldsymbol{Ap}_i\right)_2 = \sum_{j=0}^{m-1} \alpha_j \left(\boldsymbol{p}_j,\boldsymbol{Ap}_i\right)_2 = \alpha_i \underbrace{\left(\boldsymbol{p}_i,\boldsymbol{Ap}_i\right)_2}_{\neq\, 0,\ \text{since}\ \boldsymbol{A}\ \text{is pos.def.}} \ .$$

Consequently, $\alpha_i = 0$ for $i = 0,\ldots,m-1$, which proves the linear independence of the vectors. $\qquad\square$

Let $\boldsymbol{p}_0,\ldots,\boldsymbol{p}_m \in \mathbb{R}^n \backslash \{\boldsymbol{0}\}$ be given as pairwise conjugate search directions and let $\boldsymbol{x}_m$ be optimal with respect to $U_m = \operatorname{span}\{\boldsymbol{p}_0,\ldots,\boldsymbol{p}_{m-1}\}$, then we obtain the optimality of

$$\boldsymbol{x}_{m+1} = \boldsymbol{x}_m + \lambda \boldsymbol{p}_m$$

with respect to U_{m+1} if

$$0 = \left(\boldsymbol{b} - \boldsymbol{Ax}_{m+1},\boldsymbol{p}_j\right)_2 = \underbrace{\left(\boldsymbol{b} - \boldsymbol{Ax}_m,\boldsymbol{p}_j\right)_2}_{=\, 0\ \text{for}\ j\neq m} - \lambda \underbrace{\left(\boldsymbol{Ap}_m,\boldsymbol{p}_j\right)_2}_{=\, 0\ \text{for}\ j\neq m}$$

holds for $j = 0,\ldots,m$. From this, we obtain for λ the representation

$$\lambda = \frac{(\boldsymbol{r}_m,\boldsymbol{p}_m)_2}{(\boldsymbol{Ap}_m,\boldsymbol{p}_m)_2},$$

and we can write the method of conjugate directions in the following form:

Algorithm - Method of Conjugate Directions ——

<table>
<tr><td colspan="2">Choose $\boldsymbol{x}_0 \in \mathbb{R}^n$</td></tr>
<tr><td colspan="2">$\boldsymbol{r}_0 := \boldsymbol{b} - \boldsymbol{Ax}_0$</td></tr>
<tr><td colspan="2">For $m = 0,\ldots,n-1$</td></tr>
<tr><td></td><td>$\lambda_m := \dfrac{(\boldsymbol{r}_m,\boldsymbol{p}_m)_2}{(\boldsymbol{Ap}_m,\boldsymbol{p}_m)_2}$</td></tr>
<tr><td></td><td>$\boldsymbol{x}_{m+1} := \boldsymbol{x}_m + \lambda_m \boldsymbol{p}_m$</td></tr>
<tr><td></td><td>$\boldsymbol{r}_{m+1} := \boldsymbol{r}_m - \lambda_m \boldsymbol{Ap}_m$</td></tr>
</table>

If the given search directions are chosen unfavorably, it is possible that $\boldsymbol{x}_n$ is already the exact solution, even though $\boldsymbol{x}_{n-1}$ still has a very large error. In the extreme case, it can happen that $\boldsymbol{x}_m = \boldsymbol{x}_0 \neq \boldsymbol{A}^{-1}\boldsymbol{b}$ for $m = 1,2,\ldots,n-1$ and $\boldsymbol{x}_n = \boldsymbol{A}^{-1}\boldsymbol{b}$. Therefore, with fixed predetermined search directions, the method can generally only be used as a direct method. This property leads to a high computational effort for large n. A problem-adapted selection of the search directions is therefore required.

4.3.1.3 The Conjugate Gradient Method

The method of conjugate gradients (CG method), introduced by Hestenes and Stiefel [39], combines the gradient method with the method of conjugate directions. The method uses the residual vectors to define the conjugate search directions, resulting in a problem-adapted approach regarding the selection of search directions, while also preserving the optimality of the method of conjugate directions (see Figure 4.30). Analogous to the previously introduced method of conjugate directions, the conjugate gradient method can be interpreted as both a direct and an iterative method.

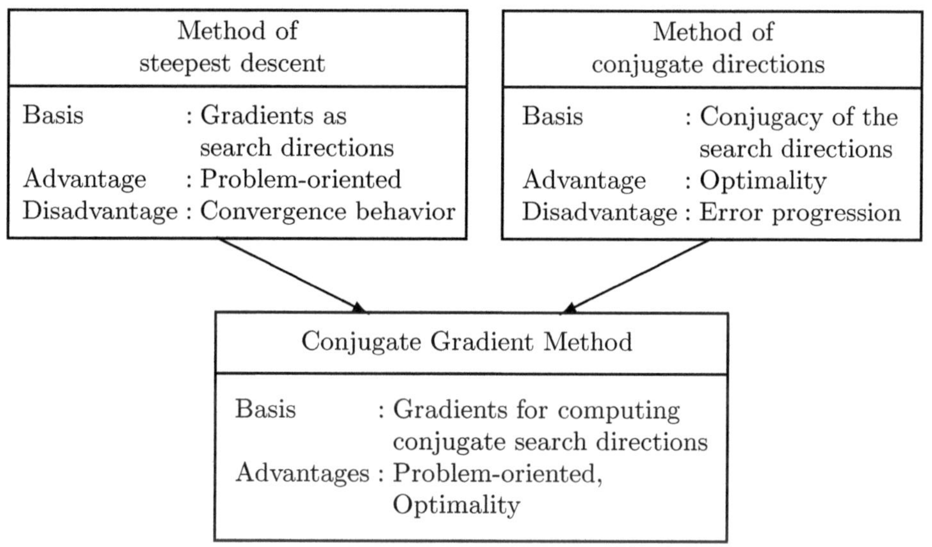

Figure 4.30 Derivation of the CG method

With the residual vectors $r_0, \ldots, r_m$, we successively determine the search directions for $m = 1, \ldots, n - 1$ according to

$$p_0 = r_0,$$

$$p_m = r_m + \sum_{j=0}^{m-1} \alpha_j p_j. \tag{4.3.17}$$

For $\alpha_j = 0$ ($j = 0, \ldots, m - 1$), we obtain a selection of search directions analogous to the method of steepest descent. By taking into account the already used search directions $p_0, \ldots, p_{m-1} \in \mathbb{R}^n \setminus \{0\}$ in the above form, there are m degrees of freedom in the choice of the coefficients α_j, which are used to ensure the conjugacy of the search directions. From the required A-orthogonality condition, it follows that

$$0 = (Ap_m, p_i)_2 = (Ar_m, p_i)_2 + \sum_{j=0}^{m-1} \alpha_j \left(Ap_j, p_i\right)_2$$

for $i = 0, \ldots, m - 1$. With

$$\left(Ap_j, p_i\right)_2 = 0 \text{ for } i, j \in \{0, \ldots, m - 1\} \text{ and } i \neq j$$

we obtain the required formula for computing the coefficients in the form

$$\alpha_i = -\frac{(A r_m, p_i)_2}{(A p_i, p_i)_2}.$$ (4.3.18)

Thus, the preliminary version of the conjugate gradient method is given in the following form:

Algorithm - Preliminary Conjugate Gradient Method —

Choose $x_0 \in \mathbb{R}^n$
$p_0 := r_0 := b - A x_0$
For $m = 0, \ldots, n-1$
$\quad \lambda_m := \dfrac{(r_m, p_m)_2}{(A p_m, p_m)_2}$
$\quad x_{m+1} := x_m + \lambda_m p_m$
$\quad r_{m+1} := r_m - \lambda_m A p_m$
$\quad p_{m+1} := r_{m+1} - \displaystyle\sum_{j=0}^{m} \frac{(A r_{m+1}, p_j)_2}{(A p_j, p_j)_2} p_j \qquad (4.3.19)$

The preliminary CG method has the crucial disadvantage that, in order to compute p_{m+1}, apparently all p_j $(j = 0, \ldots, m)$ are needed. In the worst case, we therefore require the storage space of a dense $n \times n$ matrix for the search directions. Additionally, the number of inner products, which have to calculated increases linearly with the iteration number. For large, sparse matrices A, the method is thus inefficient and possibly impractical. However, the following analysis shows that the method can be significantly improved in terms of memory requirements and computation time.

Theorem 4.77 *Assuming that the preliminary CG method does not terminate before the computation of p_k for $k > 0$, then the following holds:*

(a) p_m is conjugate to all p_j with $0 \leq j < m \leq k$,

(b) $U_{m+1} := span\{p_0, \ldots, p_m\} = span\{r_0, \ldots, r_m\}$ with $\dim U_{m+1} = m + 1$ for $m = 0, \ldots, k-1$,

(c) $r_m \perp U_m$ for $m = 1, \ldots, k$,

(d) $x_k = A^{-1} b \iff r_k = 0 \iff p_k = 0$,

(e) $U_{m+1} = span\{r_0, \ldots, A^m r_0\}$ for $m = 0, \ldots, k-1$,

(f) r_m is conjugate to all p_j with $0 \leq j < m-1 < k-1$.

Proof:

for (a): Since p_k has been computed, it holds that $p_0,\dots,p_{k-1} \in \mathbb{R}^n\backslash\{0\}$. The statement follows from the computation rules (4.3.17) and (4.3.18).

for (b): Induction on m.

For $m = 0$ we have $p_0 = r_0 \in \mathbb{R}^n\backslash\{0\}$ and thus the claim holds. Assume (b) holds for $m < k - 1$. Since $p_{m+1} \in \mathbb{R}^n\backslash\{0\}$ it follows from (a) that p_{m+1} is conjugate to all $p_0,\dots,p_m$. From Lemma 4.76 we obtain $\dim U_{m+2} = m + 2$, and

$$p_{m+1} - r_{m+1} \overset{(4.3.17)}{=} \sum_{j=0}^{m} \alpha_j p_j \in U_{m+1}$$

yields

$$U_{m+2} = \operatorname{span}\{U_{m+1}, p_{m+1}\} = \operatorname{span}\{U_{m+1}, r_{m+1}\}.$$

for (c): Induction on m:

For $m = 1$ we obtain, with $p_0 \neq 0$, the equation

$$(r_1, r_0)_2 = (r_0, r_0)_2 - \frac{(r_0, p_0)_2}{(Ap_0, p_0)_2}(Ap_0, r_0)_2 \overset{r_0 = p_0}{=} 0.$$

Assume (c) holds for $m < k$ and $\boldsymbol{\eta} \in U_m$, then it follows from part (a)

$$(r_{m+1}, \boldsymbol{\eta})_2 = \underbrace{(r_m, \boldsymbol{\eta})_2}_{=\,0} - \lambda_m \underbrace{(Ap_m, \boldsymbol{\eta})_2}_{=0} = 0.$$

Taking into account that $p_m \neq 0$, we obtain the statement by

$$(r_{m+1}, p_m)_2 = (r_m, p_m)_2 - \frac{(r_m, p_m)_2}{(Ap_m, p_m)_2}(Ap_m, p_m)_2 = 0.$$

for (d): From $r_k = b - Ax_k$ it follows directly the equivalence between $x_k = A^{-1}b$ and $r_k = 0$. If $r_k = 0$, then (4.3.19) yields the equation $p_k = 0$. If $p_k = 0$, then again by (4.3.19) $r_k \in U_k$. According to part (c), however, $r_k \perp U_k$, which implies $r_k = 0$.

for (e): Induction on m:

For $m = 0$ the statement is trivial.

Assume (e) holds for $m < k - 1$, then by (b) it follows

$$r_m \in U_{m+1} = \operatorname{span}\{r_0,\dots,r_m\} = \operatorname{span}\{r_0,\dots,A^m r_0\}$$

as well as

$$Ap_m \in AU_{m+1} = \operatorname{span}\{Ar_0,\dots,A^{m+1}r_0\}.$$

Consequently, we obtain $r_{m+1} = r_m - \lambda_m Ap_m \in \operatorname{span}\{r_0,\dots,A^{m+1}r_0\}$, so that $U_{m+2} = \operatorname{span}\{r_0,\dots,r_{m+1}\} \subset \operatorname{span}\{r_0,\dots,A^{m+1}r_0\}$ holds. Part (b) yields $\dim U_{m+2} = m + 2$, which implies $U_{m+2} = \operatorname{span}\{r_0,\dots,A^{m+1}r_0\}$.

for (f): For $0 \leq j < m - 1 \leq k - 1$ we have $\boldsymbol{p}_j \in U_{m-1}$. Thus, $\boldsymbol{A}\boldsymbol{p}_j \in U_m$, and we obtain

$$\left(\boldsymbol{A}\boldsymbol{r}_m, \boldsymbol{p}_j\right)_2 \overset{\boldsymbol{A} \text{ symm.}}{=\!=\!=} \left(\boldsymbol{r}_m, \boldsymbol{A}\boldsymbol{p}_j\right)_2 \overset{(c)}{=\!=} 0.$$

$\square$

From the above theorem, we can extract three essential statements for improving the method:

- From statement (f) it follows that

$$\boldsymbol{p}_m = \boldsymbol{r}_m - \sum_{j=0}^{m-1} \frac{\left(\boldsymbol{A}\boldsymbol{r}_m, \boldsymbol{p}_j\right)_2}{\left(\boldsymbol{A}\boldsymbol{p}_j, \boldsymbol{p}_j\right)_2} \boldsymbol{p}_j = \boldsymbol{r}_m - \frac{\left(\boldsymbol{A}\boldsymbol{r}_m, \boldsymbol{p}_{m-1}\right)_2}{\left(\boldsymbol{A}\boldsymbol{p}_{m-1}, \boldsymbol{p}_{m-1}\right)_2} \boldsymbol{p}_{m-1}. \qquad (4.3.20)$$

  Thus, the memory requirement is independent of the number of iterations.

- The method terminates in the $k+1$-th iteration if and only if $\boldsymbol{p}_k = \boldsymbol{0}$. By (d), $\boldsymbol{x}_k$ in this case already provides the exact solution, so that $\boldsymbol{p}_k = \boldsymbol{0}$ can be used as a stopping criterion. In the following diagram, we also present a variant that uses the residual as a stopping criterion without additional computational effort.

- From $\boldsymbol{r}_{m+1} = \boldsymbol{r}_m - \lambda_m \boldsymbol{A}\boldsymbol{p}_m$ we obtain, due to property (c), the equation $(\boldsymbol{r}_m - \lambda_m \boldsymbol{A}\boldsymbol{p}_m, \boldsymbol{r}_m)_2 = 0$, which allows the scalar quantity to be written in the form

$$\lambda_m = \frac{(\boldsymbol{r}_m, \boldsymbol{p}_m)_2}{(\boldsymbol{A}\boldsymbol{p}_m, \boldsymbol{p}_m)_2} = \frac{(\boldsymbol{r}_m, \boldsymbol{r}_m)_2}{(\boldsymbol{A}\boldsymbol{p}_m, \boldsymbol{r}_m)_2}. \qquad (4.3.21)$$

Using equation (4.3.20) reveals

$$(\boldsymbol{A}\boldsymbol{p}_m, \boldsymbol{r}_m)_2 = \left(\boldsymbol{A}\boldsymbol{p}_m, \boldsymbol{p}_m + \frac{(\boldsymbol{A}\boldsymbol{r}_m, \boldsymbol{p}_{m-1})_2}{(\boldsymbol{A}\boldsymbol{p}_{m-1}, \boldsymbol{p}_{m-1})_2} \boldsymbol{p}_{m-1}\right)_2 = (\boldsymbol{A}\boldsymbol{p}_m, \boldsymbol{p}_m)_2,$$

so that (4.3.21) yields the property

$$(\boldsymbol{r}_m, \boldsymbol{r}_m)_2 = (\boldsymbol{r}_m, \boldsymbol{p}_m)_2. \qquad (4.3.22)$$

Furthermore, in the case $\lambda_m \neq 0$, the preliminary CG method always yields

$$\boldsymbol{A}\boldsymbol{p}_m = -\frac{1}{\lambda_m}(\boldsymbol{r}_{m+1} - \boldsymbol{r}_m), \qquad (4.3.23)$$

so that

$$\frac{(\boldsymbol{A}\boldsymbol{r}_{m+1}, \boldsymbol{p}_m)_2}{(\boldsymbol{A}\boldsymbol{p}_m, \boldsymbol{p}_m)_2} \overset{\boldsymbol{A} \text{ symm.}}{=\!=} \frac{(\boldsymbol{A}\boldsymbol{p}_m, \boldsymbol{r}_{m+1})_2}{(\boldsymbol{A}\boldsymbol{p}_m, \boldsymbol{p}_m)_2} \overset{(4.3.23)}{=\!=} \frac{(\boldsymbol{r}_{m+1} - \boldsymbol{r}_m, \boldsymbol{r}_{m+1})_2}{(\boldsymbol{r}_{m+1} - \boldsymbol{r}_m, \boldsymbol{p}_m)_2}$$

$$\overset{(b),(c)}{=\!=} -\frac{(\boldsymbol{r}_{m+1}, \boldsymbol{r}_{m+1})_2}{(\boldsymbol{r}_m, \boldsymbol{p}_m)_2} \overset{(4.3.22)}{=\!=} -\frac{(\boldsymbol{r}_{m+1}, \boldsymbol{r}_{m+1})_2}{(\boldsymbol{r}_m, \boldsymbol{r}_m)_2}$$

follows, and thus within each iteration a matrix-vector multiplication is avoided.

These considerations yield the CG method in the following form. An implementation of the method can be found in Appendix A.

Algorithm - Conjugate Gradient Method   —

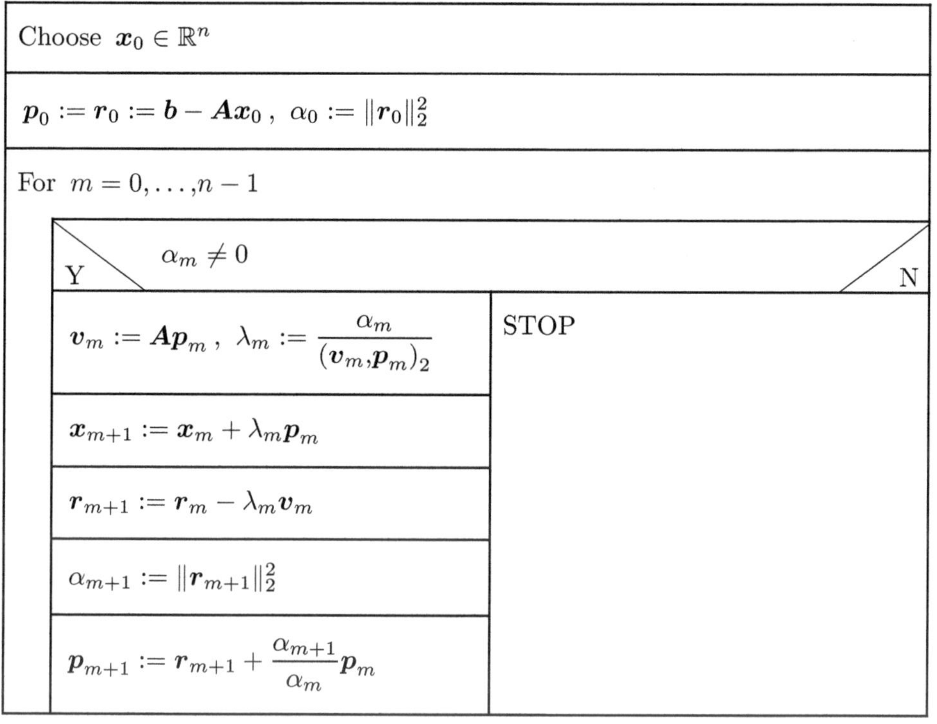

Choose $x_0 \in \mathbb{R}^n$

$p_0 := r_0 := b - Ax_0$, $\alpha_0 := \|r_0\|_2^2$

For $m = 0, \dots, n-1$

$\alpha_m \neq 0$ Y / N

$v_m := Ap_m$, $\lambda_m := \dfrac{\alpha_m}{(v_m, p_m)_2}$ STOP

$x_{m+1} := x_m + \lambda_m p_m$

$r_{m+1} := r_m - \lambda_m v_m$

$\alpha_{m+1} := \|r_{m+1}\|_2^2$

$p_{m+1} := r_{m+1} + \dfrac{\alpha_{m+1}}{\alpha_m} p_m$

Example 4.78 Let us consider the one-dimensional boundary value problem (4.2.1) with $h_\ell = \frac{1}{8}$ (that is, $N_\ell = 7$), which yields $A_\ell u^\ell = f^\ell$ with

$$\mathbb{R}^{N_\ell \times N_\ell} \ni A_\ell = \text{tridiag}\{-64, 128, -64\}.$$

Furthermore, let the right-hand side be given by

$$f^\ell = (128, -448, 704, -832, 512, 128, 320)^T.$$

With the initial vector $u_0 = 0$, we obtain the convergence history presented in the following table:

\multicolumn{9}{c}{Conjugate Gradient Method (CG method)}								
m	$u_{m,1}$	$u_{m,2}$	$u_{m,3}$	$u_{m,4}$	$u_{m,5}$	$u_{m,6}$	$u_{m,7}$	$\|r_m\|_2$
0	0.00	0.00	0.00	0.00	0.00	0.00	0.00	1336.36
1	0.58	-2.04	3.21	-3.79	2.33	0.58	1.46	363.57
2	-0.39	-1.72	2.81	-4.57	3.00	4.99	4.26	252.76
3	-0.01	-2.38	2.06	-3.53	4.87	6.07	6.25	153.30
4	-0.14	-2.88	2.57	-2.13	6.50	7.48	5.93	117.64
5	-0.70	-2.18	3.53	-1.12	7.65	7.81	6.27	103.52
6	0.13	-1.14	5.40	0.54	8.23	8.54	6.98	89.70
7	1.00	0.00	6.00	1.00	9.00	9.00	7.00	0.00

This simple example confirms the theoretical property of the method, namely that after at most N_ℓ iterations the exact solution is obtained. In general, however, the numerical algorithm exhibits Inaccuracies due to rounding errors occur, so that in a system of equations (4.3.1), the result obtained after n iterations generally deviates from the desired solution. Therefore, in practice, for small n, additional iterations are performed until a sufficiently accurate solution has been computed. An Algol program for a variant of this algorithm can be found in [78], and a comprehensive report on numerical experiences in [59].

Many systems of equations arising from technical applications have a large, sparse matrix $A \in \mathbb{R}^{n \times n}$. There are two main reasons why one is not interested in an exact solution to such a system: First, even with exact arithmetic, the potentially required n iterations would result in extremely long computation times and thus lead to an impractical method. Furthermore, the exact solution of the system of equations usually only provides an approximation to the solution of the underlying problem. Therefore, one is only interested in an error that lies within the range of the approximation quality of the discretization method. These arguments show that even a theoretically direct algorithm such as the CG method is usually used only as an iterative method.

The quotients that appear in the CG algorithm illustrate the fundamental dependence of the method on the positive definiteness of the matrix. Solvers for systems of equations that exploit the symmetry of the matrix but do not require positive definiteness are discussed in detail by Fischer [26].

Example 4.79 As another example, let us consider the Poisson equation (see Example 1.1), where $N = 200$ is chosen, resulting in 40,000 unknowns. We also use $\varphi \equiv 0$ and $f(x,y) = 2x + 2y$. The following table shows that the CG algorithm reaches a typical limit of machine precision after only 641 iterations. This again emphatically demonstrates that the method should be used as an iterative method for large systems of equations. In Example 4.16, we have already shown that the matrix of the system under consideration satisfies the conditions of Theorem 4.15 and thus the Jacobi method converges. The residual progression of this splitting method is also included in the table and highlights the immense efficiency gain achieved by the CG method.

	CG method	Jacobi method
Iterations	Residual progression $\|r_m\|_2 = \|b - Ar_m\|_2$	
0	140.348	140.348
150	1.83245	134.735
300	2.40822e-05	131.221
450	1.77391e-12	128.135
600	1.88161e-15	125.292
641	8.91038e-17	124.547

Remark:

For the CG method, it always holds that

$$
x_m \in x_0 + \underbrace{\mathrm{span}\left\{p_0, \ldots, p_{m-1}\right\}}_{=\mathrm{span}\left\{r_0, \ldots, A^{m-1}r_0\right\}=K_m},
$$

and x_m is optimal with respect to K_m, since

$$
r_m \perp K_m
$$

holds. Thus, the CG method is an orthogonal Krylov subspace method. Furthermore,

$$
x_m = \arg\min_{x \in x_0 + K_m} F(x). \tag{4.3.24}
$$

We have already been able to derive a convergence statement for the method of steepest descent based on the condition number of the matrix of the linear system. We will now formulate an analogous statement for the CG method.

Theorem 4.80 *Let $A \in \mathbb{R}^{n \times n}$ be symmetric and positive definite, and let $\{x_m\}_{m \in \mathbb{N}_0}$ be the sequence of approximate solutions generated by the conjugate gradient method. Then the error vector $e_m = x_m - A^{-1}b$ corresponding to x_m satisfies the inequality*

$$
\|e_m\|_A \le 2 \left(\frac{\sqrt{\mathrm{cond}_2(A)} - 1}{\sqrt{\mathrm{cond}_2(A)} + 1} \right)^m \|e_0\|_A.
$$

Proof:

Again, we use the set $\mathcal{P}_m^1$ of all polynomials p of degree at most m that satisfy the side condition $p(0) = I$. Thus, due to the equation

$$
e_m = x_m - A^{-1}b = \underbrace{x_0 - A^{-1}b}_{=e_0} - \sum_{i=1}^{m} c_i \underbrace{A^{i-1}r_0}_{=-A^i e_0}
$$

a polynomial $p \in \mathcal{P}_m^1$ with

$$
e_m = p(A)e_0. \tag{4.3.25}
$$

Equation (4.3.24) yields $\|x_m - A^{-1}b\|_A = \min_{x \in x_0 + K_m}\|x - A^{-1}b\|_A$, which leads to

$$
\|e_m\|_A = \min_{p \in \mathcal{P}_m^1} \|p(A)e_0\|_A. \tag{4.3.26}
$$

As a symmetric and positive definite matrix, A has real and positive eigenvalues $\lambda_n \geq \ldots \geq \lambda_1 > 0$ and the corresponding eigenvectors $v_1, \ldots, v_n$ can be chosen as an orthonormal basis of $\mathbb{R}^n$. Thus, there exists a representation of the error vector e_0 in the form

$$e_0 = \sum_{i=1}^{n} \alpha_i v_i$$

with $\alpha_i \in \mathbb{R}$ for $i = 1, \ldots, n$. From this we obtain

$$\|e_0\|_A^2 = \left(\sum_{i=1}^{n} \alpha_i v_i, \sum_{i=1}^{n} \alpha_i \lambda_i v_i \right)_2 = \sum_{i=1}^{n} \alpha_i^2 \lambda_i$$

and analogously

$$\|p(A)e_0\|_A^2 = \sum_{i=1}^{n} p(\lambda_i)^2 \alpha_i^2 \lambda_i.$$

Consequently, by using equation (4.3.26), we obtain the inequality

$$
\begin{aligned}
\|e_m\|_A &= \min_{p \in \mathcal{P}_m^1} \|p(A)e_0\|_A \\
&= \min_{p \in \mathcal{P}_m^1} \left(\sum_{i=1}^{n} p(\lambda_i)^2 \alpha_i^2 \lambda_i \right)^{1/2} \\
&\leq \min_{p \in \mathcal{P}_m^1} \max_{j=1,\ldots,n} |p(\lambda_j)| \left(\sum_{i=1}^{n} \alpha_i^2 \lambda_i \right)^{1/2} \\
&\leq \min_{p \in \mathcal{P}_m^1} \max_{\lambda \in [\lambda_1, \lambda_n]} |p(\lambda)| \|e_0\|_A.
\end{aligned}
\tag{4.3.27}
$$

The further task in proving the assertion consists in specifying a particular polynomial. For the Chebyshev polynomials $T_m : [-1,1] \mapsto [-1,1]$ with

$$T_m(x) = \cos(m \arccos x), \quad m \in \mathbb{N}_0$$

one can use induction, together with $T_0(x) = 1$ and $T_1(x) = x$, to derive the recursion formula

$$T_{m+1}(x) = 2x T_m(x) - T_{m-1}(x) \quad \text{for } m \in \mathbb{N}$$

from the cosine trigonometric identity. This representation shows $T_m \in \mathcal{P}_m$ and is also used to extend the Chebyshev polynomials to $\mathbb{R}$. Furthermore, by complete induction, one obtains the equation

$$T_m \left(\frac{1}{2} \left(x + \frac{1}{x} \right) \right) = \frac{1}{2} \left(x^m + \frac{1}{x^m} \right) \quad \text{for } m \in \mathbb{N}_0. \tag{4.3.28}$$

Let us first consider the case $\lambda_1 \neq \lambda_n$. Under this assumption, the polynomial

$$p_m(\lambda) = \frac{T_m \left(\frac{2\lambda - (\lambda_n + \lambda_1)}{\lambda_1 - \lambda_n} \right)}{T_m \left(\frac{\lambda_n + \lambda_1}{\lambda_n - \lambda_1} \right)}$$

is well-defined and $p_m \in \mathcal{P}_m^1$. Starting from the inequality (4.3.27), we obtain with

$$\left| T_m \left(\frac{2\lambda - (\lambda_n + \lambda_1)}{\lambda_1 - \lambda_n} \right) \right| \le 1 \quad \forall \lambda \in [\lambda_1, \lambda_n]$$

and

$$\frac{\lambda_n + \lambda_1}{\lambda_n - \lambda_1} = \frac{\mathrm{cond}_2(\boldsymbol{A}) + 1}{\mathrm{cond}_2(\boldsymbol{A}) - 1} = \frac{1}{2} \left(\frac{\sqrt{\mathrm{cond}_2(\boldsymbol{A})} + 1}{\sqrt{\mathrm{cond}_2(\boldsymbol{A})} - 1} + \frac{\sqrt{\mathrm{cond}_2(\boldsymbol{A})} - 1}{\sqrt{\mathrm{cond}_2(\boldsymbol{A})} + 1} \right) \tag{4.3.29}$$

the estimate

$$
\begin{aligned}
\|\boldsymbol{e}_m\|_{\boldsymbol{A}} \quad &\le \quad \max_{\lambda \in [\lambda_1, \lambda_n]} |p_m(\lambda)| \|\boldsymbol{e}_0\|_{\boldsymbol{A}} \\[2mm]
&\le \quad \frac{1}{\left| T_m \left(\frac{\lambda_n + \lambda_1}{\lambda_n - \lambda_1} \right) \right|} \|\boldsymbol{e}_0\|_{\boldsymbol{A}} \\[2mm]
\overset{(4.3.29),(4.3.28)}{=} \quad &\left| 2 \left(\underbrace{\left(\frac{\sqrt{\mathrm{cond}_2(\boldsymbol{A})} + 1}{\sqrt{\mathrm{cond}_2(\boldsymbol{A})} - 1} \right)^m}_{>0} + \underbrace{\left(\frac{\sqrt{\mathrm{cond}_2(\boldsymbol{A})} - 1}{\sqrt{\mathrm{cond}_2(\boldsymbol{A})} + 1} \right)^m}_{>0} \right)^{-1} \right| \|\boldsymbol{e}_0\|_{\boldsymbol{A}} \\[2mm]
&\le \quad 2 \left(\frac{\sqrt{\mathrm{cond}_2(\boldsymbol{A})} - 1}{\sqrt{\mathrm{cond}_2(\boldsymbol{A})} + 1} \right)^m \|\boldsymbol{e}_0\|_{\boldsymbol{A}}.
\end{aligned}
$$

For the special case $\lambda_1 = \lambda_n > 0$ we use $p_m \in \mathcal{P}_m^1$ with $p_m(\lambda) = 1 - \frac{\lambda}{\lambda_n}$ and obtain, with $\mathrm{cond}_2(\boldsymbol{A}) = \frac{\lambda_n}{\lambda_1} = 1$, the claimed inequality

$$\|\boldsymbol{e}_m\|_{\boldsymbol{A}} \le \underbrace{\max_{\lambda \in [\lambda_1, \lambda_n]} |p_m(\lambda)|}_{=0} \|\boldsymbol{e}_0\|_{\boldsymbol{A}} = 0 = 2 \left(\frac{\sqrt{\mathrm{cond}_2(\boldsymbol{A})} - 1}{\sqrt{\mathrm{cond}_2(\boldsymbol{A})} + 1} \right)^m \|\boldsymbol{e}_0\|_{\boldsymbol{A}}.$$

$\square$

Example 4.81 The derivation of the conjugate gradient method is fundamentally based on the assumption that the matrix $\boldsymbol{A} \in \mathbb{R}^{n \times n}$ of the considered system of equations $\boldsymbol{A}\boldsymbol{x} = \boldsymbol{b}$ is symmetric and positive definite. In this example, we want to investigate how the loss of these properties affects the convergence behavior of the method and also make a comparison with the steepest descent method. For this, we use the discretization of the convection-diffusion equation presented in Example 1.4 for different parameters a and ε. We use the step size $h = 1/(N + 1)$ with $N = 100$ in both coordinate directions, so that the matrix $\boldsymbol{A} \in \mathbb{R}^{10,000 \times 10,000}$. The following table lists the properties of the matrix $\boldsymbol{A}$ for various combinations of parameter values.

	a	ϵ	Properties of the matrix $\boldsymbol{A}$
Test 1	0	1	Symmetric, positive definite
Test 2	0.1	1	Nonsymmetric, non-singular
Test 3	1	0.1	Nonsymmetric, non-singular

The first test case leads to a system of equations with a symmetric, positive definite matrix $\boldsymbol{A}$. In this context, by Theorem 4.68, since

$$0 \le \frac{\mathrm{cond}_2(\boldsymbol{A}) - 1}{\mathrm{cond}_2(\boldsymbol{A}) + 1} < 1$$

the convergence of the steepest descent method is established. Likewise,

$$0 \le \frac{\sqrt{\mathrm{cond}_2(\boldsymbol{A})} - 1}{\sqrt{\mathrm{cond}_2(\boldsymbol{A})} + 1} < 1,$$

so that Theorem 4.80 guarantees the convergence of the conjugate gradient method. The convergence behaviors for the mentioned methods are shown in Figures 4.31 and 4.32 for the initial vector $\boldsymbol{x}_0 = \boldsymbol{0}$. As expected, both algorithms yield convergence of the respective sequence of approximate solutions to the uniquely determined solution of the system of equations. It is also evident that the CG method is significantly more efficient compared to the steepest descent method. The conjugate gradient method requires only 335 iterations to reduce the Euclidean norm of the residual vector $\|\boldsymbol{r}_m\|_2 = \|\boldsymbol{b} - \boldsymbol{A}\boldsymbol{r}_m\|_2$ below the required accuracy of 10^{-10}, whereas the steepest descent method requires 43,994 iterations for this. This also means that the steepest descent method requires more than 100 times the computation time of the CG method.

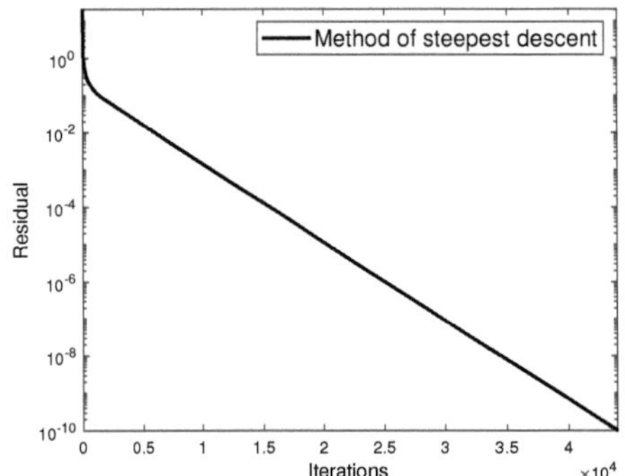

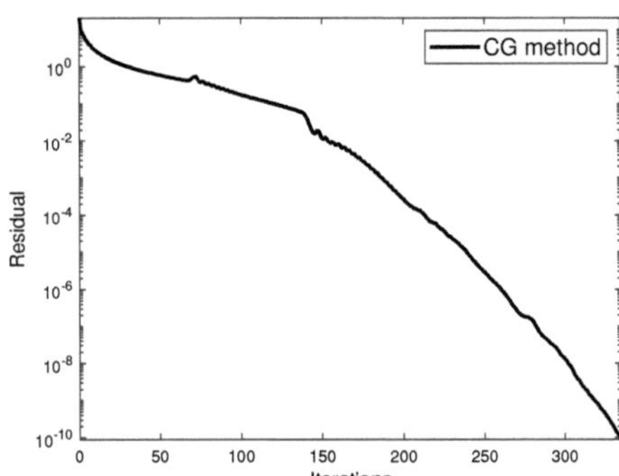

Figure 4.31　Convergence behavior of the steepest descent method for test case 1

Figure 4.32　Convergence behavior of the CG method for test case 1

We now turn to the second test case, where $a = 0.1$ and $\varepsilon = 1$ hold. For these parameter values, one also speaks of a diffusion-dominated flow. Although it is obvious that the matrix $\boldsymbol{A}$ is not symmetric, it can be assumed—due to the convection parameter a being small in relation to the diffusion parameter ε—that the convexity of the mapping $F(\boldsymbol{x}) = \frac{1}{2}(\boldsymbol{A}\boldsymbol{x},\boldsymbol{x})_2 - (\boldsymbol{b},\boldsymbol{x})_2$ is only slightly disturbed, and therefore no local minima are created and the sequence of approximate solutions still converges to the solution of the system of equations. This assumption is confirmed for both the steepest descent method and the CG method with the initial vector given above, as shown in Figures 4.33 and 4.34. Although the number of iterations required by the CG method to achieve the prescribed accuracy approximately doubles in contrast to the steepest descent method, a significant efficiency advantage of the CG method is still evident here.

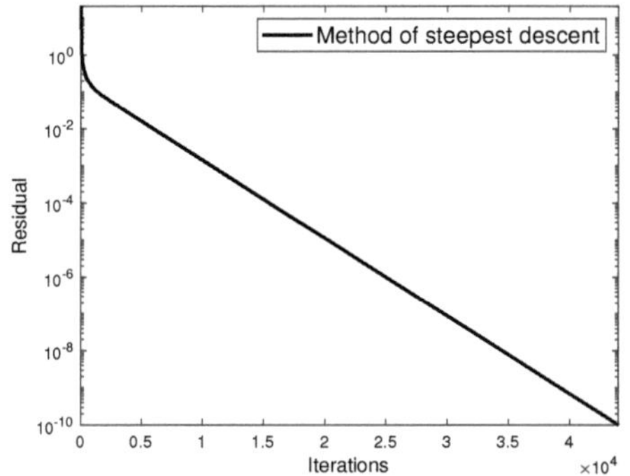

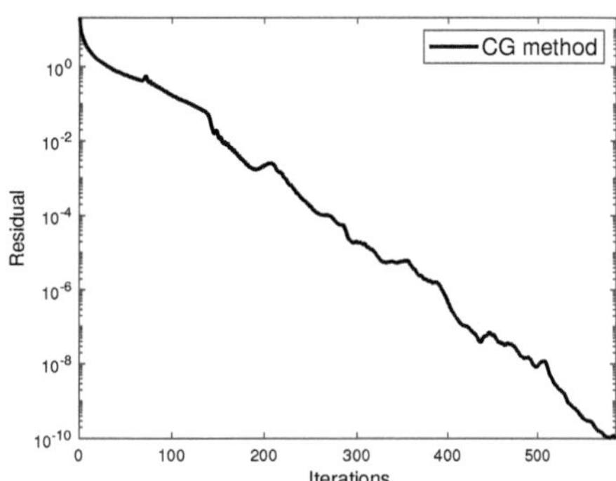

Figure 4.33 Convergence behavior of the steepest descent method for test case 2

Figure 4.34 Convergence behavior of the CG method for test case 2

Based on the results obtained in the case of a diffusion-dominated flow, the question arises as to whether convergence also occurs in the context of a convection-dominated flow. For this, we use the third test case with the diffusion parameter $\varepsilon = 0.1$ and a convection parameter $a = 1$, while keeping the initial vector unchanged. From Figure 4.36, the divergence of the CG method can be clearly observed, while the steepest descent method is still able to determine the solution of the system of equations.

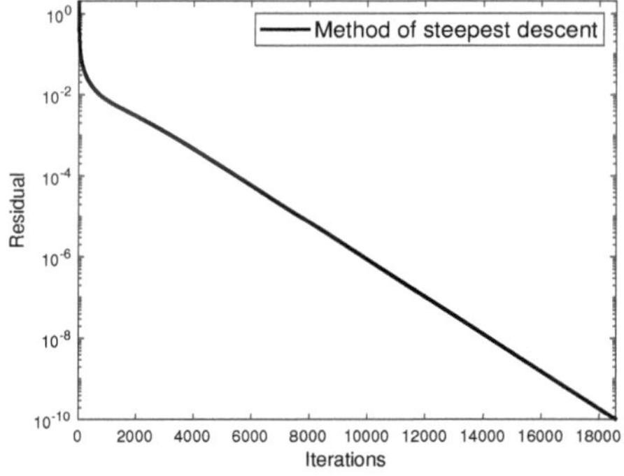

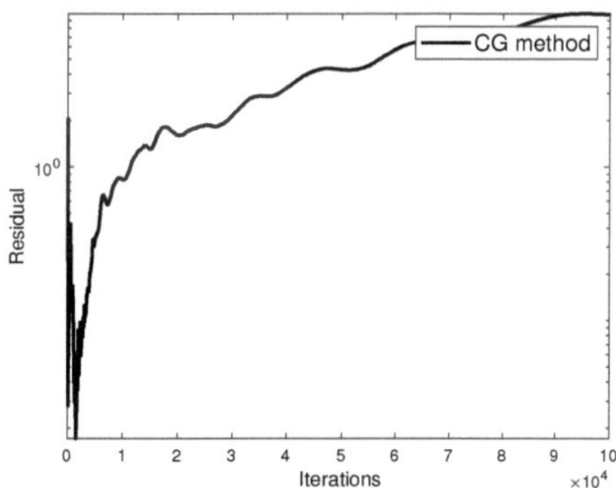

Figure 4.35 Convergence behavior of the steepest descent method for test case 3

Figure 4.36 Convergence behavior of the CG method for test case 3

From the previous investigations, it can be concluded that the CG method should always be preferred over the steepest descent method for linear systems with a symmetric and positive definite matrix, and also in the case of a slight violation of the matrix properties. If there is a significant disturbance of the matrix properties, the convergence of the sequences of approximate solutions generated by the two methods to the desired solution of the system of equations can no longer be guaranteed. Even though the steepest descent method behaves more robustly than the CG method in the convection-dominated flows considered, in such cases a method should fundamentally be used that, apart from invertibility, imposes as few additional requirements on the matrix as possible.

4.3.2   Methods for Non-singular Matrices

In the context of complex applications, it is often not possible to make statements about the properties of the resulting systems of equations. As an example, the discretization of the convection-diffusion equation (Example 1.4) with a nonzero advection parameter a shows that, even for simple fundamental equations, the symmetry of the matrix cannot be guaranteed. Therefore, in this section, we will focus on methods that, apart from the non-singularity of the matrix, initially do not impose any further requirements on the system of equations under consideration. For the GMRES method, this condition proves to be theoretically sufficient for the convergence and stability of the method. The other iterative solvers for systems of equations are based directly or indirectly on the Bi-Lanczos algorithm and, consequently, inherit its possible premature breakdowns in the case of an indefinite matrix. Despite this issue, the algorithms discussed are very common in scientific computing and have proven to be robust, stable, and efficient.

4.3.2.1   The Arnoldi Algorithm and the FOM

In 1951, Arnoldi developed a method for the successive transformation of dense matrices into upper Hessenberg form. This algorithm was later also used for the computation of eigenvalues and, in the following form, serves to compute an orthonormal basis $\mathcal{V}_m = \{v_1, \ldots, v_m\}$ of the Krylov subspace K_m, which is of crucial importance for the GMRES methods.

To derive the Arnoldi algorithm, we assume that $\{v_1, \ldots, v_j\}$ is an orthogonal basis of $K_j = \operatorname{span}\{r_0, \ldots, A^{j-1} r_0\}$ for $j = 1, \ldots, m$. Since

$$AK_m = \operatorname{span}\{Ar_0, \ldots, A^m r_0\} \subset K_{m+1}$$

the idea suggests itself to define v_{m+1} in the form

$$v_{m+1} = Av_m + \xi \text{ with } \xi \in \operatorname{span}\{v_1, \ldots, v_m\} = K_m$$

With $\xi = -\sum_{j=1}^{m} \alpha_j v_j$ it follows that

$$(v_{m+1}, v_j)_2 = (Av_m, v_j)_2 - \alpha_j (v_j, v_j)_2,$$

which, due to the orthogonality condition, yields

$$\alpha_j = \frac{(Av_m, v_j)_2}{(v_j, v_j)_2}$$

for $j = 1, \ldots, m$. If we also consider only normalized basis vectors, the computation of the coefficients simplifies to

$$\alpha_j = (v_j, Av_m)_2$$

and, under the assumption $r_0 \neq 0$, we obtain the following procedure.

Algorithm - Arnoldi —

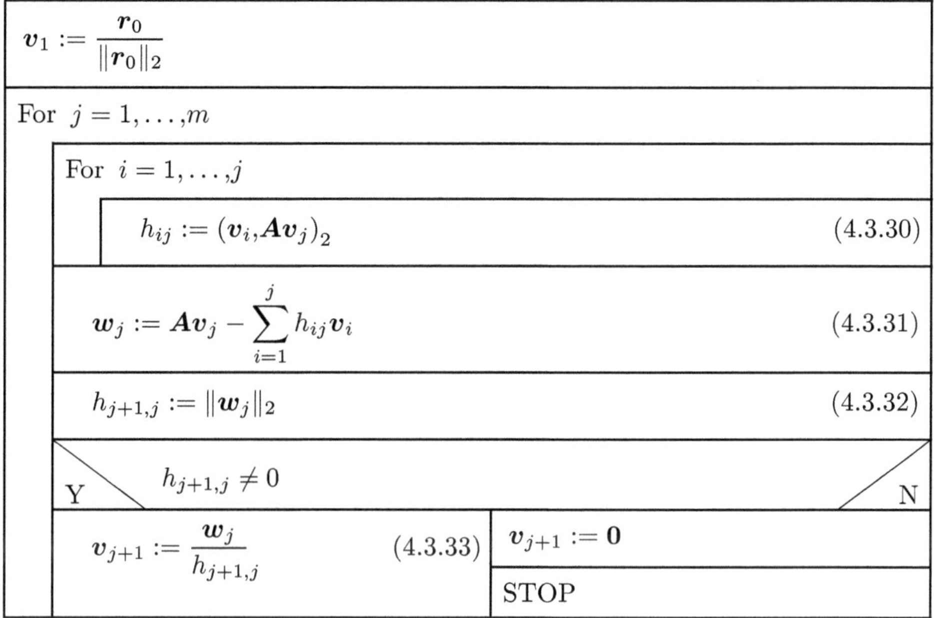

$$v_1 := \frac{r_0}{\|r_0\|_2}$$

For $j = 1, \ldots, m$

  For $i = 1, \ldots, j$

$$h_{ij} := (v_i, Av_j)_2 \tag{4.3.30}$$

$$w_j := Av_j - \sum_{i=1}^{j} h_{ij} v_i \tag{4.3.31}$$

$$h_{j+1,j} := \|w_j\|_2 \tag{4.3.32}$$

Y $h_{j+1,j} \neq 0$ N

$$v_{j+1} := \frac{w_j}{h_{j+1,j}} \tag{4.3.33} \qquad v_{j+1} := 0$$

STOP

An implementation of this procedure in the form of a MATLAB code can be found in Appendix A.

Theorem 4.82 *Provided that the Arnoldi algorithm does not terminate before the computation of $v_m \neq 0$, then $V_j = \{v_1, \ldots, v_j\}$ forms an orthonormal basis of the j-th Krylov subspace K_j for $j = 1, \ldots, m$.*

Proof:
We first show that the vectors $v_1, \ldots, v_j$ for all $j = 1, \ldots, m$ represent an orthonormal system (ONS). To this end, we proceed by induction on j.

For $j = 1$, the statement is trivial since $r_0 \neq 0$. Assume that V_k for $k = 1, \ldots, j < m$ is an ONS, then, using the assumption $v_{j+1} \neq 0$, we obtain the equation

$$
(v_{j+1}, v_k)_2 \overset{\substack{(4.3.31) \\ (4.3.33)}}{=} \frac{1}{h_{j+1,j}} \left(Av_j - \sum_{i=1}^{j} h_{ij} v_i, v_k \right)_2
$$

$$
= \frac{1}{h_{j+1,j}} \left((Av_j, v_k)_2 - h_{kj} \right)
$$

$$
\overset{(4.3.30)}{=} \frac{1}{h_{j+1,j}} \left((Av_j, v_k)_2 - (v_k, Av_j)_2 \right)
$$

$$
= 0.
$$

The normalization (4.3.33) yields the statement.

We now proceed to prove the basis property and again use induction on j.

For $j = 1$, the statement is trivial. Let $\mathcal{V}_k$ for $k = 1,\ldots,j < m$ be a basis of K_k, then it follows that

$$w_j = \boldsymbol{A} \underbrace{v_j}_{\in K_j} - \underbrace{\sum_{i=1}^{j} h_{ij} v_i}_{\in K_j} \in K_{j+1}.$$

Thus, $\mathrm{span}\{v_1,\ldots,v_{j+1}\} \subset K_{j+1}$ holds. Due to the proven orthonormality of the vectors $v_1,\ldots,v_{j+1} \in \mathbb{R}^n \setminus \{0\}$, we have $\dim \mathrm{span}\{v_1,\ldots,v_{j+1}\} = j + 1$, which implies $\mathrm{span}\{v_1,\ldots,v_{j+1}\} = K_{j+1}$. $\qquad\square$

Theorem 4.83 *Provided that the Arnoldi algorithm does not terminate before the computation of $v_m \neq 0$, then using $\boldsymbol{V}_m = (v_1 \ldots v_m) \in \mathbb{R}^{n \times m}$ with*

$$\boldsymbol{H}_m := \boldsymbol{V}_m^T \boldsymbol{A} \boldsymbol{V}_m \in \mathbb{R}^{m \times m} \tag{4.3.34}$$

we obtain an upper Hessenberg matrix, for which

$$(\boldsymbol{H}_m)_{ij} = \begin{cases} h_{ij} & \text{from the Arnoldi algorithm} \quad \text{for } i \leq j + 1, \\ 0 & \text{for } i > j + 1 \end{cases}$$

holds.

Proof:
Let us denote the elements of the matrix $\boldsymbol{H}_m$ by $\tilde{h}_{ij}$, then from (4.3.34) it follows that

$$\tilde{h}_{ij} = (v_i, \boldsymbol{A} v_j)_2 \overset{(4.3.30)}{=} h_{ij} \quad \text{for} \quad i \leq j.$$

Let $j \in \{1,\ldots,m-1\}$ be arbitrary but fixed, then for $k \in \{1,\ldots,m-j\}$ we obtain the representation

$$\begin{aligned}
\tilde{h}_{j+k,j} \quad &= \quad (v_{j+k}, \boldsymbol{A} v_j)_2 \\[2mm]
&\overset{(4.3.31)}{=} \quad (v_{j+k}, w_j)_2 + \sum_{i=1}^{j} h_{ij} \underbrace{(v_{j+k}, v_i)_2}_{=0} \\[2mm]
&\overset{(4.3.33)}{=} \quad h_{j+1,j}(v_{j+k}, v_{j+1})_2 = \begin{cases} h_{j+1,j} & \text{for} \quad k = 1 \\ 0 & \text{for} \quad k = 2,\ldots,m-j. \end{cases}
\end{aligned}$$

$$\square$$

Theorem 4.84 *Provided that the Arnoldi algorithm does not terminate before the computation of v_{m+1}, then*

$$\boldsymbol{A} \boldsymbol{V}_m = \boldsymbol{V}_{m+1} \overline{\boldsymbol{H}}_m, \tag{4.3.35}$$

where $\overline{\boldsymbol{H}}_m \in \mathbb{R}^{(m+1) \times m}$ is given by

$$\overline{\boldsymbol{H}}_m = \begin{pmatrix} \boldsymbol{H}_m \\ 0 \ldots 0 \; h_{m+1,m} \end{pmatrix}. \tag{4.3.36}$$

Proof:
From (4.3.31) and (4.3.33) it follows that

$$\boldsymbol{A}\boldsymbol{v}_j = h_{j+1,j}\boldsymbol{v}_{j+1} + \sum_{i=1}^{j} h_{ij}\boldsymbol{v}_i \text{ for } j = 1,\ldots,m$$

and thus (4.3.35). $\hspace{1cm}$ $\square$

With the Arnoldi algorithm, an iterative method for solving the equation $\boldsymbol{A}\boldsymbol{x} = \boldsymbol{b}$ can be defined directly. First, every vector $\boldsymbol{x}_m \in \boldsymbol{x}_0 + K_m$ can be written in the form

$$\boldsymbol{x}_m = \boldsymbol{x}_0 + \boldsymbol{V}_m\boldsymbol{\alpha}_m \quad \text{with} \quad \boldsymbol{\alpha}_m \in \mathbb{R}^m \tag{4.3.37}$$

since the columns of the matrix $\boldsymbol{V}_m$ form a basis of the Krylov subspace K_m.

If we make the ansatz of an orthogonal Krylov subspace method, then using the first unit vector $\boldsymbol{e}_1 = (1,0,\ldots,0)^T \in \mathbb{R}^m$, we obtain the condition

$$\begin{aligned}
\boldsymbol{r}_m \quad &= \quad \boldsymbol{b} - \boldsymbol{A}\boldsymbol{x}_m \perp K_m \\[4pt]
\Longleftrightarrow (\boldsymbol{b} - \boldsymbol{A}\boldsymbol{x}_m, \boldsymbol{v}_j)_2 \quad &= \quad 0 \text{ for } j = 1,\ldots,m \\[4pt]
\Longleftrightarrow \qquad \mathbb{R}^m \ni \boldsymbol{0} \quad &= \quad \boldsymbol{V}_m^T(\boldsymbol{b} - \boldsymbol{A}\boldsymbol{x}_m) \\[4pt]
\stackrel{(4.3.37)}{=} \quad & \quad \boldsymbol{V}_m^T(\boldsymbol{r}_0 - \boldsymbol{A}\boldsymbol{V}_m\boldsymbol{\alpha}_m) \\[4pt]
&= \quad \|\boldsymbol{r}_0\|_2\boldsymbol{e}_1 - \underbrace{\boldsymbol{V}_m^T\boldsymbol{A}\boldsymbol{V}_m}_{=\boldsymbol{H}_m}\boldsymbol{\alpha}_m.
\end{aligned} \tag{4.3.38}$$

With this, we obtain the orthogonal Krylov subspace method known as the Full Orthogonalization Method (FOM) in the following form:

Algorithm - Full Orthogonalization Method (FOM) —

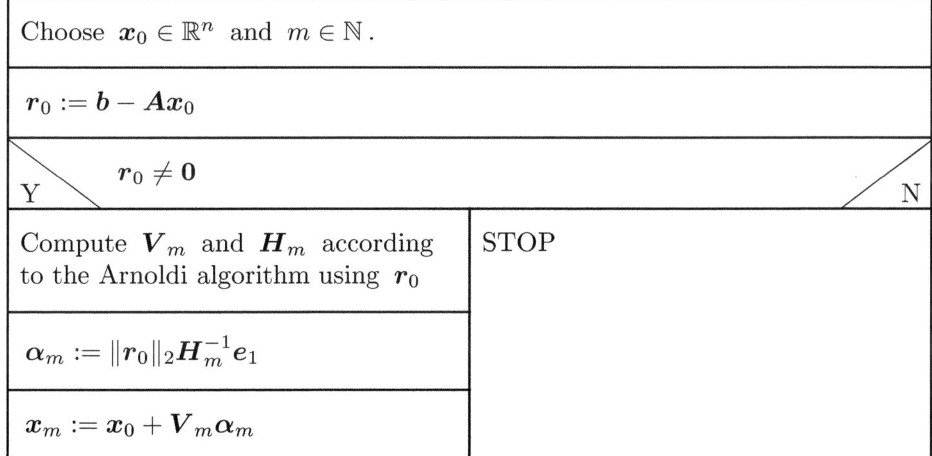

Choose $\boldsymbol{x}_0 \in \mathbb{R}^n$ and $m \in \mathbb{N}$.	
$\boldsymbol{r}_0 := \boldsymbol{b} - \boldsymbol{A}\boldsymbol{x}_0$	
Y $\quad\boldsymbol{r}_0 \neq \boldsymbol{0}\quad$ N	
Compute $\boldsymbol{V}_m$ and $\boldsymbol{H}_m$ according to the Arnoldi algorithm using $\boldsymbol{r}_0$	STOP
$\boldsymbol{\alpha}_m := \|\boldsymbol{r}_0\|_2\boldsymbol{H}_m^{-1}\boldsymbol{e}_1$	
$\boldsymbol{x}_m := \boldsymbol{x}_0 + \boldsymbol{V}_m\boldsymbol{\alpha}_m$	

The solution of the equation

$$H_m \alpha_m = \|r_0\|_2 e_1$$

can, for example, be carried out using $m-1$ Givens rotations with $G = G_{m,m-1} \cdot \ldots \cdot G_{2,1}$ and subsequent backward elimination of the remaining system

$$R \alpha_m = \|r_0\|_2 G e_1$$

with an upper right triangular matrix R. For the residual, we obtain

$$
\begin{aligned}
\|r_m\|_2 \quad &= \quad \|b - A(x_0 + V_m \alpha_m)\|_2 \\
&\overset{\text{Theorem 4.84}}{=} \|V_{m+1}\left(\|r_0\|_2 e_1 - \overline{H}_m \alpha_m\right)\|_2 \\
&= \left\| \begin{pmatrix} G\|r_0\|_2 e_1 \\ 0 \end{pmatrix} - \begin{pmatrix} R \\ 0 \ldots 0 \; h_{m+1,m} \end{pmatrix} \alpha_m \right\|_2 = h_{m+1,m}\,|(\alpha_m)_m|,
\end{aligned}
\tag{4.3.39}
$$

where it should be noted that in (4.3.39) $e_1 \in \mathbb{R}^{m+1}$ is the unit vector, and otherwise $e_1 \in \mathbb{R}^m$. In FOM, it is often necessary to choose m very large in order to expect an acceptable solution. Therefore, checking the residual as an indicator for terminating the procedure would be desirable. One possibility is to specify an accuracy threshold with $\varepsilon > 0$. If x_m satisfies the condition $\|b - Ax_m\|_2 \leq \varepsilon$, then the procedure is terminated; otherwise, one will perform a restart with $x_0 = x_m$. This limits the memory requirement for the orthonormal basis to $n \cdot m$ real numbers.

To minimize computational effort and required memory, we thus obtain the following variant of the Full Orthogonalization Method, for which a MATLAB implementation is given in Appendix A.

Algorithm - Restarted FOM —

<table>
<tr><td>

Choose $x_0 \in \mathbb{R}^n$, $\varepsilon > 0$ and $m \in \mathbb{N}$.

</td></tr>
<tr><td>

$r_0 := b - Ax_0$

</td></tr>
<tr><td>

As long as $\|r_0\|_2 > \varepsilon$

Compute V_m and H_m according to the Arnoldi algorithm using r_0
$\alpha_m := \|r_0\|_2 H_m^{-1} e_1$
$x_m := x_0 + V_m \alpha_m$
$x_0 := x_m$
$r_0 := b - Ax_0$

</td></tr>
</table>

A problem with the presented algorithm is that the orthogonality of the vectors can be lost after only a few iterations. An improvement is provided by the so-called Arnoldi-modified Gram-Schmidt algorithm. This results from the Arnoldi process by replacing (4.3.30) and (4.3.31)

Current loop (Arnoldi) —

<table>
<tr><td>For $i = 1, \ldots, j$</td></tr>
<tr><td>$\quad h_{ij} := (\boldsymbol{v}_i, \boldsymbol{A}\boldsymbol{v}_j)_2$</td></tr>
<tr><td>$\boldsymbol{w}_j := \boldsymbol{A}\boldsymbol{v}_j - \sum_{i=1}^{j} h_{ij} \boldsymbol{v}_i$</td></tr>
</table>

by

Modified loop (Arnoldi) —

<table>
<tr><td>$\boldsymbol{w}_j := \boldsymbol{A}\boldsymbol{v}_j$</td></tr>
<tr><td>For $i = 1, \ldots, j$</td></tr>
<tr><td>$\quad h_{ij} := (\boldsymbol{v}_i, \boldsymbol{w}_j)_2$</td></tr>
<tr><td>$\quad \boldsymbol{w}_j := \boldsymbol{w}_j - h_{ij} \boldsymbol{v}_i$</td></tr>
</table>

$$(4.3.40)$$

If a maximum bandwidth of the matrix $\boldsymbol{H}_m$ is specified and all matrix entries outside of it are subsequently neglected, the IOM (Incomplete Orthogonalization Method) results, which, analogous to DIOM (Direct IOM), represents a skew projection method. DIOM and IOM differ only in that, in the direct version, an LU decomposition of the matrix $\boldsymbol{H}_m$ is performed, resulting in an algorithm that is easy to implement. A specific implementation of IOM is presented in [60].

4.3.2.2 The Lanczos Algorithm and the D-Lanczos Method

With the goal of determining eigenvectors and eigenvalues, Lanczos [46] already introduced in 1950 a method for transforming symmetric matrices into tridiagonal form. As we will see, the method can be regarded as a simplification of the Arnoldi algorithm for symmetric matrices. Let us consider a symmetric matrix $\boldsymbol{A} \in \mathbb{R}^{n \times n}$, then from (4.3.34) it follows that

$$\boldsymbol{H}_m = \boldsymbol{V}_m^T \boldsymbol{A} \boldsymbol{V}_m = \boldsymbol{V}_m^T \boldsymbol{A}^T \boldsymbol{V}_m = (\boldsymbol{V}_m^T \boldsymbol{A} \boldsymbol{V}_m)^T = \boldsymbol{H}_m^T.$$

With this, we obtain the Lanczos algorithm from the Arnoldi algorithm, since

(a) the matrix element $h_{j-1,j}$ has already been computed as $h_{j,j-1}$ in the previous step, and thus the loop (4.3.30) degenerates to $h_{jj} = (\boldsymbol{v}_j, \boldsymbol{A}\boldsymbol{v}_j)_2$ and

(b) (4.3.31) reduces to $\boldsymbol{w}_j = \boldsymbol{A}\boldsymbol{v}_j - h_{j-1,j}\boldsymbol{v}_{j-1} - h_{jj}\boldsymbol{v}_j$.

If we choose the notations $a_j = h_{jj}$ and $c_j = h_{j-1,j}$, then $\boldsymbol{H}_m$ has the form

$$\boldsymbol{H}_m = \begin{pmatrix} a_1 & c_2 & & & \\ c_2 & a_2 & \ddots & & \\ & \ddots & \ddots & \ddots & \\ & & \ddots & a_{m-1} & c_m \\ & & & c_m & a_m \end{pmatrix}.$$

From the modified variant of the Arnoldi algorithm (see (4.3.40)), the Lanczos algorithm results, using $h_{j-1,j} = h_{j,j-1}$, in the form

Algorithm - Lanczos —

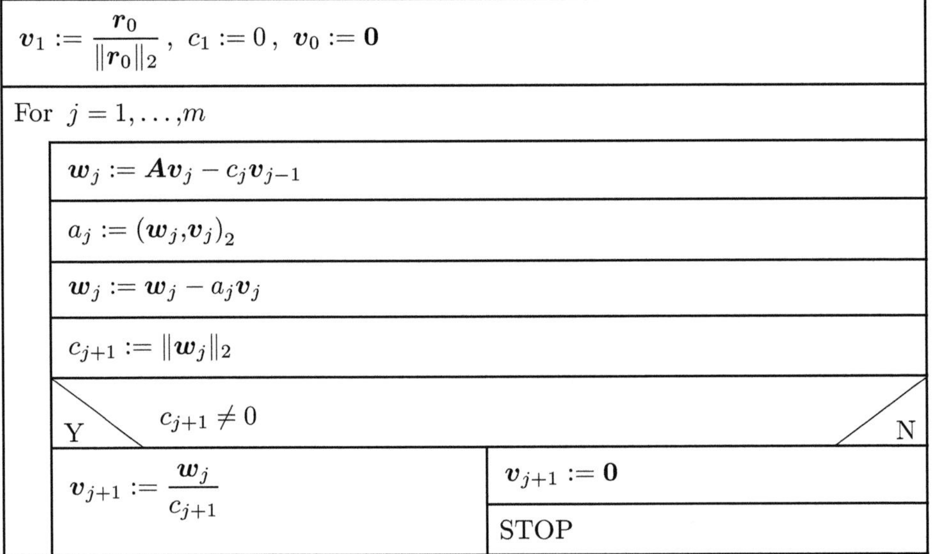

Analogous to the Arnoldi algorithm, the Lanczos method can also be used to solve the linear system $\boldsymbol{Ax} = \boldsymbol{b}$. Provided that the matrix $\boldsymbol{H}_m$ can be decomposed in the form

$$\boldsymbol{H}_m = \underbrace{\begin{pmatrix} 1 & & & & \\ \gamma_2 & \ddots & & 0 & \\ & \ddots & \ddots & & \\ 0 & & \ddots & \ddots & \\ & & & \gamma_m & 1 \end{pmatrix}}_{=:\ \boldsymbol{L}_m} \underbrace{\begin{pmatrix} \beta_1 & \delta_2 & & & \\ & \ddots & \ddots & 0 & \\ & & \ddots & \ddots & \\ 0 & & & \ddots & \delta_m \\ & & & & \beta_m \end{pmatrix}}_{=:\ \boldsymbol{U}_m} \qquad (4.3.41)$$

then it follows that

$$\delta_j = c_j \text{ for } j = 2,\ldots,m$$

$$\beta_1 = a_1, \tag{4.3.42}$$

$$\gamma_j = \frac{c_j}{\beta_{j-1}} \text{ for } j = 2,\ldots,m, \tag{4.3.43}$$

$$\beta_j = a_j - \gamma_j c_j \text{ for } j = 2,\ldots,m. \tag{4.3.44}$$

Analogous to FOM, we write the approximate solution using the first unit vector $\boldsymbol{e}_1 \in \mathbb{R}^m$ in the form

$$\boldsymbol{x}_m = \boldsymbol{x}_0 + \boldsymbol{V}_m \boldsymbol{H}_m^{-1} \underbrace{\|\boldsymbol{r}_0\|_2 \boldsymbol{e}_1}_{=:\; \tilde{\boldsymbol{e}}_1} \overset{(4.3.41)}{=} \boldsymbol{x}_0 + \boldsymbol{V}_m \boldsymbol{U}_m^{-1} \boldsymbol{L}_m^{-1} \tilde{\boldsymbol{e}}_1$$

$$= \boldsymbol{x}_0 + \boldsymbol{P}_m \boldsymbol{z}_m$$

with $\boldsymbol{P}_m := \boldsymbol{V}_m \boldsymbol{U}_m^{-1} \in \mathbb{R}^{n \times m}$ and $\boldsymbol{z}_m := \boldsymbol{L}_m^{-1} \tilde{\boldsymbol{e}}_1$. Let $\boldsymbol{p}_1, \ldots, \boldsymbol{p}_m \in \mathbb{R}^n$ be the column vectors of $\boldsymbol{P}_m$, then it follows with

$$\boldsymbol{P}_m \begin{pmatrix} \beta_1 & \delta_2 & & \\ & \ddots & \ddots & \\ & & \ddots & \delta_m \\ & & & \beta_m \end{pmatrix} = (\boldsymbol{v}_1 \ldots \boldsymbol{v}_m)$$

the equation $\delta_m \boldsymbol{p}_{m-1} + \beta_m \boldsymbol{p}_m = \boldsymbol{v}_m$, which gives the rule for computing the m-th column as

$$\boldsymbol{p}_m = \frac{1}{\beta_m}(\boldsymbol{v}_m - \delta_m \boldsymbol{p}_{m-1}) \tag{4.3.45}$$

taking into account the definition $\delta_1 := 0$ and $\boldsymbol{p}_0 := \boldsymbol{0}$. Let's assume that the relation $\boldsymbol{P}_{m-1} \boldsymbol{U}_{m-1} = \boldsymbol{V}_{m-1}$ holds, then from (4.3.45) it follows that $\boldsymbol{P}_m \boldsymbol{U}_m = \boldsymbol{V}_m$, which yields $\boldsymbol{P}_m = \boldsymbol{V}_m \boldsymbol{U}_m^{-1}$. Furthermore, let $\boldsymbol{z}_{m-1} = (\xi_1, \ldots, \xi_{m-1})^T \in \mathbb{R}^{m-1}$ be the solution of $\boldsymbol{L}_{m-1} \boldsymbol{z}_{m-1} = \tilde{\boldsymbol{e}}_1 \in \mathbb{R}^{m-1}$, then by incrementing the space dimension we obtain

$$\mathbb{R}^m \ni \tilde{\boldsymbol{e}}_1 = \boldsymbol{L}_m \boldsymbol{z}_m = \begin{pmatrix} & & & 0 \\ & \boldsymbol{L}_{m-1} & & \vdots \\ & & & 0 \\ 0 & \ldots & 0 & \gamma_m & 1 \end{pmatrix} \begin{pmatrix} \boldsymbol{z}_{m-1} \\ \xi_m \end{pmatrix} = \begin{pmatrix} \tilde{\boldsymbol{e}}_1 \\ \gamma_m \xi_{m-1} + \xi_m \end{pmatrix}$$

and thus

$$\xi_m = -\gamma_m \xi_{m-1}. \tag{4.3.46}$$

Here, the first equation directly gives $\xi_1 = 1$. For the approximate solution, it follows that

$$\boldsymbol{x}_m = \boldsymbol{x}_0 + (\boldsymbol{P}_{m-1}, \boldsymbol{p}_m) \begin{pmatrix} \boldsymbol{z}_{m-1} \\ \xi_m \end{pmatrix} = \boldsymbol{x}_0 + \boldsymbol{P}_{m-1} \boldsymbol{z}_{m-1} + \xi_m \boldsymbol{p}_m$$

$$= \boldsymbol{x}_{m-1} + \xi_m \boldsymbol{p}_m \tag{4.3.47}$$

and we thus obtain the D(Direct)-Lanczos method in summary:

D-Lanczos-Algorithmus —

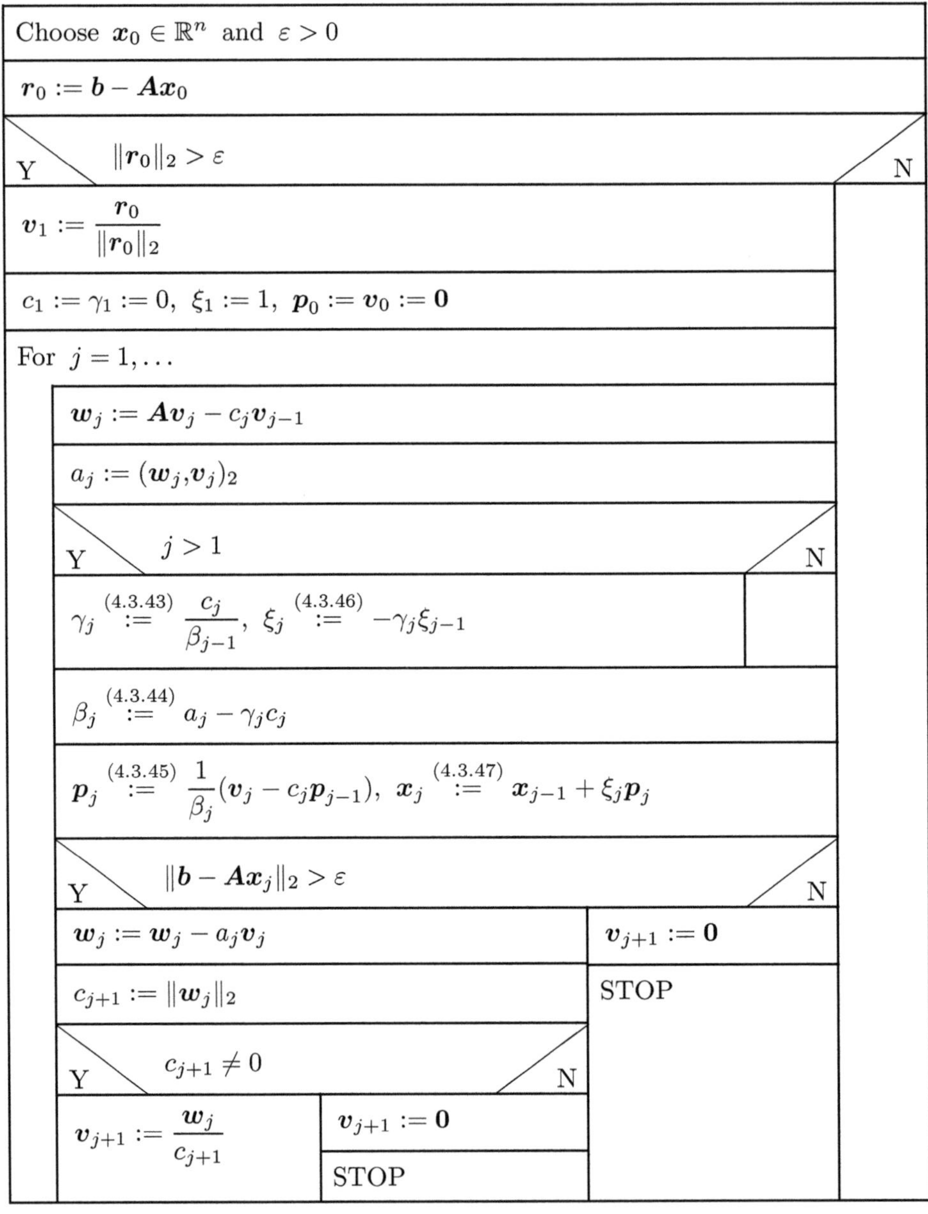

The application of an LQ decomposition instead of an LU decomposition leads to the SYMMLQ method, which was introduced in 1975 by Paige and Saunders [56]. The obvious idea of using a QR decomposition is also presented in [26].

The D-Lanczos algorithm was derived based on the transformation (4.3.38) and thus, analogous to the FOM, represents an orthogonal Krylov subspace method. Let us now consider the special case of a positive definite and symmetric matrix A. Since V_m has orthonormal column vectors, we have $V_m x \neq 0$ for all $x \in \mathbb{R}^n \setminus \{0\}$, which yields

$$(x, H_m x)_2 = (V_m x, A V_m x)_2$$

so that, in addition to symmetry, positive definiteness is also transferred to the matrix $H_m \in \mathbb{R}^{m \times m}$. With H_m, all leading principal submatrices $H_m[k]$ for $k = 1, \ldots, m$ are also positive definite matrices and therefore non-singular matrices. From this, Theorem 3.8 implies the existence and uniqueness of the decomposition (4.3.41). In this case, the algorithm is well-defined and, as an orthogonal Krylov subspace method, is also equivalent to the CG method. Thus, we can view the D-Lanczos algorithm as a generalization of the CG method to indefinite symmetric matrices.

4.3.2.3 The Bi-Lanczos Algorithm

The iterative methods FOM and D-Lanczos, which are derived from the Arnoldi and Lanczos algorithms, exhibit different advantages and disadvantages.

The advantage of the Full Orthogonalization Method lies in its applicability to arbitrary non-singular matrices $A \in \mathbb{R}^{n \times n}$, whereas the D-Lanczos algorithm is characterized by a low memory requirement for the orthonormal basis $\{v_1, \ldots, v_m\}$ of the Krylov subspace K_m. Considering the disadvantages of both algorithms, FOM has a high memory requirement for the orthonormal basis, which often does not allow n iterations for sparse matrices $A \in \mathbb{R}^{n \times n}$. A remedy is provided by the "restarted version", which, however, does not guarantee convergence even theoretically after n steps. The D-Lanczos method, on the other hand, is restricted to solving systems of equations with a symmetric matrix A and can be interpreted as a special case of FOM for symmetric matrices. Both methods are formally equivalent to the conjugate gradient method for symmetric positive definite matrices.

The goal of the Bi-Lanczos algorithm is to determine a basis of the Krylov subspace K_m in such a way that iterative methods based on it combine the advantage of low memory usage with applicability to nonsymmetric matrices.

Let $A \in \mathbb{R}^{n \times n}$ be an arbitrary non-singular matrix. In addition to

$$K_m = K_m(A, r_0) = \text{span}\left\{r_0, A r_0, \ldots, A^{m-1} r_0\right\}$$

we define the *transposed Krylov subspace* corresponding to K_m

$$K_m^T = K_m\left(A^T, r_0\right) = \text{span}\left\{r_0, A^T r_0, \ldots, \left(A^T\right)^{m-1} r_0\right\}.$$

Definition 4.85 Let $v_1, \ldots, v_m, w_1, \ldots, w_m \in \mathbb{R}^n$, then the two sets $\{v_1, \ldots, v_m\}$ and $\{w_1, \ldots, w_m\}$ are called bi-orthogonal if

$$(v_i, w_j)_2 = \delta_{ij} c_{ij} \text{ with } c_{ij} \in \mathbb{R} \backslash \{0\} \text{ for } i, j = 1, \ldots, m \tag{4.3.48}$$

holds. In the case $c_{ii} = 1$ for $i = 1, \ldots, m$, the sets are called bi-orthonormal.

An essential property of bi-orthogonal sets is described by the following lemma.

Lemma 4.86 *If the sets* $\{v_1,\ldots,v_m\}$ *and* $\{w_1,\ldots,w_m\}$ *are bi-orthogonal, then*

$$\dim \operatorname{span}\{v_1,\ldots,v_m\} = m = \dim \operatorname{span}\{w_1,\ldots,w_m\}.$$

Proof:
For reasons of symmetry, it is sufficient to prove the statement for the vectors $v_1,\ldots,v_m$. Suppose $0 = \sum_{i=1}^{m}\lambda_i v_i$, then for $j = 1,\ldots,m$ we obtain

$$\lambda_j = \frac{1}{c_{jj}}\sum_{i=1}^{m}\lambda_i c_{ij}\delta_{ij} = \frac{1}{c_{jj}}\sum_{i=1}^{m}\lambda_i\,(v_i,w_j)_2 = \frac{1}{c_{jj}}\left(\sum_{i=1}^{m}\lambda_i v_i,w_j\right)_2 = \frac{1}{c_{jj}}\,(0,w_j)_2 = 0.$$

$\square$

The Bi-Lanczos algorithm is based on the following idea: Instead of determining an orthonormal basis of K_m , a basis $\{v_1,\ldots,v_m\}$ of K_m and a basis $\{w_1,\ldots,w_m\}$ of K_m^T are computed simultaneously, which satisfy the bi-orthogonality condition (4.3.48).

Bi-Lanczos Algorithm —

<table>
<tr><td colspan="2">$h_{1,0} = h_{0,1} := 0$</td></tr>
<tr><td colspan="2">$v_0 = w_0 := 0$</td></tr>
<tr><td>$v_1 = w_1 := \dfrac{r_0}{\|r_0\|_2}$</td><td>(4.3.49)</td></tr>
<tr><td colspan="2">for $j = 1,\ldots,m$</td></tr>
<tr><td>$h_{jj} := (w_j, A\,v_j)_2$</td><td>(4.3.50)</td></tr>
<tr><td>$v_{j+1}^* := A\,v_j - h_{jj}v_j - h_{j-1,j}v_{j-1}$</td><td>(4.3.51)</td></tr>
<tr><td>$w_{j+1}^* := A^T w_j - h_{jj}w_j - h_{j,j-1}w_{j-1}$</td><td>(4.3.52)</td></tr>
<tr><td colspan="2">$h_{j+1,j} := |(v_{j+1}^*,w_{j+1}^*)_2|^{1/2}$</td></tr>
<tr><td colspan="2">Y $h_{j+1,j} \neq 0$ N</td></tr>
<tr><td>
$h_{j,j+1} := \dfrac{(v_{j+1}^*,w_{j+1}^*)_2}{h_{j+1,j}}$ (4.3.53)

$v_{j+1} := \dfrac{v_{j+1}^*}{h_{j+1,j}}$ (4.3.54)

$w_{j+1} := \dfrac{w_{j+1}^*}{h_{j,j+1}}$ (4.3.55)
</td><td>
$h_{j,j+1} := 0$

$v_{j+1} := 0$

$w_{j+1} := 0$

STOP
</td></tr>
</table>

With the next theorem, we will show that the present Bi-Lanczos algorithm indeed generates bi-orthonormal bases of the Krylov subspaces K_m and K_m^T .

Theorem 4.87 *Assuming that the Bi-Lanczos algorithm does not terminate before the computation of $\boldsymbol{w}_m \neq \boldsymbol{0}$, then $\mathcal{V}_j = \{\boldsymbol{v}_1,\ldots,\boldsymbol{v}_j\}$ and $\mathcal{W}_j = \{\boldsymbol{w}_1,\ldots,\boldsymbol{w}_j\}$ form a basis of the Krylov subspace K_j and K_j^T, respectively, for $j = 1,\ldots,m$. Moreover, the basis vectors satisfy the bi-orthogonality condition (4.3.48) with $c_{ii} = 1$ for $i = 1,\ldots,m$.*

Proof:
The proof of bi-orthonormality is obtained by complete induction on j.

For $j = 1$ it follows that

$$(\boldsymbol{w}_1,\boldsymbol{v}_1)_2 = \frac{(\boldsymbol{r}_0,\boldsymbol{r}_0)_2}{\|\boldsymbol{r}_0\|_2^2} = 1.$$

Assume the condition holds for $\ell = 1,\ldots,j < m$, then we obtain

$$(\boldsymbol{v}_{j+1},\boldsymbol{w}_j)_2 \overset{\substack{(4.3.51)\\(4.3.54)}}{=} \frac{1}{h_{j+1,j}}\left(\underbrace{(\boldsymbol{Av}_j,\boldsymbol{w}_j)_2}_{=\,h_{jj}} - h_{jj}\underbrace{(\boldsymbol{v}_j,\boldsymbol{w}_j)_2}_{=\,1} - h_{j-1,j}\underbrace{(\boldsymbol{v}_{j-1},\boldsymbol{w}_j)_2}_{=\,0}\right)$$

$$= \quad 0,$$

where in the case $j = 1$ the property $(\boldsymbol{v}_{j-1},\boldsymbol{w}_j)_2 = (\boldsymbol{v}_0,\boldsymbol{w}_1)_2 = 0$ follows from the definition of $\boldsymbol{v}_0 := \boldsymbol{0}$. Furthermore, for $0 < i < j$ we have

$$(\boldsymbol{Av}_j,\boldsymbol{w}_i)_2 \quad = \quad \left(\boldsymbol{v}_j,\boldsymbol{A}^T\boldsymbol{w}_i\right)_2$$

$$\overset{\substack{(4.3.52)\\(4.3.55)}}{=} \quad h_{i,i+1}(\boldsymbol{v}_j,\boldsymbol{w}_{i+1})_2 + \underbrace{(\boldsymbol{v}_j,h_{ii}\boldsymbol{w}_i + h_{i,i-1}\boldsymbol{w}_{i-1})_2}_{=\,0}$$

$$= \quad \begin{cases} h_{j-1,j} & \text{for } i = j-1, \\ 0 & \text{for } i < j-1. \end{cases}$$

The use of the two statements above yields, for $0 < i < j$, the equation

$$(\boldsymbol{v}_{j+1},\boldsymbol{w}_i)_2 \overset{\substack{(4.3.51)\\(4.3.54)}}{=} \frac{1}{h_{j+1,j}}\left(\underbrace{(\boldsymbol{Av}_j,\boldsymbol{w}_i)_2}_{=\,\begin{cases}h_{j-1,j},\ i=j-1\\0,\qquad j<j-1\end{cases}} - h_{jj}\underbrace{(\boldsymbol{v}_j,\boldsymbol{w}_i)_2}_{=\,0} - h_{j-1,j}\underbrace{(\boldsymbol{v}_{j-1},\boldsymbol{w}_i)_2}_{=\,\begin{cases}1,\ i=j-1\\0,\ i<j-1\end{cases}}\right)$$

$$= \quad 0.$$

Analogously, we obtain $(\boldsymbol{w}_{j+1},\boldsymbol{v}_i)_2 = 0$ for $i \leq j$. With

$$(\boldsymbol{v}_{j+1},\boldsymbol{w}_{j+1})_2 \quad = \quad \left(\frac{\boldsymbol{v}_{j+1}^*}{h_{j+1,j}},\frac{\boldsymbol{w}_{j+1}^*}{h_{j,j+1}}\right)_2$$

$$\overset{(4.3.53)}{=} \quad \frac{1}{h_{j+1,j}} \cdot \frac{h_{j+1,j}}{(\boldsymbol{v}_{j+1}^*,\boldsymbol{w}_{j+1}^*)_2} \cdot (\boldsymbol{v}_{j+1}^*,\boldsymbol{w}_{j+1}^*)_2$$

$$= \quad 1$$

we finally obtain the claimed bi-orthonormality of the vectors.

Let $q_0,\ldots,q_{m-1}$ be arbitrary polynomials with $\deg(q_i) = i$ for $i = 0,\ldots,m-1$ and $q_0 \not\equiv 0$, then every $\boldsymbol{x} \in K_m$ can be written in the form

$$\boldsymbol{x} = \sum_{i=0}^{m-1} \alpha_i \boldsymbol{A}^i \boldsymbol{r}_0 = \sum_{i=0}^{m-1} \beta_i q_i(\boldsymbol{A}) \boldsymbol{r}_0,$$

which implies

$$\operatorname{span}\{q_0(\boldsymbol{A})\boldsymbol{r}_0,\ldots,q_{m-1}(\boldsymbol{A})\boldsymbol{r}_0\} = K_m. \tag{4.3.56}$$

Based on equation (4.3.56), we obtain the proof of the generating property by complete induction on j. For $j = 1$ it follows that

$$\boldsymbol{v}_1 = \frac{\boldsymbol{r}_0}{\|\boldsymbol{r}_0\|_2} = q_0(\boldsymbol{A})\boldsymbol{r}_0 \quad \text{with} \quad q_0(\boldsymbol{A}) = \frac{1}{\|\boldsymbol{r}_0\|_2}\boldsymbol{A}^0.$$

With $\deg(q_0) = 0$ and $q_0(\boldsymbol{A}) \neq \boldsymbol{0}$ we thus obtain $\operatorname{span}\{\boldsymbol{v}_1\} = K_1$.

For $j = 2$, from (4.3.50) and (4.3.54), taking into account $\boldsymbol{v}_0 = \boldsymbol{0}$ and $h_{j+1,j} \neq 0$, we obtain the representation

$$\boldsymbol{v}_2 = \frac{1}{h_{2,1}}\left(\boldsymbol{A}\boldsymbol{v}_1 - h_{2,2}\boldsymbol{v}_1 - h_{1,2}\boldsymbol{v}_0\right) = \frac{1}{h_{2,1}}\underbrace{\left(\left(\boldsymbol{A} - h_{2,2}\boldsymbol{I}\right)q_0(\boldsymbol{A})\right)}_{q_1(\boldsymbol{A}):=}\boldsymbol{r}_0$$

with $\deg(q_1) = 1$. Suppose the generating property holds for $\ell = 1,\ldots,j < m$ with $j \geq 2$, then for the scalar quantities $h_{j+1,j}$ computed in the Bi-Lanczos algorithm, due to the assumed non-breakdown of the method and taking into account $j < m$, we have $h_{j+1,j} \neq 0$. This yields

$$\boldsymbol{v}_{j+1} \overset{\substack{(4.3.51)\\(4.3.54)}}{=} \frac{1}{h_{j+1,j}}\left(\boldsymbol{A}\boldsymbol{v}_j - h_{jj}\boldsymbol{v}_j - h_{j-1,j}\boldsymbol{v}_{j-1}\right) = q_j(\boldsymbol{A})\boldsymbol{r}_0$$

with

$$q_j(\boldsymbol{A}) = \frac{1}{h_{j+1,j}}\left(\left(\boldsymbol{A} - h_{jj}\boldsymbol{I}\right)q_{j-1}(\boldsymbol{A}) - h_{j-1,j}q_{j-2}(\boldsymbol{A})\right).$$

Thus, we obtain $\deg(q_j) = j$, which gives

$$\operatorname{span}\{\boldsymbol{v}_1,\ldots,\boldsymbol{v}_{j+1}\} = \operatorname{span}\{q_0(\boldsymbol{A})\boldsymbol{r}_0,\ldots,q_j(\boldsymbol{A})\boldsymbol{r}_0\} = K_{j+1}.$$

The proof for K_m^T is analogous. The linear independence follows from Lemma 4.86 due to the bi-orthonormality of the vectors. $\qquad\square$

The properties of the basis vectors described in Theorems 4.83 and 4.84 have already proven useful in the description of the FOM and will also be of crucial importance in the derivation of the GMRES method. Analogous and fundamental statements regarding the present bi-orthonormal basis vectors for further methods are provided by the following theorem.

Theorem 4.88 *Suppose the Bi-Lanczos algorithm does not terminate before the computation of* $\boldsymbol{w}_m \neq \boldsymbol{0}$, *then let* $\boldsymbol{V}_m = (\boldsymbol{v}_1 \ldots \boldsymbol{v}_m) \in \mathbb{R}^{n \times m}$ *and* $\boldsymbol{W}_m = (\boldsymbol{w}_1 \ldots \boldsymbol{w}_m) \in \mathbb{R}^{n \times m}$ *be the matrices formed from the determined basis vectors of* K_m *and* K_m^T. *Furthermore, let* h_{ij} *be the scalar quantities appearing in the algorithm, then the following equations hold:*

$$\boldsymbol{T}_m = \boldsymbol{W}_m^T \boldsymbol{A} \boldsymbol{V}_m \tag{4.3.57}$$

$$\boldsymbol{A} \boldsymbol{V}_m = \boldsymbol{V}_m \boldsymbol{T}_m + h_{m+1,m} \boldsymbol{v}_{m+1} \boldsymbol{e}_m^T \tag{4.3.58}$$

$$\boldsymbol{A}^T \boldsymbol{W}_m = \boldsymbol{W}_m \boldsymbol{T}_m^T + h_{m,m+1} \boldsymbol{w}_{m+1} \boldsymbol{e}_m^T \tag{4.3.59}$$

with $\boldsymbol{e}_m = (0, \ldots, 0, 1)^T \in \mathbb{R}^m$ *and the tridiagonal matrix* $\boldsymbol{T}_m \in \mathbb{R}^{m \times m}$ *given by*

$$(\boldsymbol{T}_m)_{ij} := \begin{cases} h_{ij} & \text{for } j - 1 \leq i \leq j + 1, \\ 0 & \text{otherwise} \end{cases}$$

and the matrices $\boldsymbol{V}_m$ *and* $\boldsymbol{W}_m$ *have full rank.*

Proof:
From (4.3.50) it follows directly that

$$(\boldsymbol{T}_m)_{jj} = h_{jj} = (\boldsymbol{w}_j, \boldsymbol{A} \boldsymbol{v}_j)_2 .$$

The definition of the vector $\boldsymbol{v}_{j+1}^*$ within the Bi-Lanczos process yields, together with (4.3.54), for $i > j$ the equation

$$(\boldsymbol{w}_i, \boldsymbol{A} \boldsymbol{v}_j)_2 = \underbrace{(\boldsymbol{w}_i, \boldsymbol{v}_{j+1}^*)_2}_{= h_{j+1,j}(\boldsymbol{w}_i, \boldsymbol{v}_{j+1})_2} + h_{jj} \underbrace{(\boldsymbol{w}_i, \boldsymbol{v}_j)_2}_{= 0} + h_{j-1,j} \underbrace{(\boldsymbol{w}_i, \boldsymbol{v}_{j-1})_2}_{= 0}$$

$$= \begin{cases} h_{j+1,j} & \text{for } i = j + 1, \\ 0 & \text{for } i > j + 1. \end{cases}$$

Analogously, with the definition of $\boldsymbol{w}_{j+1}^*$ and (4.3.52), for $i < j$ we obtain

$$(\boldsymbol{w}_i, \boldsymbol{A} \boldsymbol{v}_j)_2 = \left(\boldsymbol{A}^T \boldsymbol{w}_i, \boldsymbol{v}_j \right)_2 = \begin{cases} h_{j-1,j} & \text{for } i = j - 1, \\ 0 & \text{for } i < j - 1, \end{cases}$$

which shows the claimed representation of the matrix $\boldsymbol{T}_m$ according to (4.3.57).

Taking into account the structure of the matrix $\boldsymbol{T}_m$, the equations (4.3.58) and (4.3.59) represent the matrix forms of equations (4.3.51) and (4.3.52) using (4.3.54) and (4.3.55), respectively.

The proof that $\mathrm{rang}(\boldsymbol{V}_m) = \mathrm{rang}(\boldsymbol{W}_m) = m$ follows directly from the basis property of the corresponding column vectors. $\square$

The Bi-Lanczos algorithm, in the form presented, terminates in the case $h_{j+1,j} = 0$. Such a situation occurs when one of the two basis vectors vanishes identically. If $\boldsymbol{v}_{j+1}^* = \boldsymbol{0}$, then K_j is an $\boldsymbol{A}$-invariant Krylov subspace, i.e., $K_j = K_{j+1}$, and both orthogonal a skew Krylov subspace methods based on this yield the exact solution of the system of equations, see [63]. An analogous statement holds for $\boldsymbol{w}_{j+1}^* = \boldsymbol{0}$. These two cases are described as regular breakdowns of the method. For orthogonal vectors $\boldsymbol{v}^*_{j+1} \neq \boldsymbol{0}$ and $\boldsymbol{w}^*_{j+1} \neq \boldsymbol{0}$, a so-called serious breakdown occurs, since projection methods based on this typically terminate without computing an acceptable approximate solution. This issue can often be resolved by a look-ahead variant of the method (look-ahead Lanczos algorithm). Here, it is exploited that the vectors $\boldsymbol{v}^*_{j+2}$ and $\boldsymbol{w}^*_{j+2}$ can often be defined without explicit use of $\boldsymbol{v}^*_{j+1}$ and $\boldsymbol{w}^*_{j+1}$. If $\boldsymbol{v}^*_{j+2}$ and $\boldsymbol{w}^*_{j+2}$ also cannot be determined, the algorithm attempts to compute the vectors $\boldsymbol{v}^*_{j+3}$ and $\boldsymbol{w}^*_{j+3}$, and so on. If the computation of such a vector pair succeeds, the algorithm can be continued in its original form. Iterative methods based on the Bi-Lanczos algorithm have proven to be stable in practical applications, even when using the described Lanczos variant [50, 52]. We therefore refrain from providing a detailed description of the look-ahead variant and refer the interested reader to the works of Brezinski, Zaglia, Sadok [13] and Parlett, Taylor, Liu [57], the latter of which also discusses so-called incurable breakdowns.

4.3.2.4   The GMRES Method

The GMRES method (Generalized Minimal Residual), introduced by Saad and Schultz [64] in 1986, is applicable to arbitrary non-singular matrices. Analogous to the CG algorithm, the method can formally be regarded as both a direct and an iterative method. However, the use of the GMRES method in a direct form is generally not practical due to the required memory space. Therefore, the algorithm is usually used in a restarted version for practically relevant problems, as was already introduced with the FOM. The method can be viewed in two different ways.

One approach is to consider GMRES as a Krylov subspace method, in which the projection is given by a Petrov-Galerkin condition with $L_m = \boldsymbol{A}K_m$, and thus the method represents a skew projection method.

The second possibility for deriving the method consists in reformulating the linear system of equations as a minimization problem. In contrast to the conjugate gradient method, we define the function F by

$$\begin{aligned} F : \mathbb{R}^n &\to \mathbb{R} \\ \boldsymbol{x} &\mapsto \|\boldsymbol{b} - \boldsymbol{A}\boldsymbol{x}\|_2^2. \end{aligned} \qquad (4.3.60)$$

Lemma 4.89 *Let $\boldsymbol{A} \in \mathbb{R}^{n \times n}$ be invertible and $\boldsymbol{b} \in \mathbb{R}^n$, then for the function F given by (4.3.60) we have*

$$\hat{\boldsymbol{x}} = \arg \min_{\boldsymbol{x} \in \mathbb{R}^n} F(\boldsymbol{x})$$

if and only if $\hat{\boldsymbol{x}} = \boldsymbol{A}^{-1}\boldsymbol{b}$ holds.

Proof:
We have $F(\boldsymbol{x}) = \|\boldsymbol{b} - \boldsymbol{A}\boldsymbol{x}\|_2^2 \geq 0$ for all $\boldsymbol{x} \in \mathbb{R}^n$. Thus, we obtain

$$F(\hat{\boldsymbol{x}}) = 0 \quad \Leftrightarrow \quad \boldsymbol{A}\hat{\boldsymbol{x}} = \boldsymbol{b},$$

which yields the statement

$$\hat{\boldsymbol{x}} = \arg \min_{\boldsymbol{x} \in \mathbb{R}^n} F(\boldsymbol{x}) = \boldsymbol{A}^{-1}\boldsymbol{b}.$$

$\square$

To conclude this brief classification of the GMRES method, we still need to show that both derivations lead to the same method, that is, that the approximate solutions obtained under the different conditions coincide. This statement is provided by the following lemma.

Lemma 4.90 *Let* $F : \mathbb{R}^n \to \mathbb{R}$ *be given by (4.3.60) and let* $x_0 \in \mathbb{R}^n$ *be arbitrary. Then*

$$\widetilde{x} = \arg \min_{x \in x_0 + K_m} F(x) \tag{4.3.61}$$

if and only if

$$b - A\widetilde{x} \perp L_m = AK_m \tag{4.3.62}$$

holds.

Proof:
Let $x_m, \widetilde{x} \in x_0 + K_m$ with $\widetilde{x} = x_0 + z$ and $x_m = x_0 + z_m$, then we obtain

$$\begin{aligned}
& F(x_m) - F(\widetilde{x}) \\[2mm]
= \; & (Ax_m - b, Ax_m - b)_2 - (A\widetilde{x} - b, A\widetilde{x} - b)_2 \\[2mm]
= \; & (A(x_0 + z_m) - b, A(x_0 + z_m) - b)_2 - (A(x_0 + z) - b, A(x_0 + z) - b)_2 \\[2mm]
= \; & (Az_m, Az_m)_2 + 2\,(Ax_0 - b, Az_m)_2 + (Ax_0 - b, Ax_0 - b)_2 \\[2mm]
& - (Az, Az)_2 - 2\,(Ax_0 - b, Az)_2 - (Ax_0 - b, Ax_0 - b)_2 \\[2mm]
= \; & \underbrace{(Az_m, Az_m)_2 - 2\,(Az, Az_m)_2 + (Az, Az)_2}_{=\,(A(z_m - z), A(z_m - z))_2} \\[2mm]
& + 2\underbrace{\Big\{(Ax_0 - b, Az_m)_2 + (Az, Az_m)_2 - (Ax_0 - b, Az)_2 - (Az, Az)_2\Big\}}_{=\,(A\widetilde{x} - b, A(z_m - z))_2} \\[2mm]
= \; & (A(z_m - z), A(z_m - z))_2 + 2\,(A\widetilde{x} - b, A(z_m - z))_2 . \tag{4.3.63}
\end{aligned}$$

"$(4.3.62) \Rightarrow (4.3.61)$"

Suppose $b - A\widetilde{x} \perp AK_m$.

Then it follows that $(b - A\widetilde{x}, Az)_2 = 0$ for all $z \in K_m$, and we obtain

$$F(x_m) - F(\widetilde{x}) \overset{(4.3.63)}{=} \|A(z_m - z)\|_2^2 .$$

Taking into account the invertibility of the matrix A, it follows that

$$F(x_m) > F(\widetilde{x}) \quad \text{for all} \quad x_m \in \{x_0 + K_m\} \backslash \{\widetilde{x}\}.$$

"$(4.3.61) \Rightarrow (4.3.62)$"

Suppose $\widetilde{x} = \arg \min\limits_{x \in x_0 + K_m} F(x)$.

We proceed by contradiction and assume that there exists a $z_m \in K_m$ such that

$$(b - A\widetilde{x}, Az_m)_2 = \varepsilon \neq 0$$

holds. Without loss of generality, we may assume $\varepsilon > 0$. By the invertibility of the matrix A, we have $z_m \neq 0$, and we define

$$\eta := (Az_m, Az_m)_2 > 0.$$

Furthermore, for a given $\xi \in \mathbb{R}$ with $0 < \xi < \dfrac{2\varepsilon}{\eta}$, we consider the vectors

$$z_m^{\xi} := \xi z_m + z \in K_m \tag{4.3.64}$$

and

$$x_m^{\xi} := x_0 + z_m^{\xi}.$$

Thus, it follows that

$$
\begin{aligned}
F\left(x_m^{\xi}\right) - F(\widetilde{x}) \;\; &\overset{(4.3.63)}{=}\;\; \left(A\left(z_m^{\xi} - z\right), A\left(z_m^{\xi} - z\right)\right)_2 + 2\left(A\widetilde{x} - b, A\left(z_m^{\xi} - z\right)\right)_2 \\[2mm]
&\overset{(4.3.64)}{=}\;\; \left(A\left(\xi z_m\right), A\left(\xi z_m\right)\right)_2 + 2\left(A\widetilde{x} - b, A\left(\xi z_m\right)\right)_2 \\[2mm]
&\;\;=\;\; \xi^2 \eta - 2\xi\varepsilon \\[2mm]
&\;\;=\;\; \xi(\xi\eta - 2\varepsilon) \\[2mm]
&\;\;<\;\; 0.
\end{aligned}
$$

This leads to a contradiction to the minimality property of $\widetilde{x}$. $\qquad\square$

The GMRES method is based on the orthonormal basis $\{v_1, \ldots, v_m\}$ of K_m computed by the Arnoldi algorithm. In principle, the determination of the orthonormal basis can also be carried out by other methods, such as a Householder transformation. Such an implementation is described in [75]. In the derivation, we start from the formulation of the method as a minimization problem. Let $V_m = (v_1 \ldots v_m) \in \mathbb{R}^{n \times m}$, then every $x_m \in x_0 + K_m$ can be written as

$$x_m = x_0 + V_m \alpha_m \text{ with } \alpha_m \in \mathbb{R}^m.$$

With

$$
\begin{aligned}
J_m : \mathbb{R}^m &\;\rightarrow\; \mathbb{R} \\[2mm]
\alpha &\;\mapsto\; \|b - A\left(x_0 + V_m\alpha\right)\|_2
\end{aligned}
\tag{4.3.65}
$$

the minimization according to (4.3.61) is equivalent to

$$\boldsymbol{\alpha}_m \;:=\; \arg\min_{\boldsymbol{\alpha}\in\mathbb{R}^m} J_m(\boldsymbol{\alpha}), \tag{4.3.66}$$

$$\boldsymbol{x}_m \;:=\; \boldsymbol{x}_0 + \boldsymbol{V}_m\boldsymbol{\alpha}_m. \tag{4.3.67}$$

Two central goals of the further investigation are now to find a way to calculate $\boldsymbol{\alpha}_m$ that is as simple as possible and to only have to compute $\boldsymbol{\alpha}_m$ when

$$\|\boldsymbol{b} - \boldsymbol{A}\boldsymbol{x}_m\|_2 \leq \varepsilon$$

holds for a given accuracy threshold $\varepsilon > 0$.

With the residual vector $\boldsymbol{r}_0 = \boldsymbol{b} - \boldsymbol{A}\boldsymbol{x}_0$, we write, using the first unit vector $\boldsymbol{e}_1 = (1,0,\ldots,0)^T \in \mathbb{R}^{m+1}$,

$$
\begin{aligned}
J_m(\boldsymbol{\alpha}) \;&=\; \|\boldsymbol{b} - \boldsymbol{A}\,(\boldsymbol{x}_0 + \boldsymbol{V}_m\boldsymbol{\alpha})\|_2 \\[2mm]
&=\; \|\boldsymbol{r}_0 - \boldsymbol{A}\boldsymbol{V}_m\boldsymbol{\alpha}\|_2 \\[2mm]
&=\; \big\|\,\|\boldsymbol{r}_0\|_2\,\boldsymbol{v}_1 - \boldsymbol{A}\boldsymbol{V}_m\boldsymbol{\alpha}\,\big\|_2 \\[2mm]
&\overset{\text{Theorem } 4.84}{=}\; \big\|\,\|\boldsymbol{r}_0\|_2\,\boldsymbol{v}_1 - \boldsymbol{V}_{m+1}\overline{\boldsymbol{H}}_m\boldsymbol{\alpha}\,\big\|_2 \\[2mm]
&=\; \big\|\boldsymbol{V}_{m+1}\left(\|\boldsymbol{r}_0\|_2\,\boldsymbol{e}_1 - \overline{\boldsymbol{H}}_m\boldsymbol{\alpha}\right)\big\|_2 .
\end{aligned}
\tag{4.3.68}
$$

Here, $\overline{\boldsymbol{H}}_m$ is the matrix

$$\overline{\boldsymbol{H}}_m = \begin{pmatrix} & \boldsymbol{H}_m & \\ 0 \ldots 0 & h_{m+1,m} \end{pmatrix} \in \mathbb{R}^{(m+1)\times m}$$

with an upper right Hessenberg matrix $\boldsymbol{H}_m$.

The purpose of the performed equivalent transformation of the minimization problem lies in the special structure of the $(m+1)\times m$ matrix $\overline{\boldsymbol{H}}_m$. The first formulation of the minimization problem has the disadvantage that, for a given accuracy, we must explicitly compute the sequence

$$\boldsymbol{x}_m = \arg\min_{\boldsymbol{x}\in\boldsymbol{x}_0+K_m} F(\boldsymbol{x}), \quad m = 1,\ldots$$

successively until the accuracy threshold is reached. As we will show below, the formulation of the problem given by (4.3.66) and (4.3.67) allows us to compute the minimal error

$$\min_{\boldsymbol{x}\in\boldsymbol{x}_0+K_m} F(\boldsymbol{x})$$

without having to explicitly determine $\boldsymbol{x}_m$. This gives us the possibility to only determine $\boldsymbol{x}_m$ when the expected error meets the required accuracy.

Lemma 4.91 *Suppose that the Arnoldi algorithm does not break down before the computation of $\boldsymbol{v}_{m+1}$ and that the matrices $\boldsymbol{G}_{i+1,i} \in \mathbb{R}^{(m+1)\times(m+1)}$ for $i = 1,\ldots,m$ are given by*

$$
\boldsymbol{G}_{i+1,i} := \begin{pmatrix} 1 & & & & & & & \\ & \ddots & & & & & & \\ & & 1 & & & & & \\ & & & c_i & s_i & & & \\ & & & -s_i & c_i & & & \\ & & & & & 1 & & \\ & & & & & & \ddots & \\ & & & & & & & 1 \end{pmatrix}
$$

where c_i and s_i are defined by

$$
c_i := \frac{a}{\sqrt{a^2+b^2}} \quad and \quad s_i := \frac{b}{\sqrt{a^2+b^2}} \tag{4.3.69}
$$

with

$$
a := \left(\boldsymbol{G}_{i,i-1} \cdot \ldots \cdot \boldsymbol{G}_{3,2} \cdot \boldsymbol{G}_{2,1} \overline{\boldsymbol{H}}_m\right)_{i,i}
$$

and

$$
b := \left(\boldsymbol{G}_{i,i-1} \cdot \ldots \cdot \boldsymbol{G}_{3,2} \cdot \boldsymbol{G}_{2,1} \overline{\boldsymbol{H}}_m\right)_{i+1,i}.
$$

Then

$$
\boldsymbol{Q}_m := \boldsymbol{G}_{m+1,m} \cdot \ldots \cdot \boldsymbol{G}_{2,1}
$$

is an orthogonal matrix for which

$$
\boldsymbol{Q}_m \overline{\boldsymbol{H}}_m = \overline{\boldsymbol{R}}_m
$$

with

$$
\overline{\boldsymbol{R}}_m = \begin{pmatrix} \overline{r}_{11} & \cdots & \cdots & \overline{r}_{1m} \\ 0 & \ddots & & \vdots \\ \vdots & \ddots & \ddots & \vdots \\ \vdots & & \ddots & \overline{r}_{mm} \\ 0 & \cdots & \cdots & 0 \end{pmatrix} =: \begin{pmatrix} \boldsymbol{R}_m \\ 0 \ldots 0 \end{pmatrix} \in \mathbb{R}^{(m+1)\times m} \tag{4.3.70}
$$

holds and $\boldsymbol{R}_m$ is invertible.

Proof:

If $\boldsymbol{v}_{m+1} \neq \boldsymbol{0}$ holds, then it follows that $h_{j+1,j} \neq 0$ for $j = 1,\ldots,m$, which means that all column vectors of the matrix $\overline{\boldsymbol{H}}_m$ are linearly independent and thus $\operatorname{rank}\overline{\boldsymbol{H}}_m = m$ holds. In the case where $\boldsymbol{v}_{m+1} = \boldsymbol{0}$, we have $h_{m+1,m} = 0$, so that with (4.3.35) $\boldsymbol{A}\boldsymbol{V}_m = \boldsymbol{V}_m\boldsymbol{H}_m$ and since $\operatorname{rank}(\boldsymbol{A}\boldsymbol{V}_m) = \operatorname{rank}(\boldsymbol{V}_m\boldsymbol{H}_m) = m$, the property $\operatorname{rank}\overline{\boldsymbol{H}}_m = \operatorname{rank}\boldsymbol{H}_m = m$ follows.

We formulate the proof by means of a complete induction on i.

For $i = 1$, we obtain directly from $\operatorname{rank}\overline{\boldsymbol{H}}_m = m$

$$
h_{11}^2 + h_{21}^2 \neq 0
$$

and thus the well-definedness of the rotation matrix $\boldsymbol{G}_{2,1}$. By straightforward calculation, it is easy to see that an orthogonal transformation of the matrix $\overline{\boldsymbol{H}}_m$ by $\boldsymbol{G}_{2,1}$ yields

$$
\boldsymbol{G}_{2,1}\overline{\boldsymbol{H}}_m =
\begin{pmatrix}
h_{11}^{(1)} & h_{12}^{(1)} & \cdots & \cdots & h_{1m}^{(1)} \\
0 & h_{22}^{(1)} & \cdots & \cdots & h_{2m}^{(1)} \\
0 & h_{32} & & & \vdots \\
0 & 0 & \ddots & & \vdots \\
\vdots & \vdots & & \ddots & \vdots \\
0 & 0 & \cdots & 0 & h_{m+1,m}
\end{pmatrix}.
$$

Now assume for $i < m$

$$
\boldsymbol{G}_{i+1,i} \cdot \ldots \cdot \boldsymbol{G}_{2,1}\overline{\boldsymbol{H}}_m =
\begin{pmatrix}
h_{11}^{(i)} & \cdots & \cdots & \cdots & \cdots & \cdots & h_{1m}^{(i)} \\
0 & \ddots & & & & & \vdots \\
\vdots & \ddots & \ddots & & & & \vdots \\
\vdots & & 0 & h_{i+1,i+1}^{(i)} & \cdots & \cdots & h_{i+1,m}^{(i)} \\
\vdots & & & h_{i+2,i+1} & \cdots & \cdots & h_{i+2,m} \\
\vdots & & & 0 & \ddots & & \vdots \\
\vdots & & \vdots & \vdots & \ddots & \ddots & \vdots \\
0 & \cdots & 0 & 0 & \cdots & 0 & h_{m+1,m}
\end{pmatrix}.
$$

Since all Givens rotations $\boldsymbol{G}_{j+1,j}$, $j = 1,\ldots,i$ are orthogonal rotation matrices, it follows that

$$
\operatorname{rank}\left(\boldsymbol{G}_{i+1,i} \cdot \ldots \cdot \boldsymbol{G}_{2,1}\overline{\boldsymbol{H}}_m\right) = m,
$$

which yields

$$
\left(h_{i+1,i+1}^{(i)}\right)^2 + \left(h_{i+2,i+1}\right)^2 \neq 0.
$$

Consequently, the matrix $\boldsymbol{G}_{i+2,i+1}$ is also well-defined and it follows that

$$
\boldsymbol{G}_{i+2,i+1} \cdot \ldots \cdot \boldsymbol{G}_{2,1}\overline{\boldsymbol{H}}_m =
\begin{pmatrix}
h_{11}^{(i+1)} & \cdots & \cdots & \cdots & \cdots & \cdots & h_{1m}^{(i+1)} \\
0 & \ddots & & & & & \vdots \\
\vdots & \ddots & \ddots & & & & \vdots \\
\vdots & & 0 & h_{i+2,i+2}^{(i+1)} & \cdots & \cdots & h_{i+2,m}^{(i+1)} \\
\vdots & & & h_{i+3,i+2} & \cdots & \cdots & h_{i+3,m} \\
\vdots & & & 0 & \ddots & & \vdots \\
\vdots & & \vdots & \vdots & \ddots & \ddots & \vdots \\
0 & \cdots & 0 & 0 & \cdots & 0 & h_{m+1,m}
\end{pmatrix}.
$$

The matrix $\boldsymbol{Q}_m = \boldsymbol{G}_{m+1,m} \cdot \ldots \cdot \boldsymbol{G}_{21} \in \mathbb{R}^{(m+1)\times(m+1)}$ is, by Lemma 2.27, orthogonal and it holds that

$$
\boldsymbol{Q}_m\overline{\boldsymbol{H}}_m = \overline{\boldsymbol{R}}_m \quad \text{with} \quad \overline{r}_{ij} = h_{ij}^{(m)}
$$

for $i = 1,\ldots,m$, $j = 1,\ldots,m$. From $\operatorname{rank}\overline{\boldsymbol{R}}_m = \operatorname{rank}(\boldsymbol{Q}_m\overline{\boldsymbol{H}}_m) = \operatorname{rank}\overline{\boldsymbol{H}}_m = m$ it finally follows that the matrix $\boldsymbol{R}_m$ is non-singular. $\qquad\square$

Let us define, with the matrix $\boldsymbol{Q}_m \in \mathbb{R}^{(m+1)\times(m+1)}$ given by Lemma 4.91, and using $\boldsymbol{e}_1 = (1,0,\ldots,0)^T \in \mathbb{R}^{m+1}$, the vector

$$\overline{\boldsymbol{g}}_m := \|\boldsymbol{r}_0\|_2 \boldsymbol{Q}_m \boldsymbol{e}_1 = \left(\gamma_1^{(m)},\ldots,\gamma_m^{(m)},\gamma_{m+1}\right)^T = \left(\boldsymbol{g}_m^T,\gamma_{m+1}\right)^T \in \mathbb{R}^{m+1}, \qquad (4.3.71)$$

then, using (4.3.68) in the case $\boldsymbol{v}_{m+1} \neq \boldsymbol{0}$, we obtain the representation

$$\begin{aligned}
\min_{\boldsymbol{\alpha}\in\mathbb{R}^m} J_m(\boldsymbol{\alpha}) \quad &= \quad \min_{\boldsymbol{\alpha}\in\mathbb{R}^m} \left\|\boldsymbol{V}_{m+1}\left(\|\boldsymbol{r}_0\|_2 \boldsymbol{e}_1 - \overline{\boldsymbol{H}}_m\boldsymbol{\alpha}\right)\right\|_2 \\[2mm]
&= \quad \min_{\boldsymbol{\alpha}\in\mathbb{R}^m} \left\|\,\|\boldsymbol{r}_0\|_2 \boldsymbol{e}_1 - \overline{\boldsymbol{H}}_m\boldsymbol{\alpha}\,\right\|_2 \\[2mm]
&= \quad \min_{\boldsymbol{\alpha}\in\mathbb{R}^m} \left\|\boldsymbol{Q}_m\left(\|\boldsymbol{r}_0\|_2 \boldsymbol{e}_1 - \overline{\boldsymbol{H}}_m\boldsymbol{\alpha}\right)\right\|_2 \\[2mm]
&\overset{\text{Lemma 4.91}}{=} \quad \min_{\boldsymbol{\alpha}\in\mathbb{R}^m} \left\|\overline{\boldsymbol{g}}_m - \overline{\boldsymbol{R}}_m\boldsymbol{\alpha}\right\|_2 \\[2mm]
&= \quad \min_{\boldsymbol{\alpha}\in\mathbb{R}^m} \sqrt{|\gamma_{m+1}|^2 + \|\boldsymbol{g}_m - \boldsymbol{R}_m\boldsymbol{\alpha}\|_2^2}. \qquad (4.3.72)
\end{aligned}$$

Due to the invertibility of the matrix $\boldsymbol{R}_m$, it follows that

$$\min_{\boldsymbol{\alpha}\in\mathbb{R}^m} J_m(\boldsymbol{\alpha}) = |\gamma_{m+1}|. \qquad (4.3.73)$$

If $\boldsymbol{v}_{m+1} = \boldsymbol{0}$, we obtain

$$\min_{\boldsymbol{\alpha}\in\mathbb{R}^m} J_m(\boldsymbol{\alpha}) = \min_{\boldsymbol{\alpha}\in\mathbb{R}^m} \left\|\boldsymbol{V}_m\left(\|\boldsymbol{r}_0\|_2 \boldsymbol{e}_1 - \boldsymbol{H}_m\boldsymbol{\alpha}\right)\right\|_2 = \min_{\boldsymbol{\alpha}\in\mathbb{R}^m} \|\boldsymbol{g}_m - \boldsymbol{R}_m\boldsymbol{\alpha}\|_2 = 0.$$

Remark:
For the error quantities $\gamma_1,\ldots,\gamma_{m+1}$ given by (4.3.71), we have

$$\|\boldsymbol{r}_j\|_2 = |\gamma_{j+1}| \leq |\gamma_j| = \|\boldsymbol{r}_{j-1}\|_2$$

for $j = 1,\ldots,m$.

In summary, we obtain the GMRES algorithm in the following form. A possible implementation in MATLAB for this method can also be found in Appendix A. It should be noted here that the subsequent algorithmic description uses the basic form of the Arnoldi algorithm, while the MATLAB implementation uses the more stable and, in practice, preferable modified form of the Arnoldi process.

GMRES Algorithm —

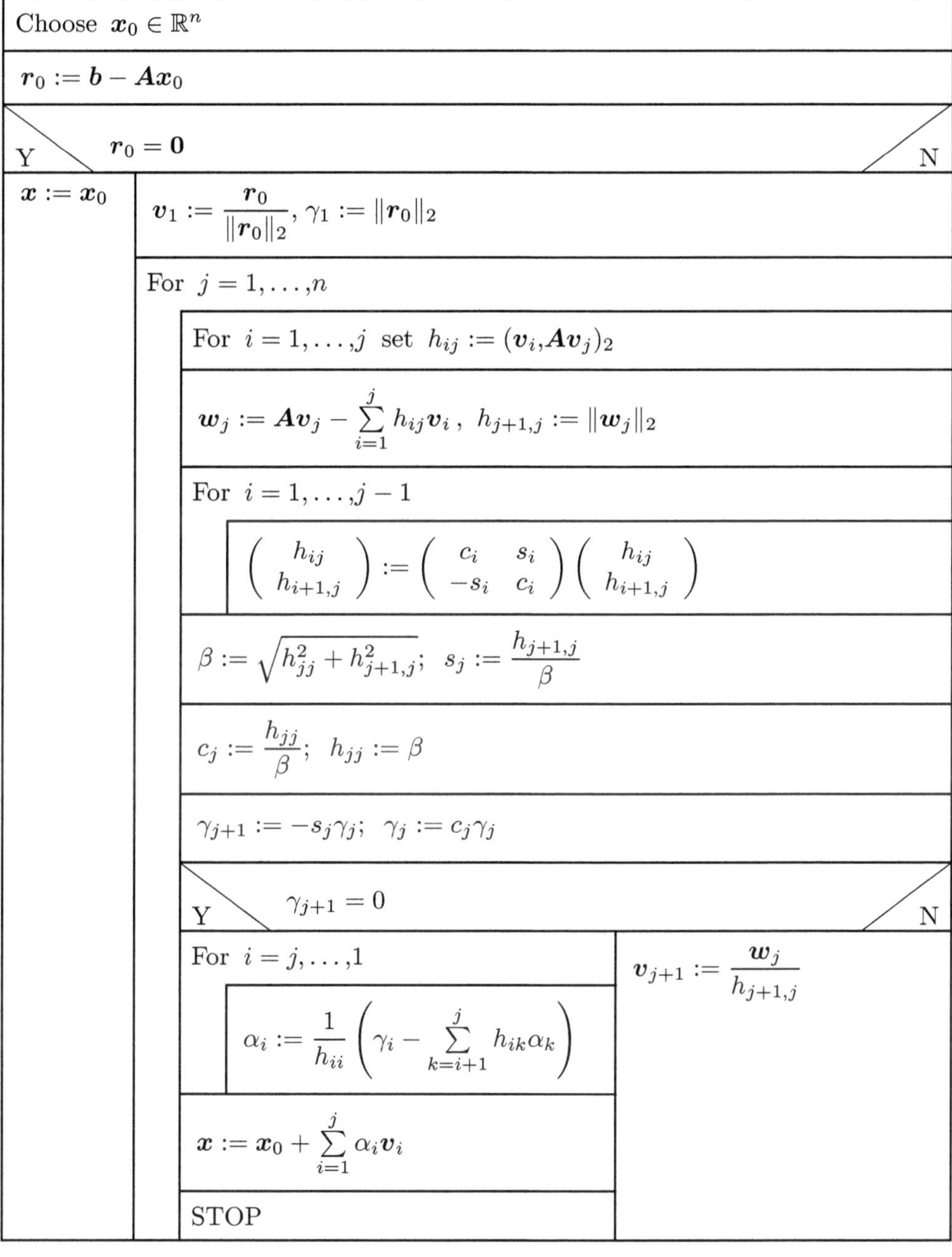

If we analyze the method, we see that the only possible breakdown is given by the vanishing of $h_{j+1,j}$. An essential feature of the method is that GMRES does not break down at this point, but instead provides the exact solution. A mathematical description of this property is given by the following theorem.

Theorem 4.92 *Let $\boldsymbol{A} \in \mathbb{R}^{n \times n}$ be a non-singular matrix, and let $h_{j+1,j}$ and $\boldsymbol{w}_j$ be given by the Arnoldi algorithm, with $j < n$. Then the following statements are equivalent:*

(1) For the sequence of Krylov subspaces, we have

$$K_1 \subset K_2 \subset \ldots \subset K_j = K_{j+1} = \ldots.$$

(2) The GMRES method yields the exact solution at the j-th step.

(3) $\boldsymbol{w}_j = \boldsymbol{0} \in \mathbb{R}^n$.

(4) $h_{j+1,j} = 0$.

Proof:
"(1) $\Leftrightarrow$ (3)"
We obtain the statement directly through the following equivalent reformulations:

$$K_{j+1} = K_j$$
$$\Leftrightarrow \quad \mathrm{span}\left\{\boldsymbol{r}_0, \ldots, \boldsymbol{A}^j \boldsymbol{r}_0\right\} = \mathrm{span}\left\{\boldsymbol{r}_0, \ldots, \boldsymbol{A}^{j-1}\boldsymbol{r}_0\right\}$$
$$\Leftrightarrow \quad \mathrm{span}\left\{\boldsymbol{v}_1, \ldots, \boldsymbol{v}_j, \boldsymbol{w}_j\right\} = \mathrm{span}\left\{\boldsymbol{v}_1, \ldots, \boldsymbol{v}_j\right\}$$
$$\Leftrightarrow \quad \boldsymbol{w}_j \in \mathrm{span}\left\{\boldsymbol{v}_1, \ldots, \boldsymbol{v}_j\right\}$$
$$\Leftrightarrow \quad \boldsymbol{w}_j = \boldsymbol{0} \in \mathbb{R}^n, \text{ since } (\boldsymbol{v}_i, \boldsymbol{w}_j)_2 = 0 \text{ for } i = 1, \ldots, j \text{ holds.}$$

"(3) $\Leftrightarrow$ (4)"
The proof follows immediately from $h_{j+1,j} = \|\boldsymbol{w}_j\|_2$.

"(2) $\Rightarrow$ (4)"
Let $\boldsymbol{x}_j$ be the exact solution of $\boldsymbol{A}\boldsymbol{x} = \boldsymbol{b}$, then from (4.3.73) it follows that $\gamma_{j+1} = 0$. Without loss of generality, we assume that $\boldsymbol{x}_{j-1} \neq \boldsymbol{x}_j$ and thus $\gamma_j \neq 0$ holds. We use the equation

$$0 = \gamma_{j+1} \;=\; \boldsymbol{e}_{j+1}^T \boldsymbol{G}_{j+1,j} \begin{pmatrix} \boldsymbol{Q}_{j-1} & \boldsymbol{0} \\ \boldsymbol{0}^T & 1 \end{pmatrix} \|\boldsymbol{r}_0\|_2 \, \boldsymbol{e}_1$$

$$= \; \boldsymbol{e}_{j+1}^T \begin{pmatrix} 1 & & & & \\ & \ddots & & & \\ & & 1 & & \\ & & & c_j & s_j \\ & & & -s_j & c_j \end{pmatrix} \begin{pmatrix} \gamma_1^{(j-1)} \\ \vdots \\ \gamma_{j-1}^{(j-1)} \\ \gamma_j \\ 0 \end{pmatrix}$$

$$= \; -s_j \gamma_j$$

and with $\gamma_j \neq 0$ we obtain the statement $s_j = 0$. Using equation (4.3.69) this yields $h_{j+1,j} = 0$.

"(4) $\Rightarrow$ (2)"

Since $h_{j+1,j} = 0$ was computed, the Arnoldi algorithm did not terminate before the computation of $v_j \neq 0$. By Lemma 4.91 we thus obtain the existence of an orthogonal matrix

$$Q_j := G_{j+1,j} \cdot \ldots \cdot G_{2,1} \in \mathbb{R}^{(j+1)\times(j+1)}$$

with

$$\overline{R}_j = \begin{pmatrix} R_j \\ 0 \ldots 0 \end{pmatrix} = Q_j \overline{H}_j, \tag{4.3.74}$$

where R_j represents a non-singular upper triangular matrix. The condition $h_{j+1,j} = 0$ yields, together with (4.3.69), $s_j = 0$, so that

$$G_{j+1,j} = \begin{pmatrix} 1 & & & & \\ & \ddots & & & \\ & & 1 & & \\ & & & c_j & \\ & & & & c_j \end{pmatrix} \tag{4.3.75}$$

follows. Furthermore, the vectors $v_1,\ldots,v_j$ determined in the Arnoldi algorithm form an orthonormal basis of K_j.

Let $V_{j+1} = (V_j, \tilde{v}_{j+1}) \in \mathbb{R}^{n\times(j+1)}$ with a normalized vector $\tilde{v}_{j+1} \in \mathbb{R}^n$ orthogonal to $v_1,\ldots,v_j$, then with $h_{j+1,j} = 0$ it follows from Theorem 4.83 that

$$AV_j = V_j H_j = V_{j+1}\overline{H}_j. \tag{4.3.76}$$

Let $\overline{g}_j$ be given according to (4.3.71) in the form

$$\overline{g}_j = \|r_0\|_2\, Q_j e_1 = \left(g_j^T, \gamma_{j+1}\right)^T \in \mathbb{R}^{j+1}$$

then we define

$$\alpha_j := R_j^{-1} g_j \in \mathbb{R}^j$$

and

$$x_j := x_0 + V_j \alpha_j \in \mathbb{R}^n. \tag{4.3.77}$$

With this, using

$$\begin{aligned}
\overline{g}_j &= \|r_0\|_2\, G_{j+1,j} \begin{pmatrix} Q_{j-1} & 0 \\ 0^T & 1 \end{pmatrix} e_1 \\[2mm]
&= G_{j+1,j} \begin{pmatrix} g_{j-1} \\ \gamma_j \\ 0 \end{pmatrix} \\[2mm]
&\overset{(4.3.75)}{=} \begin{pmatrix} g_j \\ 0 \end{pmatrix}
\end{aligned}$$

we obtain the equation

$$
\begin{aligned}
\|\boldsymbol{b} - \boldsymbol{A}\boldsymbol{x}_j\|_2 &= \|\boldsymbol{r}_0 - \boldsymbol{A}\boldsymbol{V}_j\boldsymbol{\alpha}_j\|_2 \\
&= \left\|\boldsymbol{V}_{j+1}\left(\|\boldsymbol{r}_0\|_2\, \boldsymbol{e}_1 - \overline{\boldsymbol{H}}_j\boldsymbol{\alpha}_j\right)\right\|_2 \\
&= \left\|\boldsymbol{Q}_j\left(\|\boldsymbol{r}_0\|_2\, \boldsymbol{e}_1 - \overline{\boldsymbol{H}}_j\boldsymbol{\alpha}_j\right)\right\|_2 \\
&= \left\|\begin{pmatrix} \boldsymbol{g}_j \\ 0 \end{pmatrix} - \begin{pmatrix} \boldsymbol{R}_j \\ 0 \dots 0 \end{pmatrix}\boldsymbol{\alpha}_j\right\|_2 \\
&= 0.
\end{aligned}
$$

Thus, $\boldsymbol{x}_j$ is the exact solution of the equation $\boldsymbol{A}\boldsymbol{x} = \boldsymbol{b}$. $\qquad\square$

The GMRES method has two fundamental disadvantages. The first issue is the computational effort required to determine the orthonormal basis, which increases with the dimension of the Krylov subspace. Furthermore, there is a high memory requirement for the basis vectors. In the worst case, even for a sparse matrix $\boldsymbol{A} \in \mathbb{R}^{n\times n}$, a full matrix $\boldsymbol{V}_n \in \mathbb{R}^{n\times n}$ must be stored. For practical problems, the memory requirement often exceeds the available resources. For this reason, a GMRES method with restart is often used, in which the maximum Krylov subspace dimension is limited. If the residual $\|\boldsymbol{r}_m\|_2$ does not reach a prescribed accuracy $\|\boldsymbol{r}_m\|_2 \le \varepsilon > 0$ at this upper limit, the currently optimal approximate solution $\boldsymbol{x}_m$ is nevertheless determined and used as the starting vector for a restart. The following diagram describes a GMRES version with restart and a maximum Krylov subspace dimension of m. The method is often referred to as Restarted GMRES(m). The theoretically possible interpretation of the GMRES method as a direct method is lost in this formulation, but the residual still decreases monotonically.

In the context of an implicit finite volume method for the simulation of inviscid and viscous airflows, it was found that for sparse systems with about 10,000 unknowns, a maximum Krylov subspace dimension in the range of 10 to 15 was sufficient to achieve an accuracy of $\varepsilon = 10^{-6}$ without restart, provided that suitable preconditioning was implemented. It was shown that the convergence behavior of the GMRES method depends very strongly on the chosen preconditioner. With an incomplete LU decomposition (see Section 5.4), the method proved to be robust, whereas a simple scaling (see Section 5.1), analogous to the method without preconditioning, often led to unacceptable computation times. A detailed discussion of the results mentioned can be found in [50].

Replacing the Arnoldi method with an incomplete orthogonalization procedure within the GMRES method leads to the so-called Quasi-GMRES algorithm (QGMRES). This approach can also be used in a direct version, the DQGMRES. However, both methods are not as widely used and are described in detail in [63].

GMRES(m) algorithm with a limited number of restarts —

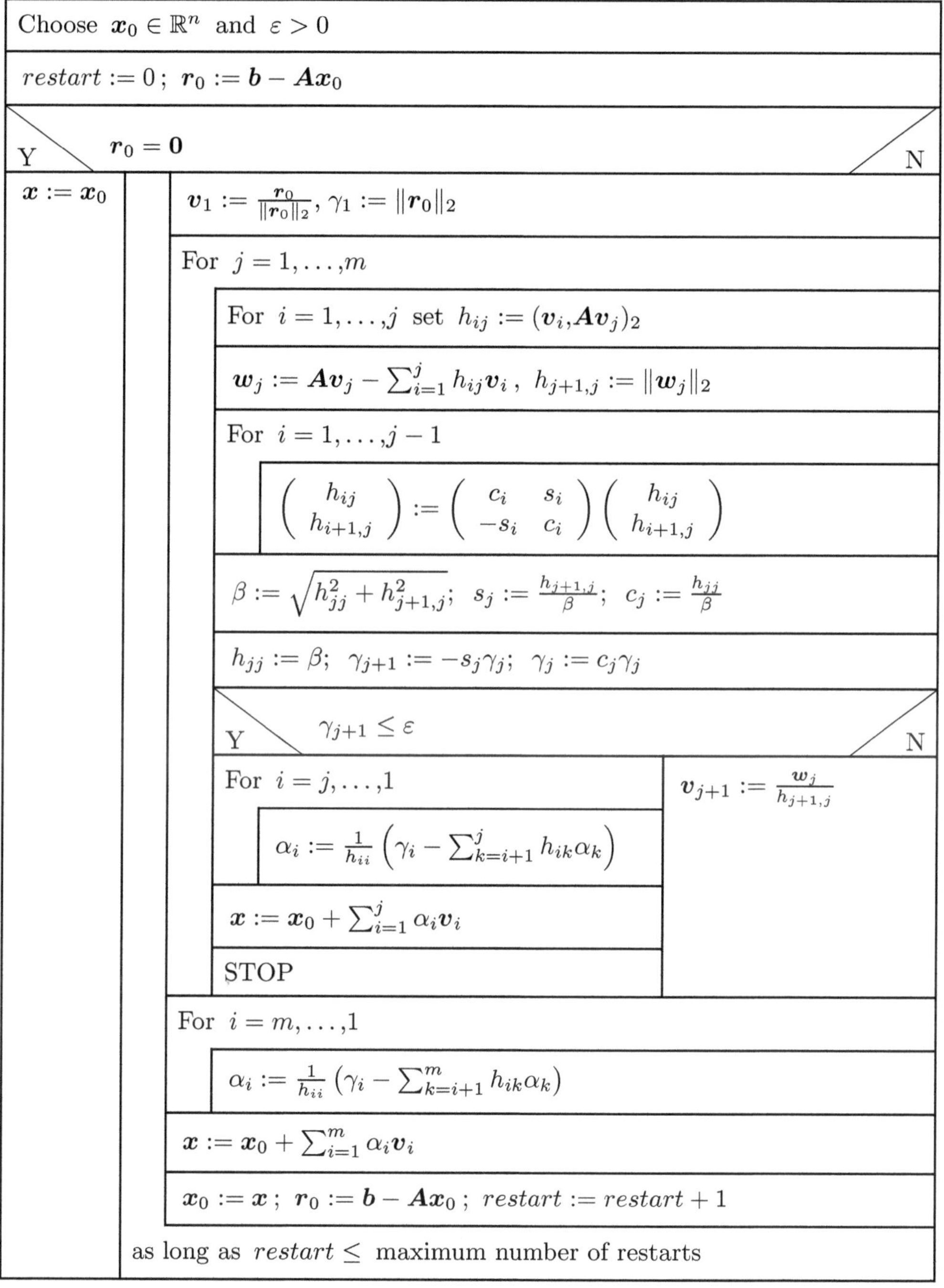

Using $\boldsymbol{W}_m = (\boldsymbol{A}\boldsymbol{v}_1 \ldots \boldsymbol{A}\boldsymbol{v}_m)$, the GMRES method satisfies the condition of Theorem 4.64. Consequently, we have

$$\|\boldsymbol{r}_m\| \le \|\boldsymbol{P}_m\| \min_{p \in \mathcal{P}_m^1} \|p(\boldsymbol{A})\boldsymbol{r}_0\|$$

with respect to any norm, where $\boldsymbol{P}_m$ denotes the projection given in (4.3.4). For projections, it always holds that $\boldsymbol{P}_m^2 = \boldsymbol{P}_m$ and thus $\|\boldsymbol{P}_m\| \ge 1$, which, in the special case of the Euclidean norm, already leads to an improvement of the statement for the approximation $\boldsymbol{x}_m$ determined by the GMRES method from the definition of the function F according to (4.3.60), namely

$$\|\boldsymbol{r}_m\|_2 = \min_{p \in \mathcal{P}_m^1} \|p(\boldsymbol{A})\boldsymbol{r}_0\|_2. \tag{4.3.78}$$

Further estimates of the residual can be obtained under additional assumptions on the matrix, as will be shown in the following statements.

For a given symmetric matrix $\boldsymbol{B} \in \mathbb{R}^{n \times n}$, let $\lambda_{\min}(\boldsymbol{B})$ and $\lambda_{\max}(\boldsymbol{B})$ always denote the smallest and largest eigenvalue of $\boldsymbol{B}$ in absolute value, respectively.

Theorem 4.93 *Let $\boldsymbol{A} \in \mathbb{R}^{n \times n}$ be positive definite and $\boldsymbol{r}_m$ the m-th residual vector determined by the GMRES method. Then the GMRES method converges, and*

$$\|\boldsymbol{r}_m\|_2 \le \left(1 - \frac{\lambda_{\min}^2\left(\frac{\boldsymbol{A}^T + \boldsymbol{A}}{2}\right)}{\lambda_{\max}\left(\boldsymbol{A}^T \boldsymbol{A}\right)}\right)^{\frac{m}{2}} \|\boldsymbol{r}_0\|_2. \tag{4.3.79}$$

Proof:
From (4.3.78) we obtain

$$\|\boldsymbol{r}_m\|_2 \le \min_{p \in \mathcal{P}_m^1} \|p(\boldsymbol{A})\|_2 \|\boldsymbol{r}_0\|_2. \tag{4.3.80}$$

We will now prove the estimate (4.3.79) by explicitly specifying a polynomial $p \in \mathcal{P}_m^1$.

For $\alpha > 0$ we define $p_1(\boldsymbol{A}) = \boldsymbol{I} - \alpha \boldsymbol{A} \in \mathcal{P}_1^1$, so that $(p_1(\boldsymbol{A}))^m \in \mathcal{P}_m^1$ and

$$\min_{p \in \mathcal{P}_m^1} \|p(\boldsymbol{A})\|_2 \le \|(p_1(\boldsymbol{A}))^m\|_2 \le \|p_1(\boldsymbol{A})\|_2^m \tag{4.3.81}$$

holds. From

$$\|p_1(\boldsymbol{A})\|_2^2 \overset{(2.2.2)}{=} \sup_{\|\boldsymbol{x}\|_2 = 1} \|(\boldsymbol{I} - \alpha \boldsymbol{A})\boldsymbol{x}\|_2^2 = \sup_{\boldsymbol{x} \ne \boldsymbol{0}} \frac{\|(\boldsymbol{I} - \alpha \boldsymbol{A})\boldsymbol{x}\|_2^2}{\|\boldsymbol{x}\|_2^2}$$

$$= \sup_{\boldsymbol{x} \ne \boldsymbol{0}} \left\{ 1 - 2\alpha \frac{(\boldsymbol{x}, \boldsymbol{A}\boldsymbol{x})_2}{(\boldsymbol{x}, \boldsymbol{x})_2} + \alpha^2 \frac{(\boldsymbol{A}\boldsymbol{x}, \boldsymbol{A}\boldsymbol{x})_2}{(\boldsymbol{x}, \boldsymbol{x})_2} \right\}$$

for $\boldsymbol{x} \ne \boldsymbol{0}$, using

$$0 < \frac{(\boldsymbol{A}\boldsymbol{x}, \boldsymbol{A}\boldsymbol{x})_2}{(\boldsymbol{x}, \boldsymbol{x})_2} = \frac{(\boldsymbol{x}, \boldsymbol{A}^T \boldsymbol{A}\boldsymbol{x})_2}{(\boldsymbol{x}, \boldsymbol{x})_2} \le \lambda_{\max}\left(\boldsymbol{A}^T \boldsymbol{A}\right)$$

and

$$\frac{(\boldsymbol{x}, \boldsymbol{A}\boldsymbol{x})_2}{(\boldsymbol{x}, \boldsymbol{x})_2} = \frac{\left(\boldsymbol{x}, \frac{\boldsymbol{A}^T + \boldsymbol{A}}{2}\boldsymbol{x}\right)_2}{(\boldsymbol{x}, \boldsymbol{x})_2} \ge \lambda_{\min}\left(\frac{\boldsymbol{A}^T + \boldsymbol{A}}{2}\right) \overset{\boldsymbol{A} \text{ pos. def.}}{>} 0$$

we obtain the inequality

$$\|p_1(\boldsymbol{A})\|_2^2 \le 1 - 2\alpha\lambda_{\min}\left(\frac{\boldsymbol{A}^T + \boldsymbol{A}}{2}\right) + \alpha^2\lambda_{\max}\left(\boldsymbol{A}^T\boldsymbol{A}\right). \tag{4.3.82}$$

The minimum of the right-hand side with respect to α is attained at the point

$$\alpha_{\min} = \frac{\lambda_{\min}\left(\frac{\boldsymbol{A}^T+\boldsymbol{A}}{2}\right)}{\lambda_{\max}\left(\boldsymbol{A}^T\boldsymbol{A}\right)} > 0.$$

By substituting the value $\alpha_{\min}$ into the estimate (4.3.82), we obtain for $p_1(\boldsymbol{A}) = \boldsymbol{I} - \alpha_{\min}\boldsymbol{A}$ the inequality

$$0 \le \|p_1(\boldsymbol{A})\|_2^2 \le 1 - \frac{\lambda_{\min}^2\left(\frac{\boldsymbol{A}^T+\boldsymbol{A}}{2}\right)}{\lambda_{\max}\left(\boldsymbol{A}^T\boldsymbol{A}\right)} < 1. \tag{4.3.83}$$

Thus, the GMRES method converges, and we have

$$\|\boldsymbol{r}_m\|_2 \overset{(4.3.80)}{\le} \min_{p\in\mathcal{P}_m^1}\|p(\boldsymbol{A})\|_2\|\boldsymbol{r}_0\|_2$$

$$\overset{(4.3.81)}{\le} \|p_1(\boldsymbol{A})\|_2^m\|\boldsymbol{r}_0\|_2$$

$$\overset{(4.3.83)}{\le} \left(1 - \frac{\lambda_{\min}^2\left(\frac{\boldsymbol{A}^T+\boldsymbol{A}}{2}\right)}{\lambda_{\max}\left(\boldsymbol{A}^T\boldsymbol{A}\right)}\right)^{\frac{m}{2}}\|\boldsymbol{r}_0\|_2.$$

$$\square$$

For a symmetric matrix, this also yields an estimate using the condition number of the matrix $\boldsymbol{A}$, which illustrates the advantage of a smaller condition number.

Corollary 4.94 *Let $\boldsymbol{A} \in \mathbb{R}^{n\times n}$ be positive definite and symmetric. Furthermore, let $\boldsymbol{r}_m$ be the m-th residual vector determined in the GMRES method, then the GMRES method converges and*

$$\|\boldsymbol{r}_m\|_2 \le \left(\frac{\mathrm{cond}_2^2(\boldsymbol{A}) - 1}{\mathrm{cond}_2^2(\boldsymbol{A})}\right)^{\frac{m}{2}}\|\boldsymbol{r}_0\|_2.$$

To conclude this section, we turn again to the GMRES(m) method. While Theorem 4.92 guarantees that the GMRES method yields the exact solution after at most n iterations, such a statement cannot be made for the GMRES(m) method ($m < n$) due to the restarts. Although the restarted version also exhibits a monotonic decrease of the residual, it is to be feared that a constant residual behavior may occur. We are therefore interested in convergence statements for the GMRES(m) method, which, from a practical point of view, should also require the smallest possible lower bound for the maximum Krylov subspace dimension m. Such a problem can be answered positively in special cases by the following theorems.

Theorem 4.95 *Let $A \in \mathbb{R}^{n \times n}$ be positive definite, then the GMRES(m) method converges for $m \geq 1$.*

Proof:
From the inequality (4.3.83) and Theorem 4.93 we always have

$$\|r_1\|_2 \leq \gamma \|r_0\|_2 \tag{4.3.84}$$

with $\gamma < 1$. Since γ is independent of the initial vector x_0 , the claim follows from (4.3.84).

$\square$

Theorem 4.96 *Let $A \in \mathbb{R}^{n \times n}$ be invertible and symmetric, then the GMRES(m) method converges for $m \geq 2$.*

Proof:
Starting from the estimate (4.3.80), we choose $p(A) = I - \alpha A^2$ with $\alpha > 0$. Analogous to the proof of Theorem 4.93, using the symmetry of the matrix A , we obtain the inequality

$$\|p(A)\|_2^2 \leq 1 - 2\alpha \lambda_{\min}\left(A^2\right) + \alpha^2 \lambda_{\max}\left(A^4\right) .$$

For the optimal parameter, we have

$$\alpha_{\min} = \frac{\lambda_{\min}\left(A^2\right)}{\lambda_{\max}\left(A^4\right)},$$

which yields, with $p_1(A) = I - \alpha_{\min} A^2$, the estimate

$$\|r_2\|_2 \leq \underbrace{\left(1 - \frac{\lambda_{\min}^2\left(A^2\right)}{\lambda_{\max}\left(A^4\right)}\right)^{\frac{1}{2}}}_{< 1} \|r_0\|_2$$

as claimed.

$\square$

Further convergence results for the GMRES method as well as for the CGN and CGS methods are derived in [54].

If A is a symmetric matrix, then the Arnoldi algorithm degenerates to the Lanczos method (see Section 4.3.2.2) and we obtain a tridiagonal matrix H_m . Due to the special structure of the matrix, the GMRES method can be written in the form of a three-term recurrence, which eliminates the aforementioned storage problem and makes it unnecessary to consider a restarted version. This special form of the GMRES algorithm was already introduced in 1975 by Paige and Saunders [56] and is called MINRES (Minimal Residual). A derivation of this method is presented in [34].

4.3.2.5 The BiCG Method

The BiCG method (Bi-Conjugate Gradient) was originally proposed as early as 1952 by Lanczos [47]. However, it only gained widespread use in the formulation presented by Fletcher [27]. The method is a Krylov subspace method based on the Bi-Lanczos algorithm, where orthogonality is ensured by a Petrov-Galerkin condition $L_m = K_m^T = \text{span}\{r_0, A^T r_0, \ldots, (A^T)^{m-1} r_0\}$. Compared to the GMRES method, the algorithm is characterized by a significantly lower memory requirement. However, the method has two major disadvantages. Inherited from the underlying Bi-Lanczos algorithm, multiplications with the matrix transposed to A are required, and as described in Section 4.3.2.3, the method can break down for a non-singular matrix without having computed the exact solution.

Theorem 4.97 *Assuming that the Bi-Lanczos algorithm does not break down before computing $v_m \neq 0$, and the tridiagonal matrix T_m defined in Theorem 4.88 is invertible, then*

$$x_m = x_0 + V_m T_m^{-1} \left(\|r_0\|_2 e_1 \right) \in x_0 + K_m \tag{4.3.85}$$

with $e_1 = (1, 0, \ldots, 0)^T \in \mathbb{R}^m$ is the uniquely determined solution of the Krylov subspace method based on the Petrov-Galerkin condition

$$b - A x_m \perp K_m^T \tag{4.3.86}$$

Proof:
Since the vectors of the matrix V_m form a basis of K_m according to Theorem 4.87, it follows directly that

$$x_m \in x_0 + K_m.$$

According to Theorem 4.87 and Theorem 4.88, we also have

$$T_m = W_m^T A V_m \tag{4.3.87}$$

as well as

$$W_m^T V_m = V_m^T W_m = I.$$

Thus, for $i = 1, \ldots, m$ we obtain

$$
\begin{aligned}
(b - A x_m, w_i)_2 &= \left(r_0 - A V_m T_m^{-1} \left(\|r_0\|_2 e_1 \right), w_i \right)_2 \\[2mm]
&= \|r_0\|_2 \left[\underbrace{\left(\frac{r_0}{\|r_0\|_2}, w_i \right)_2}_{= \delta_{i1}} - \left(A V_m T_m^{-1} e_1, w_i \right)_2 \right] \\[2mm]
&= \|r_0\|_2 \left[\delta_{i1} - \underbrace{w_i^T A V_m T_m^{-1}}_{\overset{(4.3.87)}{=} e_i^T} e_1 \right] \\[2mm]
&= 0,
\end{aligned}
$$

which implies $b - A x_m \perp w_i$ for $i = 1, \ldots, m$, and therefore

$$b - Ax_m \perp K_m^T = \mathrm{span}\{w_1, \ldots, w_m\}$$

holds.

We now turn to the proof of uniqueness of the solution. Let x_m, $\tilde{x}_m \in x_0 + K_m$ be two vectors that satisfy condition (4.3.86), then it follows that

$$x_m = x_0 + V_m \alpha_m \quad \text{and} \quad \tilde{x}_m = x_0 + V_m \tilde{\alpha}_m$$

with $\alpha_m, \tilde{\alpha}_m \in \mathbb{R}^m$. From the orthogonality condition, we obtain

$$\begin{aligned}
0 &= W_m^T \left(b - A\tilde{x}_m - b + Ax_m \right) \\
&= W_m^T A \left(x_m - \tilde{x}_m \right) \\
&= W_m^T A V_m \left(\alpha_m - \tilde{\alpha}_m \right) \\
&= T_m \left(\alpha_m - \tilde{\alpha}_m \right).
\end{aligned}$$

By assumption, $T_m \in \mathbb{R}^{m \times m}$ is non-singular, and thus we obtain $\alpha_m = \tilde{\alpha}_m$ and consequently $x_m = \tilde{x}_m$. $\qquad\square$

To derive the method, we assume that the matrix T_m can be decomposed as

$$T_m = \underbrace{\begin{pmatrix} 1 & & & \\ \ell_{21} & \ddots & & \\ & \ddots & \ddots & \\ & & \ell_{m,m-1} & 1 \end{pmatrix}}_{=:\, L_m} \underbrace{\begin{pmatrix} u_{11} & h_{12} & & \\ & \ddots & \ddots & \\ & & \ddots & h_{m-1,m} \\ & & & u_{mm} \end{pmatrix}}_{=:\, U_m}.$$

With

$$\tilde{P}_m = \left(\tilde{p}_0 \ldots \tilde{p}_{m-1} \right) := V_m U_m^{-1} \in \mathbb{R}^{n \times m}$$

and

$$z_m = (\xi_1, \ldots, \xi_m)^T := L_m^{-1} \left(\|r_0\|_2 e_1 \right) \in \mathbb{R}^m$$

it follows that

$$\begin{aligned}
x_m &= x_0 + V_m T_m^{-1} \left(\|r_0\|_2 e_1 \right) = x_0 + V_m U_m^{-1} L_m^{-1} \left(\|r_0\|_2 e_1 \right) \\
&= x_0 + \tilde{P}_m z_m
\end{aligned} \tag{4.3.88}$$

with the first unit vector $e_1 \in \mathbb{R}^m$. Since $V_m = \tilde{P}_m U_m$, and taking into account the conventions $\tilde{p}_{-1} := 0$ and $h_{0,1} := 0$, for the m-th column vector of the matrix V_m we have the representation $v_m = h_{m-1,m} \tilde{p}_{m-2} + u_{mm} \tilde{p}_{m-1}$ and thus

$$\tilde{p}_{m-1} = \frac{1}{u_{mm}} \left(v_m - h_{m-1,m} \tilde{p}_{m-2} \right). \tag{4.3.89}$$

Furthermore, from $L_m z_m = \|r_0\|_2 e_1$ for $m > 1$ we obtain the equation

$$\xi_m + \ell_{m,m-1} \xi_{m-1} = 0. \tag{4.3.90}$$

Moreover, $\xi_1 = \|r_0\|_2$, and from (4.3.88) it follows that

$$x_m = x_0 + \widetilde{P}_m z_m = x_0 + \widetilde{P}_{m-1} z_{m-1} + \tilde{p}_{m-1}\xi_m$$

$$= x_{m-1} + \xi_m \tilde{p}_{m-1} \tag{4.3.91}$$

and

$$r_m = b - A x_m = r_{m-1} - \xi_m A \tilde{p}_{m-1}. \tag{4.3.92}$$

For the residual vector we thus obtain the representation

$$
\begin{aligned}
r_m \quad &= \quad b - A x_m \\[2mm]
&\overset{(4.3.85)}{=} \quad r_0 - A V_m T_m^{-1}\left(\|r_0\|_2 e_1\right) \\[2mm]
&\overset{\text{Theorem 4.88}}{=} \quad r_0 - \left(V_m T_m + h_{m+1,m} v_{m+1} e_m^T\right) T_m^{-1}\left(\|r_0\|_2 e_1\right) \\[2mm]
&= \quad \underbrace{r_0 - V_m \|r_0\|_2 e_1}_{=\,0} - h_{m+1,m} v_{m+1} \underbrace{e_m^T T_m^{-1}\left(\|r_0\|_2 e_1\right)}_{\in\,\mathbb{R}} \\[2mm]
&= \quad \sigma_m v_{m+1} \tag{4.3.93}
\end{aligned}
$$

with $\sigma_m \in \mathbb{R}$. The vector $\tilde{p}_{m-1}$ can thus, using equation (4.3.89), be written as

$$\tilde{p}_{m-1} = \frac{1}{u_{mm}}\left(\frac{1}{\sigma_{m-1}} r_{m-1} - h_{m-1,m}\tilde{p}_{m-2}\right).$$

Analogously, for the transposed problem we obtain

$$b - A^T x_m^* \perp K_m$$

with $x_m^* \in x_0^* + K_m^T$ and $x_0^* = A^{-T} A x_0$. Assuming invertibility of the matrix T_m, the uniquely determined solution takes the form

$$x_m^* = x_0^* + W_m T_m^{-T}\left(\|r_0\|_2 e_1\right)$$

with W_m according to Theorem 4.88. For the residual vector $r_m^* = b - A^T x_m^*$ we obtain, according to equation (4.3.93), the representation

$$r_m^* = \sigma_m^* w_{m+1}$$

with $\sigma_m^* \in \mathbb{R}$. If we define

$$\widetilde{P}_m^* = \left(\tilde{p}_0^* \cdots \tilde{p}_{m-1}^*\right) := W_m L_m^{-T},$$

then it holds that

$$\tilde{p}_{m-1}^* = w_m - \ell_{m,m-1}\tilde{p}_{m-2}^* = \sigma_{m-1}^* r_{m-1}^* - \ell_{m,m-1}\tilde{p}_{m-2}^*,$$

where, analogous to the determination of the vectors $\widetilde{\boldsymbol{p}}_{m-1}$ according to (4.3.89), the convention $\widetilde{\boldsymbol{p}}^*_{-1} := \boldsymbol{0}$ and $\ell_{1,0} := 0$ must be taken into account. The approximate solution can be written in the form

$$\boldsymbol{x}^*_m = \boldsymbol{x}^*_0 + \widetilde{\boldsymbol{P}}^*_m \boldsymbol{z}^*_m$$

with

$$\boldsymbol{z}^*_m = (\xi^*_1, \dots, \xi^*_m)^T := \boldsymbol{U}_m^{-T} \left(\|\boldsymbol{r}_0\|_2 \boldsymbol{e}_1 \right).$$

Analogous to (4.3.90), for $m > 1$ it holds that

$$\xi^*_m = -\frac{h_{m-1,m}}{u_{mm}} \xi^*_{m-1}$$

and

$$\xi_1 = -\frac{\|\boldsymbol{r}_0\|_2}{u_{11}}.$$

Therefore, we obtain

$$\boldsymbol{x}^*_m = \boldsymbol{x}^*_{m-1} + \xi^*_m \tilde{\boldsymbol{p}}^*_{m-1}$$

as well as

$$\boldsymbol{r}^*_m = \boldsymbol{r}^*_{m-1} - \xi^*_m \boldsymbol{A}^T \tilde{\boldsymbol{p}}^*_{m-1}.$$

If we scale the search vectors according to

$$\boldsymbol{p}_{m-1} = \frac{\xi_m}{\alpha_{m-1}} \tilde{\boldsymbol{p}}_{m-1} \quad \text{and} \quad \boldsymbol{p}^*_{m-1} = \frac{\xi^*_m}{\alpha_{m-1}} \tilde{\boldsymbol{p}}^*_{m-1},$$

then it follows that

$$\begin{aligned}
\boldsymbol{x}_m &= \boldsymbol{x}_{m-1} + \alpha_{m-1}\boldsymbol{p}_{m-1}, \\
\boldsymbol{x}^*_m &= \boldsymbol{x}^*_{m-1} + \alpha_{m-1}\boldsymbol{p}^*_{m-1}
\end{aligned}$$

and

$$\begin{aligned}
\boldsymbol{p}_m &= \tau_m \boldsymbol{r}_m + \beta_{m-1}\boldsymbol{p}_{m-1}, \\
\boldsymbol{p}^*_m &= \tau^*_m \boldsymbol{r}^*_m + \beta^*_{m-1}\boldsymbol{p}^*_{m-1}
\end{aligned}$$

with $\tau_m, \tau^*_m, \beta_{m-1}, \beta^*_{m-1} \in \mathbb{R}$. Here, the computation of the scalar quantities remains to be specified. It is found to be clearer to first present the BiCG algorithm in the following form and then to show that in the chosen formulation, with $\boldsymbol{x}_{j+1}$ for $j \in \mathbb{N}_0$, the solution of the Krylov subspace method based on the Petrov-Galerkin condition (4.3.86) is always obtained. A directly executable program is given in Appendix A.

BiCG Algorithm —

<table>
<tr><td>

Choose $x_0 \in \mathbb{R}^n$ and $\varepsilon > 0$

</td></tr>
<tr><td>

$r_0 = r_0^* = p_0 = p_0^* := b - A x_0$

</td></tr>
<tr><td>

$j := 0$

</td></tr>
<tr><td>

As long as $\|r_j\|_2 > \varepsilon$

<table>
<tr><td>

$$\alpha_j := \frac{(r_j, r_j^*)_2}{(A p_j, p_j^*)_2}$$

</td></tr>
<tr><td>

$x_{j+1} := x_j + \alpha_j p_j$

</td></tr>
<tr><td>

$r_{j+1} := r_j - \alpha_j A p_j$

</td></tr>
<tr><td>

$r_{j+1}^* := r_j^* - \alpha_j A^T p_j^*$

</td></tr>
<tr><td>

$$\beta_j := \frac{(r_{j+1}, r_{j+1}^*)_2}{(r_j, r_j^*)_2}$$

</td></tr>
<tr><td>

$p_{j+1} := r_{j+1} + \beta_j p_j$

</td></tr>
<tr><td>

$p_{j+1}^* := r_{j+1}^* + \beta_j p_j^*$

</td></tr>
<tr><td>

$j := j + 1$

</td></tr>
</table>

</td></tr>
</table>

Lemma 4.98 *Assuming the BiCG algorithm does not terminate before the computation of* p_m^* *, then for* $j = 0, \ldots, m$ *the following holds:*

$$r_j = \sum_{i=1}^{j+1} \gamma_{i,j} v_i, \quad r_j^* = \sum_{i=1}^{j+1} \gamma_{i,j}^* w_i, \tag{4.3.94}$$

$$p_j = \sum_{i=1}^{j+1} \lambda_{i,j} v_i, \quad p_j^* = \sum_{i=1}^{j+1} \lambda_{i,j}^* w_i \tag{4.3.95}$$

with $\gamma_{j+1,j}, \gamma_{j+1,j}^*, \lambda_{j+1,j}, \lambda_{j+1,j}^* \neq 0$ *and the bi-orthonormal vectors* $v_1, \ldots, v_{j+1}$ *and* $w_1, \ldots, w_{j+1}$ *resulting from the Bi-Lanczos algorithm.*

Proof:

The proof follows by induction on j. For $j = 0$ we have $\boldsymbol{v}_1 = \boldsymbol{w}_1 = \boldsymbol{r}_0 = \boldsymbol{p}_0 = \boldsymbol{r}_0^* = \boldsymbol{p}_0^*$, so the representation with $\gamma_{1,0} = \gamma_{1,0}^* = \lambda_{1,0} = \lambda_{1,0}^* = 1$ is satisfied.

Assuming the statement holds for $\ell = 0,\ldots,j < m$, then from the BiCG method it follows that

$$\boldsymbol{r}_{j+1} \;=\; \boldsymbol{r}_j - \alpha_j \boldsymbol{A}\boldsymbol{p}_j$$

$$\overset{\substack{(4.3.94)\\(4.3.95)}}{=} \; -\alpha_j \lambda_{j+1,j} \underbrace{\boldsymbol{A}\boldsymbol{v}_{j+1}}_{\in K_{j+2}} + \underbrace{\sum_{i=1}^{j+1} \gamma_{i,j}\boldsymbol{v}_i - \alpha_j \sum_{i=1}^{j} \lambda_{i,j}\boldsymbol{A}\boldsymbol{v}_i}_{\in K_{j+1}}.$$

By assumption, $\boldsymbol{p}_{j+1}$ could be computed, which implies $\alpha_j \neq 0$, and with $\lambda_{j+1,j} \neq 0$, the residual vector can be written as

$$\boldsymbol{r}_{j+1} = \gamma_{j+2,j+1}\boldsymbol{v}_{j+2} + \underbrace{\boldsymbol{\psi}_{j+1}}_{\in K_{j+1}} \in K_{j+2}$$

with $\gamma_{j+2,j+1} \neq 0$. For the search direction vector, we similarly obtain

$$\boldsymbol{p}_{j+1} = \boldsymbol{r}_{j+1} + \underbrace{\beta_j \boldsymbol{p}_j}_{\in K_{j+1}} \overset{\substack{(4.3.94)\\(4.3.95)}}{=} \lambda_{j+2,j+1}\boldsymbol{v}_{j+2} + \underbrace{\boldsymbol{\phi}_{j+1}}_{\in K_{j+1}} \in K_{j+2}$$

with $\lambda_{j+2,j+1} \neq 0$. The statement for the remaining two vectors follows analogously. $\square$

Lemma 4.99 *Assuming the BiCG algorithm does not terminate before the computation of $\boldsymbol{p}_{m+1}^*$, then*

$$\left(\boldsymbol{r}_j,\boldsymbol{r}_i^*\right)_2 \;=\; 0 \quad for \quad i \neq j \leq m+1, \tag{4.3.96}$$

$$\left(\boldsymbol{A}\boldsymbol{p}_j,\boldsymbol{p}_i^*\right)_2 \;=\; 0 \quad for \quad i \neq j \leq m+1. \tag{4.3.97}$$

Proof:

We prove this by complete induction on m. For $m = 0$, the statement follows from

$$\left(\boldsymbol{r}_1,\boldsymbol{r}_0^*\right)_2 = \left(\boldsymbol{r}_0,\boldsymbol{r}_0^*\right)_2 - \frac{\left(\boldsymbol{r}_0,\boldsymbol{r}_0^*\right)_2}{\left(\boldsymbol{A}\boldsymbol{r}_0,\boldsymbol{r}_0^*\right)_2}\left(\boldsymbol{A}\boldsymbol{r}_0,\boldsymbol{r}_0^*\right)_2 = 0 \tag{4.3.98}$$

and

$$\left(\boldsymbol{A}\boldsymbol{p}_1,\boldsymbol{p}_0^*\right)_2 \;=\; \left(\boldsymbol{A}\boldsymbol{r}_1,\boldsymbol{p}_0^*\right)_2 + \beta_0\left(\boldsymbol{A}\boldsymbol{p}_0,\boldsymbol{p}_0^*\right)_2$$

$$=\; \left(\boldsymbol{r}_1,\boldsymbol{A}^T\boldsymbol{p}_0^*\right)_2 + \frac{\beta_0}{\alpha_0}\left(\boldsymbol{r}_0,\boldsymbol{r}_0^*\right)_2$$

$$=\; \left(\boldsymbol{r}_1,\frac{\boldsymbol{r}_0^* - \boldsymbol{r}_1^*}{\alpha_0}\right)_2 + \frac{\left(\boldsymbol{r}_1,\boldsymbol{r}_1^*\right)_2}{\alpha_0}$$

$$=\; 0. \tag{4.3.99}$$

Analogously, the claimed statement also holds for $(r_0, r_1^*)_2$ as well as $(Ap_0, p_1^*)_2$.

Assume the claim holds for $\ell = 0, \ldots, m$, then the proof follows in the next steps. For $j < m$, taking into account $\beta_{-1} := 0$ and $p_{-1}^* := 0$, we obtain the equations

$$
\begin{aligned}
\left(r_{m+1}, r_j^*\right)_2 &= \underbrace{\left(r_m, r_j^*\right)_2}_{=\,0} - \alpha_m \left(Ap_m, r_j^*\right)_2 \\[2mm]
&= -\alpha_m \left(Ap_m, p_j^* - \beta_{j-1} p_{j-1}^*\right)_2 \\[2mm]
&\overset{(4.3.97)}{=} 0
\end{aligned}
\tag{4.3.100}
$$

and

$$
\begin{aligned}
\left(r_{m+1}, r_m^*\right)_2 &= \left(r_m, r_m^*\right)_2 - \alpha_m \left(Ap_m, r_m^*\right)_2 \\[2mm]
&= \left(r_m, r_m^*\right)_2 - \alpha_m \left(Ap_m, p_m^*\right)_2 + \alpha_m \underbrace{\left(Ap_m, \beta_{m-1} p_{m-1}^*\right)_2}_{=\,0} \\[2mm]
&= \left(r_m, r_m^*\right)_2 - \frac{\left(r_m, r_m^*\right)_2}{\left(Ap_m, p_m^*\right)_2}\left(Ap_m, p_m^*\right)_2 \\[2mm]
&= 0.
\end{aligned}
\tag{4.3.101}
$$

Furthermore, for $j < m$ we obtain

$$
\begin{aligned}
\left(Ap_{m+1}, p_j^*\right)_2 &= \left(Ar_{m+1}, p_j^*\right)_2 + \beta_m \underbrace{\left(Ap_m, p_j^*\right)_2}_{=\,0} = \left(r_{m+1}, A^T p_j^*\right)_2 \\[2mm]
&= \left(r_{m+1}, \frac{r_j^* - r_{j+1}^*}{\alpha_j}\right)_2 \\[2mm]
&\overset{\substack{(4.3.100) \\ (4.3.101)}}{=} 0
\end{aligned}
$$

and

$$
\begin{aligned}
\left(Ap_{m+1}, p_m^*\right)_2 &= \left(Ar_{m+1}, p_m^*\right)_2 + \beta_m \left(Ap_m, p_m^*\right)_2 \\[2mm]
&= \left(r_{m+1}, A^T p_m^*\right)_2 + \frac{\beta_m}{\alpha_m}\left(r_m, r_m^*\right)_2 \\[2mm]
&= \left(r_{m+1}, \frac{r_m^* - r_{m+1}^*}{\alpha_m}\right)_2 + \frac{\left(r_{m+1}, r_{m+1}^*\right)_2}{\alpha_m} \\[2mm]
&= 0.
\end{aligned}
$$

An analogous procedure yields both the remaining property $\left(r_j, r_{m+1}^*\right)_2 = 0$ and the orthogonality $\left(Ap_j, p_{m+1}^*\right)_2 = 0$ for $j < m+1$. $\qquad\square$

Theorem 4.100 *Assuming that the BiCG algorithm does not break down before the computation of r_m^*, x_m is the solution of the problem*

$$b - Ax \perp K_m^T$$

with $x \in x_0 + K_m$.

Proof:
Substituting (4.3.94) into (4.3.96) yields

$$0 \overset{(4.3.96)}{=} r_m^T r_i^* \overset{(4.3.94)}{=} \sum_{j=1}^{m+1} \gamma_j v_j^T r_i^* \quad \text{for} \quad i = 0, \dots, m-1.$$

Starting from $i = 0$, we obtain, using the bi-orthogonality condition (4.3.48) and the representation (4.3.94) for r_i^*, successively the equation

$$\gamma_1 = \dots = \gamma_m = 0,$$

which yields

$$b - Ax_m = r_m = \gamma_{m+1} v_{m+1} \perp K_m^T$$

and thus, with x_m, the solution of the Krylov subspace method is obtained. $\qquad\square$

Remark:
The vector implicitly present in the BiCG method, $x_m^* = A^{-T}(b - r_m^*)$, represents, analogously to the approximate solution x_m, the solution of the transposed problem

$$b - A^T x \perp K_m$$

with $x \in x_0^* + K_m^T$, $x_0^* = A^{-T} A x_0$.

The BiCG algorithm essentially has three disadvantages. First, residual vectors $r_j \neq 0$ and $r_j^* \neq 0$ with $(r_j, r_j^*) = 0$ can occur, which lead to a breakdown of the method, even though the solution of the considered equation has not been computed. Furthermore, at each iteration, matrix-vector multiplications with A and A^T are required, and the BiCG iterates, in contrast to the GMRES iterates, are generally not characterized by any minimization property, which can result in oscillations in the convergence history [28, 30, 34, 43, 76]. In the case of a symmetric, positive definite matrix A, the Bi-Lanczos algorithm coincides with the Lanczos method, and accordingly the BiCG method coincides with the CG method.

4.3.2.6 The CGS Method

The CGS method (Conjugate Gradient Squared) was presented by Sonneveld in 1989 [69] and represents a further development of the BiCG method, in which the computation of the scalar quantities α_j and β_j is possible without using the residual vector r_m^* and the search vector p_m^*, thus eliminating the need to access A^T.

We consider $\mathcal{P}_j$ as the set of polynomials of degree at most j and write

$$\boldsymbol{r}_j \;=\; \varphi_j(\boldsymbol{A})\boldsymbol{r}_0, \qquad \boldsymbol{r}_j^* \;=\; \varphi_j(\boldsymbol{A}^T)\boldsymbol{r}_0$$

$$\boldsymbol{p}_j \;=\; \psi_j(\boldsymbol{A})\boldsymbol{r}_0, \qquad \boldsymbol{p}_j^* \;=\; \psi_j(\boldsymbol{A}^T)\boldsymbol{r}_0$$

with $\varphi_j, \psi_j \in \mathcal{P}_j$. Based on the BiCG algorithm, in our case the polynomials φ_j and ψ_j are recursively given by

$$\varphi_{j+1}(\lambda) = \varphi_j(\lambda) - \alpha_j \lambda \psi_j(\lambda) \tag{4.3.102}$$

and

$$\psi_j(\lambda) = \varphi_j(\lambda) + \beta_{j-1}\psi_{j-1}(\lambda) \tag{4.3.103}$$

with $\varphi_0(\lambda) = \psi_0(\lambda) = 1$. The fundamental idea of the CGS method lies in the use of the property

$$\left(\phi_j(\boldsymbol{A})\boldsymbol{r}_0, \phi_j(\boldsymbol{A}^T)\boldsymbol{r}_0\right)_2 = \left(\phi_j^2(\boldsymbol{A})\boldsymbol{r}_0, \boldsymbol{r}_0\right)_2,$$

which holds for all polynomials $\phi_j \in \mathcal{P}_j$. If we define

$$\hat{\boldsymbol{r}}_j := \varphi_j^2(\boldsymbol{A})\boldsymbol{r}_0 \quad \text{and} \quad \hat{\boldsymbol{p}}_j := \psi_j^2(\boldsymbol{A})\boldsymbol{r}_0 \tag{4.3.104}$$

then it follows that

$$\alpha_j = \frac{\left(\varphi_j(\boldsymbol{A})\boldsymbol{r}_0, \varphi_j(\boldsymbol{A}^T)\boldsymbol{r}_0\right)_2}{\left(\boldsymbol{A}\psi_j(\boldsymbol{A})\boldsymbol{r}_0, \psi_j(\boldsymbol{A}^T)\boldsymbol{r}_0\right)_2} = \frac{\left(\varphi_j^2(\boldsymbol{A})\boldsymbol{r}_0, \boldsymbol{r}_0\right)_2}{\left(\boldsymbol{A}\psi_j^2(\boldsymbol{A})\boldsymbol{r}_0, \boldsymbol{r}_0\right)_2}$$

$$= \frac{(\hat{\boldsymbol{r}}_j, \boldsymbol{r}_0)_2}{(\boldsymbol{A}\hat{\boldsymbol{p}}_j, \boldsymbol{r}_0)_2} \tag{4.3.105}$$

and

$$\beta_j = \frac{\left(\varphi_{j+1}(\boldsymbol{A})\boldsymbol{r}_0, \varphi_{j+1}(\boldsymbol{A}^T)\boldsymbol{r}_0\right)_2}{\left(\varphi_j(\boldsymbol{A})\boldsymbol{r}_0, \varphi_j(\boldsymbol{A}^T)\boldsymbol{r}_0\right)_2} = \frac{(\hat{\boldsymbol{r}}_{j+1}, \boldsymbol{r}_0)_2}{(\hat{\boldsymbol{r}}_j, \boldsymbol{r}_0)_2}. \tag{4.3.106}$$

With (4.3.103) and (4.3.102) we obtain

$$\psi_j^2(\lambda) = \varphi_j^2(\lambda) + 2\beta_{j-1}\varphi_j(\lambda)\psi_{j-1}(\lambda) + \beta_{j-1}^2\psi_{j-1}^2(\lambda), \tag{4.3.107}$$

$$\varphi_{j+1}^2(\lambda) = \varphi_j^2(\lambda) - \alpha_j\lambda\left(2\varphi_j^2(\lambda) + 2\beta_{j-1}\varphi_j(\lambda)\psi_{j-1}(\lambda) - \alpha_j\lambda\psi_j^2(\lambda)\right) \tag{4.3.108}$$

and for the mixed terms

$$\varphi_{j+1}(\lambda)\psi_j(\lambda) = \varphi_j^2(\lambda) + \beta_{j-1}\varphi_j(\lambda)\psi_{j-1}(\lambda) - \alpha_j\lambda\psi_j^2(\lambda). \tag{4.3.109}$$

We define

$$\hat{\boldsymbol{q}}_j := \varphi_{j+1}(\boldsymbol{A})\psi_j(\boldsymbol{A})\boldsymbol{r}_0 \tag{4.3.110}$$

and

$$\hat{u}_j \quad := \quad \varphi_j(\boldsymbol{A})\psi_j(\boldsymbol{A})\boldsymbol{r}_0$$

$$\overset{(4.3.103)}{=} \quad \varphi_j^2(\boldsymbol{A})\boldsymbol{r}_0 + \beta_{j-1}\varphi_j(\boldsymbol{A})\psi_{j-1}(\boldsymbol{A})\boldsymbol{r}_0$$

$$= \quad \hat{\boldsymbol{r}}_j + \beta_{j-1}\hat{\boldsymbol{q}}_{j-1}. \qquad (4.3.111)$$

This yields

$$\hat{\boldsymbol{q}}_j \overset{(4.3.109)}{=} \quad \varphi_j^2(\boldsymbol{A})\boldsymbol{r}_0 + \beta_{j-1}\varphi_j(\boldsymbol{A})\psi_{j-1}(\boldsymbol{A})\boldsymbol{r}_0 - \alpha_j \boldsymbol{A}\psi_j^2(\boldsymbol{A})\boldsymbol{r}_0$$

$$= \quad \hat{\boldsymbol{r}}_j + \beta_{j-1}\hat{\boldsymbol{q}}_{j-1} - \alpha_j \boldsymbol{A}\hat{\boldsymbol{p}}_j$$

$$\overset{(4.3.111)}{=} \quad \hat{\boldsymbol{u}}_j - \alpha_j \boldsymbol{A}\hat{\boldsymbol{p}}_j, \qquad (4.3.112)$$

$$\hat{\boldsymbol{r}}_{j+1} \quad = \quad \varphi_{j+1}^2(\boldsymbol{A})\boldsymbol{r}_0$$

$$\overset{(4.3.108)}{=} \quad \varphi_j^2(\boldsymbol{A})\boldsymbol{r}_0 - \alpha_j \boldsymbol{A}\left(2\varphi_j^2(\boldsymbol{A})\boldsymbol{r}_0 + 2\beta_{j-1}\varphi_j(\boldsymbol{A})\psi_{j-1}(\boldsymbol{A})\boldsymbol{r}_0 - \alpha_j \boldsymbol{A}\psi_j^2(\boldsymbol{A})\boldsymbol{r}_0\right)$$

$$= \quad \hat{\boldsymbol{r}}_j - \alpha_j \boldsymbol{A}\left(2\hat{\boldsymbol{r}}_j + 2\beta_{j-1}\hat{\boldsymbol{q}}_{j-1} - \alpha_j \boldsymbol{A}\hat{\boldsymbol{p}}_j\right)$$

$$= \quad \hat{\boldsymbol{r}}_j - \alpha_j \boldsymbol{A}\Big(\underbrace{\hat{\boldsymbol{r}}_j + \beta_{j-1}\hat{\boldsymbol{q}}_{j-1}}_{=\ \hat{\boldsymbol{u}}_j} + \underbrace{\hat{\boldsymbol{r}}_j + \beta_{j-1}\hat{\boldsymbol{q}}_{j-1} - \alpha_j \boldsymbol{A}\hat{\boldsymbol{p}}_j}_{=\ \hat{\boldsymbol{q}}_j} \Big)$$

$$= \quad \hat{\boldsymbol{r}}_j - \alpha_j \boldsymbol{A}\left(\hat{\boldsymbol{u}}_j + \hat{\boldsymbol{q}}_j\right) \qquad (4.3.113)$$

and

$$\hat{\boldsymbol{p}}_{j+1} \quad = \quad \psi_{j+1}^2(\boldsymbol{A})\boldsymbol{r}_0$$

$$\overset{(4.3.107)}{=} \quad \varphi_{j+1}^2(\boldsymbol{A})\boldsymbol{r}_0 + 2\beta_j\varphi_{j+1}(\boldsymbol{A})\psi_j(\boldsymbol{A})\boldsymbol{r}_0 + \beta_j^2\psi_j^2(\boldsymbol{A})\boldsymbol{r}_0$$

$$= \quad \hat{\boldsymbol{r}}_{j+1} + 2\beta_j\hat{\boldsymbol{q}}_j + \beta_j^2\hat{\boldsymbol{p}}_j$$

$$= \quad \hat{\boldsymbol{u}}_{j+1} + \beta_j\left(\hat{\boldsymbol{q}}_j + \beta_j\hat{\boldsymbol{p}}_j\right). \qquad (4.3.114)$$

With the residual vector $\hat{\boldsymbol{r}}_{j+1}$, we obtain, taking into account

$$\hat{\boldsymbol{r}}_{j+1} = \boldsymbol{b} - \boldsymbol{A}\hat{\boldsymbol{x}}_{j+1}$$

the corresponding approximate solution according to

$$\hat{x}_{j+1} = A^{-1}(b - \hat{r}_{j+1}) \overset{(4.3.113)}{=} A^{-1}\left(b - \hat{r}_j - \alpha_j A\left(\hat{u}_j + \hat{q}_j\right)\right)$$
$$= A^{-1}\left(b - \hat{r}_j\right) + \alpha_j A^{-1} A\left(\hat{u}_j + \hat{q}_j\right) = \hat{x}_j + \alpha_j\left(\hat{u}_j + \hat{q}_j\right). \tag{4.3.115}$$

If we neglect the superscript $\hat{\ }$, the CGS algorithm follows from (4.3.105), (4.3.106) as well as (4.3.111) to (4.3.115) in the following form. The implementation of such a pseudocode as an executable MATLAB program is given in Appendix A.

CGS Algorithm —

Choose $x_0 \in \mathbb{R}^n$ and $\varepsilon > 0$
$u_0 = r_0 = p_0 := b - A x_0$, $j := 0$
While $\|r_j\|_2 > \varepsilon$
$\quad v_j := A p_j$, $\alpha_j := \dfrac{(r_j, r_0)_2}{(v_j, r_0)_2}$
$\quad q_j := u_j - \alpha_j v_j$
$\quad x_{j+1} := x_j + \alpha_j\left(u_j + q_j\right)$
$\quad r_{j+1} := r_j - \alpha_j A\left(u_j + q_j\right)$
$\quad \beta_j := \dfrac{(r_{j+1}, r_0)_2}{(r_j, r_0)_2}$
$\quad u_{j+1} := r_{j+1} + \beta_j q_j$
$\quad p_{j+1} := u_{j+1} + \beta_j\left(q_j + \beta_j p_j\right)$, $j := j + 1$

Based on the fundamental definitions 4.60 and 4.62 of a Krylov subspace method, it always holds for the computed approximate solution x_m that $x_m \in x_0 + K_m$, respectively

$$x_m = x_0 + \sum_{i=0}^{m-1} \alpha_i A^i r_0 \tag{4.3.116}$$

with $\alpha_0, \ldots, \alpha_{m-1} \in \mathbb{R}$. Due to the squaring of the polynomial in the definition of the residual vector according to (4.3.104) and the subsequent determination of the corresponding approximate solution based on it, the question arises whether the property (4.3.116) is preserved in the context of the CGS method. To investigate this question, we first formulate the following auxiliary statement.

Lemma 4.101 *For the polynomial $\varphi_j \in \mathcal{P}_j$ defined by the BiCG method with $r_j = \varphi_j(A)r_0$, the following representation holds:*

$$\varphi_j(\lambda) = 1 + \sum_{i=1}^{j} \tilde{\alpha}_{j,i}\lambda^i \qquad\qquad (4.3.117)$$

with $\tilde{\alpha}_{j,i} \in \mathbb{R}$ for $i = 1, \ldots, j$.

Proof:
We prove this by complete induction. For $j = 0$, we obtain directly from the initialization in the BiCG method the representation $\varphi_0(\lambda) = 1$. Suppose for an arbitrary but fixed $j \in \mathbb{N}_0$ the representation of φ_j is given in the form (4.3.117), then using (4.3.102) and exploiting $\psi_j \in \mathcal{P}_j$ in the form $\psi_j(\lambda) = \sum_{i=0}^{j} \tilde{\beta}_{j,i}\lambda^i$, we obtain the representation

$$\varphi_{j+1}(\lambda) = \varphi_j(\lambda) - \alpha_j\lambda\psi_j(\lambda) = 1 + \sum_{i=1}^{j} \tilde{\alpha}_{j,i}\lambda^i - \alpha_j\lambda\sum_{i=0}^{j} \tilde{\beta}_{j,i}\lambda^i$$

$$= 1 + \sum_{i=1}^{j+1} \tilde{\alpha}_{j+1,i}\lambda^i$$

with

$$\tilde{\alpha}_{j+1,i} = \begin{cases} \tilde{\alpha}_{j,i} - \alpha_j\tilde{\beta}_{j,i-1} & \text{for } i = 1,\ldots,j \\[2mm] -\alpha_j\tilde{\beta}_{j,j} & \text{for } i = j+1. \end{cases}$$

$\square$

The squaring of the polynomial in the definition of the residual vector $\hat{\boldsymbol{r}}_j$ according to (4.3.104) yields, using (4.3.117) and neglecting the superscript $\hat{}$, the representation

$$\boldsymbol{r}_j = \varphi_j(\boldsymbol{A})^2\boldsymbol{r}_0 = \left(\boldsymbol{I} + \sum_{i=1}^{j} \tilde{\alpha}_{j,i}\boldsymbol{A}^i\right)^2 \boldsymbol{r}_0 = \left(\boldsymbol{I} + \sum_{i=1}^{2j} \gamma_{j,i}\boldsymbol{A}^i\right)\boldsymbol{r}_0.$$

with $\gamma_{j,i} \in \mathbb{R}$ for $i = 1,\ldots,2j$. For the corresponding approximate vector, we thus obtain

$$\boldsymbol{x}_j = \boldsymbol{A}^{-1}(\boldsymbol{b} - \boldsymbol{r}_j) = \boldsymbol{A}^{-1}\left(\boldsymbol{b} - \left(\boldsymbol{r}_0 + \sum_{i=1}^{2j} \gamma_{j,i}\boldsymbol{A}^i\boldsymbol{r}_0\right)\right)$$

$$= \underbrace{\boldsymbol{A}^{-1}(\boldsymbol{b} - \boldsymbol{r}_0)}_{=\boldsymbol{x}_0} - \underbrace{\sum_{i=1}^{2j} \gamma_{j,i}\boldsymbol{A}^{i-1}\boldsymbol{r}_0}_{\in\,\mathrm{span}\{\boldsymbol{r}_0,\ldots,\boldsymbol{A}^{2j-1}\boldsymbol{r}_0\}=K_{2j}} \qquad \in \boldsymbol{x}_0 + K_{2j}. \qquad (4.3.118)$$

If we also take into account that the polynomial φ_j has the j degrees of freedom $\tilde{\alpha}_{j,1},\ldots,\tilde{\alpha}_{j,j}$, it becomes apparent that the approximate solution determined by CGS lies in $\boldsymbol{x}_0 + M_j$, with M_j being a subspace of dimension at most j of the Krylov subspace K_{2j}, which has dimension at most $2j$. Naturally, this property also affects the convergence behavior. Let us denote by $\boldsymbol{x}_j^{\mathrm{GMRES}} \in \boldsymbol{x}_0 + K_j$, $\boldsymbol{x}_j^{\mathrm{BiCG}} \in \boldsymbol{x}_0 + K_j$, and $\boldsymbol{x}_j^{\mathrm{CGS}} \in \boldsymbol{x}_0 + K_{2j}$ the approximate solutions computed by the GMRES, BiCG, and CGS methods, respectively, for the same initial vector $\boldsymbol{x}_0$. Then, for the corresponding residual vectors, taking Lemma 4.90 into account, the following estimates hold:

$$\|\boldsymbol{r}_j^{\mathrm{GMRES}}\|_2 = \arg \min_{\boldsymbol{x} \in \boldsymbol{x}_0 + K_j} \|\boldsymbol{b} - \boldsymbol{A}\boldsymbol{x}\|_2 \leq \|\boldsymbol{r}_j^{\mathrm{BiCG}}\|_2 \qquad (4.3.119)$$

as well as

$$\|\boldsymbol{r}_{2j}^{\mathrm{GMRES}}\|_2 = \arg \min_{\boldsymbol{x} \in \boldsymbol{x}_0 + K_{2j}} \|\boldsymbol{b} - \boldsymbol{A}\boldsymbol{x}\|_2 \leq \|\boldsymbol{r}_j^{\mathrm{CGS}}\|_2, \qquad (4.3.120)$$

where, in contrast to the BiCG method, it is possible in the CGS method that

$$\|\boldsymbol{r}_j^{\mathrm{CGS}}\|_2 < \|\boldsymbol{r}_j^{\mathrm{GMRES}}\|_2$$

can occur. We want to illustrate these properties using a model problem. For this, we consider the system of equations for the convection-diffusion equation with coefficients $a = \varepsilon = 1$ introduced in Example 1.4. In addition, we use $N = 100$, so that, in summary, we have a nonsymmetric matrix $\boldsymbol{A} \in \mathbb{R}^{10^4 \times 10^4}$. The residual histories shown in Figure 4.37 for the initial vector $\boldsymbol{x}_0 = \boldsymbol{0}$ exhibit exactly the behavior expected from the preceding analysis. The residual $\|\boldsymbol{r}_j^{\mathrm{GMRES}}\|_2$, as formulated in (4.3.119), serves as a lower bound for $\|\boldsymbol{r}_j^{\mathrm{BiCG}}\|_2$, while the Euclidean norm of the residual vector $\boldsymbol{r}_j^{\mathrm{CGS}}$ is in some cases smaller than that of the vector $\boldsymbol{r}_j^{\mathrm{GMRES}}$. The line corresponding to $\mathrm{GMRES}_{\mathrm{subseq}}$ shows the value $\|\boldsymbol{r}_{2j}^{\mathrm{GMRES}}\|_2$ plotted over j, and the corresponding convergence history is, according to (4.3.120), not undercut by $\|\boldsymbol{r}_j^{\mathrm{CGS}}\|_2$.

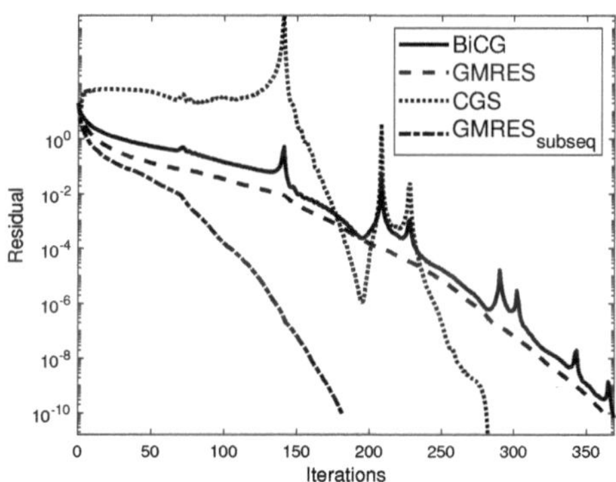

Figure 4.37   Convergence histories of the BiCG method, the GMRES method, and the CGS method

4.3.2.7 The BiCGSTAB Method

The CGS algorithm presented above stands out compared to the BiCG method due to two advantages. On the one hand, the method does not require any operations with $\boldsymbol{A}^T$, and on the other hand, due to the squaring of the polynomial in the definition of the residual vector, it exhibits faster convergence behavior with the same computational effort. However, the CGS method also contains the problem of premature breakdowns already present in the Bi-Lanczos algorithm, and sometimes oscillations appear in the residual progression [29, 32, 52, 54, 72]. The occurring premature breakdowns can be partially remedied analogously to the look-ahead approach [14].

With the BiCGSTAB method (BiCG Stabilized), van der Vorst [73] introduced a variant of the CGS method that exhibits a significantly smoother convergence behavior. In contrast to the CGS algorithm, different polynomials are deliberately considered in the definition of the search directions and residual vectors, providing an additional degree of freedom in each iteration that is used to minimize the residual.

First, the residual vector $\boldsymbol{r}_j$ and the search vector $\boldsymbol{p}_j$ are defined according to the CGS method, where the required polynomials φ_j and ψ_j are given by (4.3.103) and (4.3.102). We further set

$$\tilde{\boldsymbol{p}}_j^* := \tilde{\boldsymbol{r}}_j^* := \phi_j\left(\boldsymbol{A}^T\right)\boldsymbol{r}_0,$$

where the polynomial is defined by the simple recursion formula

$$\phi_{j+1}\left(\lambda\right) := \left(1 - \omega_j\lambda\right)\phi_j\left(\lambda\right) \tag{4.3.121}$$

with $\phi_0\left(\lambda\right) := 1$. As a consequence of Theorem 4.100 we obtain

$$\varphi_j\left(\boldsymbol{A}\right)\boldsymbol{r}_0 = \boldsymbol{r}_j \perp K_j^T$$

and thus

$$\left(\varphi_j\left(\boldsymbol{A}\right)\boldsymbol{r}_0, \pi_{j-1}\left(\boldsymbol{A}^T\right)\boldsymbol{r}_0\right)_2 = 0, \tag{4.3.122}$$

for all $\pi_{j-1} \in \mathcal{P}_{j-1}$. This yields the representation

$$
\begin{aligned}
\alpha_j &= \alpha_j \frac{\left(\varphi_j\left(\boldsymbol{A}\right)\boldsymbol{r}_0, \phi_j\left(\boldsymbol{A}^T\right)\boldsymbol{r}_0\right)_2}{\left(\varphi_j\left(\boldsymbol{A}\right)\boldsymbol{r}_0, \phi_j\left(\boldsymbol{A}^T\right)\boldsymbol{r}_0\right)_2} \\[2ex]
&\stackrel{(4.3.122)}{=} \frac{\left(\phi_j\left(\boldsymbol{A}\right)\varphi_j\left(\boldsymbol{A}\right)\boldsymbol{r}_0, \boldsymbol{r}_0\right)_2}{\left(\dfrac{\varphi_j\left(\boldsymbol{A}\right) - \varphi_{j+1}\left(\boldsymbol{A}\right)}{\alpha_j}\boldsymbol{r}_0, \phi_j\left(\boldsymbol{A}^T\right)\boldsymbol{r}_0\right)_2} \\[2ex]
&\stackrel{(4.3.102)}{=} \frac{\left(\phi_j\left(\boldsymbol{A}\right)\varphi_j\left(\boldsymbol{A}\right)\boldsymbol{r}_0, \boldsymbol{r}_0\right)_2}{\left(\phi_j\left(\boldsymbol{A}\right)\boldsymbol{A}\psi_j\left(\boldsymbol{A}\right)\boldsymbol{r}_0, \boldsymbol{r}_0\right)_2},
\end{aligned}
$$

and we therefore define

$$\hat{\boldsymbol{r}}_j := \phi_j\left(\boldsymbol{A}\right)\varphi_j\left(\boldsymbol{A}\right)\boldsymbol{r}_0 \tag{4.3.123}$$

as well as

$$\hat{\boldsymbol{p}}_j := \phi_j\left(\boldsymbol{A}\right)\psi_j\left(\boldsymbol{A}\right)\boldsymbol{r}_0. \tag{4.3.124}$$

This leads to

$$
\begin{aligned}
\hat{\boldsymbol{s}}_j &:= \phi_j\left(\boldsymbol{A}\right)\varphi_{j+1}\left(\boldsymbol{A}\right)\boldsymbol{r}_0 \\[2ex]
&\stackrel{(4.3.102)}{=} \phi_j\left(\boldsymbol{A}\right)\varphi_j\left(\boldsymbol{A}\right)\boldsymbol{r}_0 - \alpha_j\boldsymbol{A}\phi_j\left(\boldsymbol{A}\right)\psi_j\left(\boldsymbol{A}\right)\boldsymbol{r}_0 \\[2ex]
&= \hat{\boldsymbol{r}}_j - \alpha_j\boldsymbol{A}\hat{\boldsymbol{p}}_j \tag{4.3.125}
\end{aligned}
$$

and we obtain the recursion formulas

$$
\begin{aligned}
\hat{\boldsymbol{r}}_{j+1} &= \phi_{j+1}\left(\boldsymbol{A}\right)\varphi_{j+1}\left(\boldsymbol{A}\right)\boldsymbol{r}_0 \\
&= \left(\boldsymbol{I}-\omega_j\boldsymbol{A}\right)\phi_j\left(\boldsymbol{A}\right)\varphi_{j+1}\left(\boldsymbol{A}\right)\boldsymbol{r}_0 \\
&= \left(\boldsymbol{I}-\omega_j\boldsymbol{A}\right)\hat{\boldsymbol{s}}_j
\end{aligned}
\tag{4.3.126}
$$

and

$$
\begin{aligned}
\hat{\boldsymbol{p}}_{j+1} &= \phi_{j+1}\left(\boldsymbol{A}\right)\psi_{j+1}\left(\boldsymbol{A}\right)\boldsymbol{r}_0 \\
&\overset{(4.3.103)}{=} \phi_{j+1}\left(\boldsymbol{A}\right)\varphi_{j+1}\left(\boldsymbol{A}\right)\boldsymbol{r}_0 + \beta_j\phi_{j+1}\left(\boldsymbol{A}\right)\psi_j\left(\boldsymbol{A}\right)\boldsymbol{r}_0 \\
&= \hat{\boldsymbol{r}}_{j+1} + \beta_j\left(\boldsymbol{I}-\omega_j\boldsymbol{A}\right)\phi_j\left(\boldsymbol{A}\right)\psi_j\left(\boldsymbol{A}\right)\boldsymbol{r}_0 \\
&= \hat{\boldsymbol{r}}_{j+1} + \beta_j\left(\boldsymbol{I}-\omega_j\boldsymbol{A}\right)\hat{\boldsymbol{p}}_j.
\end{aligned}
\tag{4.3.127}
$$

In order to obtain a complete description of the method, we finally need to explain the computation of the scalar quantities β_j and ω_j. We write

$$
\begin{aligned}
\varphi_j\left(\lambda\right) &\overset{\substack{(4.3.103)\\(4.3.102)}}{=} -\alpha_{j-1}\lambda\varphi_{j-1}\left(\lambda\right) + \pi_{j-1}\left(\lambda\right) \\
&= \alpha_{j-1}\alpha_{j-2}\lambda^2\varphi_{j-2}\left(\lambda\right) + \tilde{\pi}_{j-1}\left(\lambda\right) \\
&= (-1)^j\alpha_{j-1}\ldots\alpha_0\lambda^j + \overline{\pi}_{j-1}\left(\lambda\right)
\end{aligned}
\tag{4.3.128}
$$

with $\pi_{j-1}, \tilde{\pi}_{j-1}, \overline{\pi}_{j-1} \in \mathcal{P}_{j-1}$. Analogously, we obtain the representation

$$
\phi_j(\lambda) = (-1)^j\omega_{j-1}\ldots\omega_0\lambda^j + \hat{\pi}_{j-1}(\lambda)
\tag{4.3.129}
$$

with $\hat{\pi}_{j-1} \in \mathcal{P}_{j-1}$, so that with $\boldsymbol{r}_j^* = \varphi_j\left(\boldsymbol{A}^T\right)\boldsymbol{r}_0$ the equation

$$
\begin{aligned}
\frac{\left(\boldsymbol{r}_j,\boldsymbol{r}_j^*\right)_2}{\left(\boldsymbol{r}_j,\tilde{\boldsymbol{r}}_j^*\right)_2}
&= \frac{\left(\varphi_j\left(\boldsymbol{A}\right)\boldsymbol{r}_0,\varphi_j\left(\boldsymbol{A}^T\right)\boldsymbol{r}_0\right)_2}{\left(\varphi_j\left(\boldsymbol{A}\right)\boldsymbol{r}_0,\phi_j\left(\boldsymbol{A}^T\right)\boldsymbol{r}_0\right)_2} \\[2mm]
&\overset{\substack{(4.3.128)\\(4.3.129)}}{=} \frac{\left(\varphi_j\left(\boldsymbol{A}\right)\boldsymbol{r}_0,(-1)^j\alpha_{j-1}\ldots\alpha_0\left(\boldsymbol{A}^T\right)^j\boldsymbol{r}_0 + \overline{\pi}_{j-1}\left(\boldsymbol{A}^T\right)\boldsymbol{r}_0\right)_2}{\left(\varphi_j\left(\boldsymbol{A}\right)\boldsymbol{r}_0,(-1)^j\omega_{j-1}\ldots\omega_0\left(\boldsymbol{A}^T\right)^j\boldsymbol{r}_0 + \hat{\pi}_{j-1}\left(\boldsymbol{A}^T\right)\boldsymbol{r}_0\right)_2} \\[2mm]
&\overset{(4.3.122)}{=} \frac{\alpha_{j-1}\ldots\alpha_0}{\omega_{j-1}\ldots\omega_0}\frac{\left(\varphi_j\left(\boldsymbol{A}\right)\boldsymbol{r}_0,\left(\boldsymbol{A}^T\right)^j\boldsymbol{r}_0\right)_2}{\left(\varphi_j\left(\boldsymbol{A}\right)\boldsymbol{r}_0,\left(\boldsymbol{A}^T\right)^j\boldsymbol{r}_0\right)_2} \\[2mm]
&= \frac{\alpha_{j-1}\ldots\alpha_0}{\omega_{j-1}\ldots\omega_0}
\end{aligned}
$$

follows. Thus, we have

$$\beta_j \;=\; \frac{\left(\boldsymbol{r}_{j+1},\boldsymbol{r}_{j+1}^{*}\right)_2}{\left(\boldsymbol{r}_j,\boldsymbol{r}_j^{*}\right)_2}$$

$$=\; \frac{\left(\boldsymbol{r}_{j+1},\boldsymbol{r}_{j+1}^{*}\right)_2}{\left(\boldsymbol{r}_{j+1},\tilde{\boldsymbol{r}}_{j+1}^{*}\right)_2}\,\frac{\left(\boldsymbol{r}_{j+1},\tilde{\boldsymbol{r}}_{j+1}^{*}\right)_2}{\left(\boldsymbol{r}_j,\tilde{\boldsymbol{r}}_j^{*}\right)_2}\,\frac{\left(\boldsymbol{r}_j,\tilde{\boldsymbol{r}}_j^{*}\right)_2}{\left(\boldsymbol{r}_j,\boldsymbol{r}_j^{*}\right)_2}$$

$$=\; \frac{\alpha_j}{\omega_j}\,\frac{\left(\boldsymbol{r}_{j+1},\tilde{\boldsymbol{r}}_{j+1}^{*}\right)_2}{\left(\boldsymbol{r}_j,\tilde{\boldsymbol{r}}_j^{*}\right)_2}$$

$$=\; \frac{\alpha_j}{\omega_j}\,\frac{\left(\varphi_{j+1}\left(\boldsymbol{A}\right)\boldsymbol{r}_0,\phi_{j+1}\left(\boldsymbol{A}^T\right)\boldsymbol{r}_0\right)_2}{\left(\varphi_j\left(\boldsymbol{A}\right)\boldsymbol{r}_0,\phi_j\left(\boldsymbol{A}^T\right)\boldsymbol{r}_0\right)_2}$$

$$=\; \frac{\alpha_j}{\omega_j}\,\frac{\left(\hat{\boldsymbol{r}}_{j+1},\boldsymbol{r}_0\right)_2}{\left(\hat{\boldsymbol{r}}_j,\boldsymbol{r}_0\right)_2}. \tag{4.3.130}$$

With ω_j, we have a parameter available for the targeted minimization of the residual. We consider the function $f_j : \mathbb{R} \to \mathbb{R}$

$$f_j(\omega) := \|\left(\boldsymbol{I} - \omega\boldsymbol{A}\right)\hat{\boldsymbol{s}}_j\|_2^2.$$

For $\hat{\boldsymbol{s}}_j \neq \boldsymbol{0}$, we obtain $f_j''(\omega) = 2\|\boldsymbol{A}\hat{\boldsymbol{s}}_j\|_2^2 > 0$, which means that the function is strictly convex in this case. As we will see below, in the case $\hat{\boldsymbol{s}}_j = \boldsymbol{0}$, the parameter ω_j can be chosen arbitrarily and the method yields the exact solution. We therefore restrict ourselves to the case $\hat{\boldsymbol{s}}_j \neq \boldsymbol{0}$, so that we can determine

$$\omega_j := \arg\min_{\omega \in \mathbb{R}} f_j(\omega)$$

by

$$0 = f_j'(\omega_j) = -2\left(\boldsymbol{A}\hat{\boldsymbol{s}}_j,\hat{\boldsymbol{s}}_j\right)_2 + 2\,\omega_j\left(\boldsymbol{A}\hat{\boldsymbol{s}}_j,\boldsymbol{A}\hat{\boldsymbol{s}}_j\right)_2$$

in the form

$$\omega_j = \frac{\left(\boldsymbol{A}\hat{\boldsymbol{s}}_j,\hat{\boldsymbol{s}}_j\right)_2}{\left(\boldsymbol{A}\hat{\boldsymbol{s}}_j,\boldsymbol{A}\hat{\boldsymbol{s}}_j\right)_2}. \tag{4.3.131}$$

The BiCGSTAB iterate can thus be written according to

$$\hat{\boldsymbol{x}}_{j+1} \;=\; \boldsymbol{A}^{-1}\left(\boldsymbol{b} - \hat{\boldsymbol{r}}_{j+1}\right)$$

$$\overset{(4.3.126)}{=}\; \boldsymbol{A}^{-1}\left(\boldsymbol{b} - \hat{\boldsymbol{s}}_j + \omega_j \boldsymbol{A}\hat{\boldsymbol{s}}_j\right)$$

$$\overset{(4.3.125)}{=}\; \boldsymbol{A}^{-1}\left(\boldsymbol{b} - \hat{\boldsymbol{r}}_j + \alpha_j \boldsymbol{A}\hat{\boldsymbol{p}}_j + \omega_j \boldsymbol{A}\hat{\boldsymbol{s}}_j\right)$$

$$=\; \hat{\boldsymbol{x}}_j + \alpha_j \hat{\boldsymbol{p}}_j + \omega_j \hat{\boldsymbol{s}}_j. \tag{4.3.132}$$

If we again neglect the superscript $\hat{}$, then, using (4.3.125) to (4.3.127), (4.3.130) to (4.3.132), and the definition of the scalar quantity α_j, we obtain the BiCGSTAB algorithm in the following form. For implementation, see Appendix A.

BiCGSTAB Algorithm —

<table>
<tr><td>Choose $\boldsymbol{x}_0 \in \mathbb{R}^n$ and $\varepsilon > 0$</td></tr>
<tr><td>$\boldsymbol{r}_0 := \boldsymbol{p}_0 := \boldsymbol{b} - \boldsymbol{A}\,\boldsymbol{x}_0$, $\rho_0 := (\boldsymbol{r}_0,\boldsymbol{r}_0)_2$, $j := 0$</td></tr>
<tr><td>While $\|\boldsymbol{r}_j\|_2 > \varepsilon$

$\boldsymbol{v}_j := \boldsymbol{A}\,\boldsymbol{p}_j$, $\alpha_j := \dfrac{\rho_j}{(\boldsymbol{v}_j,\boldsymbol{r}_0)_2}$

$\boldsymbol{s}_j := \boldsymbol{r}_j - \alpha_j \boldsymbol{v}_j$, $\boldsymbol{t}_j := \boldsymbol{A}\,\boldsymbol{s}_j$

$\omega_j := \dfrac{(\boldsymbol{t}_j,\boldsymbol{s}_j)_2}{(\boldsymbol{t}_j,\boldsymbol{t}_j)_2}$

$\boldsymbol{x}_{j+1} := \boldsymbol{x}_j + \alpha_j \boldsymbol{p}_j + \omega_j \boldsymbol{s}_j$

$\boldsymbol{r}_{j+1} := \boldsymbol{s}_j - \omega_j \boldsymbol{t}_j$

$\rho_{j+1} := (\boldsymbol{r}_{j+1},\boldsymbol{r}_0)_2$, $\beta_j := \dfrac{\alpha_j}{\omega_j}\dfrac{\rho_{j+1}}{\rho_j}$

$\boldsymbol{p}_{j+1} := \boldsymbol{r}_{j+1} + \beta_j \left(\boldsymbol{p}_j - \omega_j \boldsymbol{v}_j\right)$, $j := j + 1$</td></tr>
</table>

Analogous to the CGS method, the approximate vector $\boldsymbol{x}_j$ in the above method is formally determined based on the residual vector $\boldsymbol{r}_j$ according to (4.3.132). For the latter, according to (4.3.123), the representation

$$\boldsymbol{r}_j = \phi_j\left(\boldsymbol{A}\right)\varphi_j\left(\boldsymbol{A}\right)\boldsymbol{r}_0,$$

holds, where φ_j represents the polynomial originating from the BiCG algorithm and ϕ_j is defined recursively by (4.3.121) and $\phi_0(\lambda) = 1$. From this definition, it follows directly that

$$\phi_j\left(\boldsymbol{A}\right) = \boldsymbol{I} + \sum_{i=1}^{j} \tilde{\beta}_{j,i} \boldsymbol{A}^i$$

so that, using Lemma 4.101, the residual vector can be written as

$$\boldsymbol{r}_j = \left(\boldsymbol{I} + \sum_{i=1}^{j} \tilde{\beta}_{j,i} \boldsymbol{A}^i\right)\left(\boldsymbol{I} + \sum_{i=1}^{j} \tilde{\alpha}_{j,i} \boldsymbol{A}^i\right)\boldsymbol{r}_0 = \left(\boldsymbol{I} + \sum_{i=1}^{2j} \tilde{\gamma}_{j,i} \boldsymbol{A}^i\right)\boldsymbol{r}_0$$

with $\tilde{\gamma}_{j,i} \in \mathbb{R}$ for $i = 1,\ldots,2j$. As already shown in (4.3.118), in the context of the BiCGSTAB method, the approximate vector, analogous to the CGS algorithm, satisfies the property

$$\boldsymbol{x}_j \in \boldsymbol{x}_0 + K_{2j}.$$

Thus, for the residual vectors $\boldsymbol{r}_j^{\mathrm{BiCGSTAB}}$ of the BiCGSTAB method and $\boldsymbol{r}_{2j}^{\mathrm{GMRES}}$ of the GMRES method, we obtain the estimate

$$\|\boldsymbol{r}_{2j}^{\mathrm{GMRES}}\|_2 \leq \|\boldsymbol{r}_{j}^{\mathrm{BiCGSTAB}}\|_2$$

and it is also possible that

$$\|\boldsymbol{r}_{j}^{\mathrm{BiCGSTAB}}\|_2 < \|\boldsymbol{r}_{j}^{\mathrm{GMRES}}\|_2$$

occurs. For practical investigation, we use the model problem of the convection-diffusion equation already considered in Section 4.3.2.6. The properties described above are impressively demonstrated by the convergence history presented in Figure 4.38.

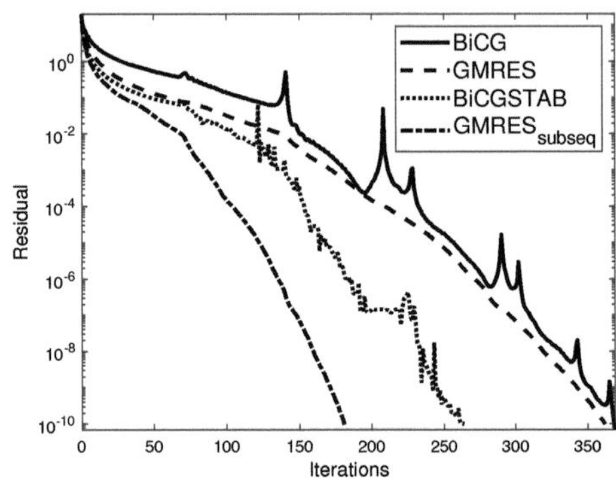

Figure 4.38   Convergence histories of the BiCG method, the GMRES method, and the BiCGSTAB method

A generalization of the presented method in the sense of an ℓ-dimensional instead of a one-dimensional minimization is given by the BiCGSTAB(ℓ) method, which was derived in 1993 by Sleipjen and Fokkema [66]. The algorithm can be interpreted as a combination of the GMRES(ℓ) method and the BiCG method and coincides with the BiCGSTAB method in the case $\ell = 1$. In this approach, ℓ BiCG residual vectors are first computed implicitly, and then a minimization is performed over the ℓ-dimensional subspace thus defined.

We have already discussed the possible breakdowns of the Bi-Lanczos algorithm in Section 4.3.2.3. With the following variant of the BiCGSTAB method, we will present a stabilization of the method with respect to this issue. First, before computing the scalar α_j, we check the magnitude of the denominator that arises, in order to avoid division by very small values in the case of large but nearly orthogonal vectors $\boldsymbol{v}_j$ and $\boldsymbol{r}_0$. If the absolute value of the denominator relative to the length of the vectors $\boldsymbol{v}_j$ and $\boldsymbol{r}_0$ falls below a given threshold, a restart of the method is performed. Another possible division by zero occurs when computing ω_j. To address this issue, [8] recommends monitoring the vector $\boldsymbol{s}_j$, which corresponds to the residual vector in the BiCG method. If $\|\boldsymbol{s}_j\|_2 \leq \varepsilon$, then we have

$$\boldsymbol{x}_{j+1} = \boldsymbol{x}_j + \alpha_j \boldsymbol{p}_j$$

and thus

$$\|\boldsymbol{r}_{j+1}\|_2 = \|\boldsymbol{r}_j - \alpha_j \boldsymbol{v}_j\|_2 = \|\boldsymbol{s}_j\|_2 \leq \varepsilon,$$

so that there is no need to compute the vector $\boldsymbol{t}_j$ and thus to calculate the scalar quantity ω_j. Taking into account the above improvements, the following stabilization of the BiCGSTAB method results.

Stabilized BiCGSTAB Algorithm —

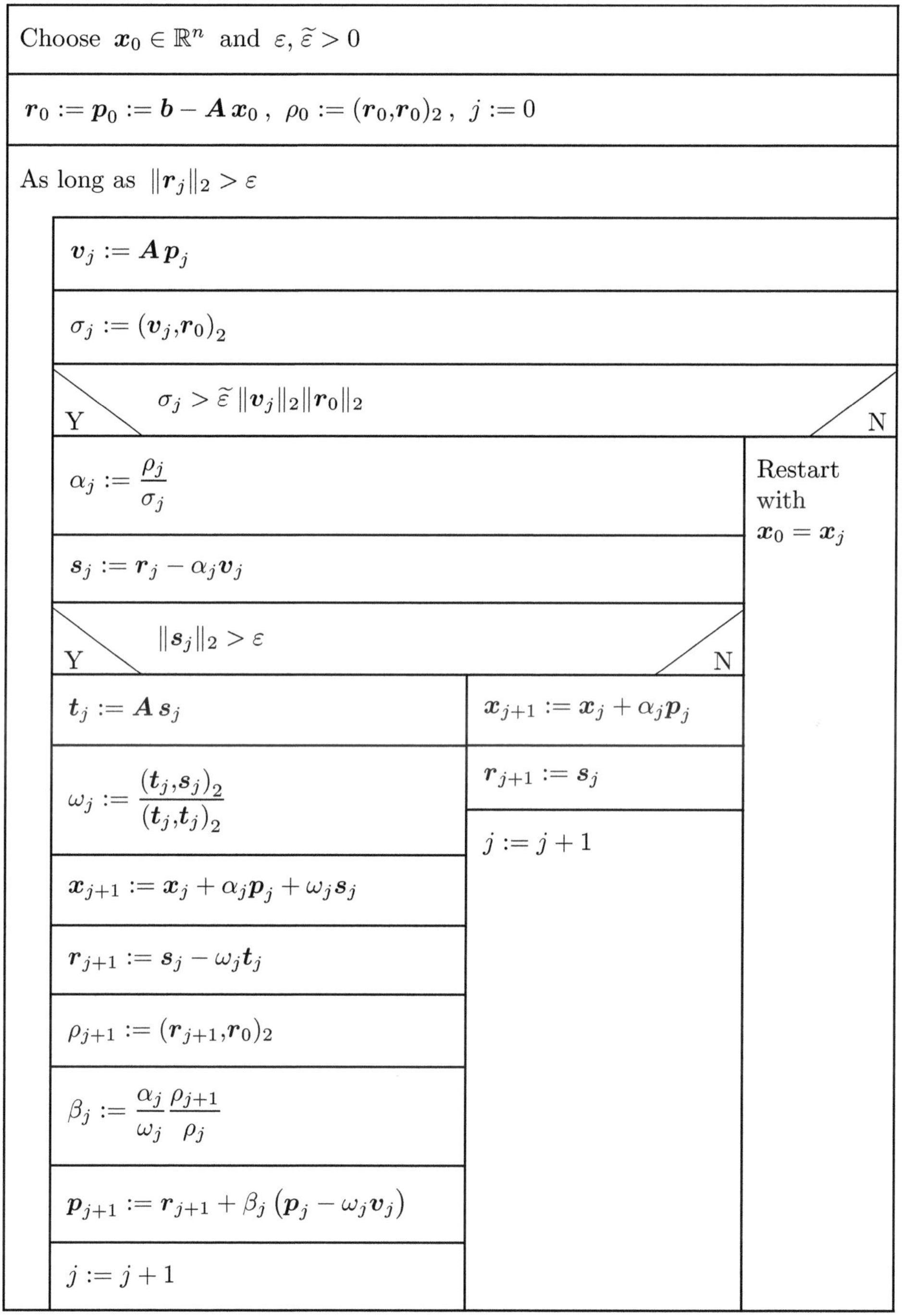

<table>
<tr><td colspan="3">Choose $\boldsymbol{x}_0 \in \mathbb{R}^n$ and $\varepsilon, \widetilde{\varepsilon} > 0$</td></tr>
<tr><td colspan="3">$\boldsymbol{r}_0 := \boldsymbol{p}_0 := \boldsymbol{b} - \boldsymbol{A}\boldsymbol{x}_0\,,\ \rho_0 := (\boldsymbol{r}_0, \boldsymbol{r}_0)_2\,,\ j := 0$</td></tr>
<tr><td colspan="3">As long as $\|\boldsymbol{r}_j\|_2 > \varepsilon$</td></tr>
<tr><td colspan="2">$\boldsymbol{v}_j := \boldsymbol{A}\boldsymbol{p}_j$</td><td></td></tr>
<tr><td colspan="2">$\sigma_j := (\boldsymbol{v}_j, \boldsymbol{r}_0)_2$</td><td></td></tr>
<tr><td colspan="2">Y $\sigma_j > \widetilde{\varepsilon}\,\|\boldsymbol{v}_j\|_2 \|\boldsymbol{r}_0\|_2$ N</td><td></td></tr>
<tr><td colspan="2">$\alpha_j := \dfrac{\rho_j}{\sigma_j}$</td><td rowspan="3">Restart with $\boldsymbol{x}_0 = \boldsymbol{x}_j$</td></tr>
<tr><td colspan="2">$\boldsymbol{s}_j := \boldsymbol{r}_j - \alpha_j \boldsymbol{v}_j$</td></tr>
<tr><td colspan="2">Y $\|\boldsymbol{s}_j\|_2 > \varepsilon$ N</td></tr>
<tr><td>$\boldsymbol{t}_j := \boldsymbol{A}\boldsymbol{s}_j$</td><td>$\boldsymbol{x}_{j+1} := \boldsymbol{x}_j + \alpha_j \boldsymbol{p}_j$</td><td></td></tr>
<tr><td rowspan="2">$\omega_j := \dfrac{(\boldsymbol{t}_j, \boldsymbol{s}_j)_2}{(\boldsymbol{t}_j, \boldsymbol{t}_j)_2}$</td><td>$\boldsymbol{r}_{j+1} := \boldsymbol{s}_j$</td><td></td></tr>
<tr><td>$j := j + 1$</td><td></td></tr>
<tr><td>$\boldsymbol{x}_{j+1} := \boldsymbol{x}_j + \alpha_j \boldsymbol{p}_j + \omega_j \boldsymbol{s}_j$</td><td></td><td></td></tr>
<tr><td>$\boldsymbol{r}_{j+1} := \boldsymbol{s}_j - \omega_j \boldsymbol{t}_j$</td><td></td><td></td></tr>
<tr><td>$\rho_{j+1} := (\boldsymbol{r}_{j+1}, \boldsymbol{r}_0)_2$</td><td></td><td></td></tr>
<tr><td>$\beta_j := \dfrac{\alpha_j}{\omega_j}\dfrac{\rho_{j+1}}{\rho_j}$</td><td></td><td></td></tr>
<tr><td>$\boldsymbol{p}_{j+1} := \boldsymbol{r}_{j+1} + \beta_j\left(\boldsymbol{p}_j - \omega_j \boldsymbol{v}_j\right)$</td><td></td><td></td></tr>
<tr><td>$j := j + 1$</td><td></td><td></td></tr>
</table>

Of course, further modifications can also contribute to a better behavior of the method. One possibility is to control the variation of the relevant quantities. Because

$$\frac{\|r_{j+1} - r_j\|_2}{\|r_j\|_2} \leq \frac{\|s_j - r_j\|_2}{\|r_j\|_2} = \frac{|\alpha_j|\|v_j\|_2}{\|r_j\|_2}$$

in the case

$$\frac{|\alpha_j|\|v_j\|_2}{\|r_j\|_2} \leq \gamma$$

the relative change of the residual is below a chosen threshold γ, so that a restart appears reasonable. To control rounding errors, the residual vector r_{j+1} computed in the algorithm can be compared at regular intervals with the actual residual vector $b - Ax_{j+1}$. Larger deviations indicate rounding errors, in which case a restart is also appropriate. A transfer of the look-ahead strategy to the BiCGSTAB method is presented in [12].

4.3.2.8   The TFQMR Method

With the QMR method (Quasi Minimal Residual), Freund and Nachtigal [30] presented a novel BiCG-like method that can be seen as a combination of the GMRES method and the BiCG method. The method combines the advantages of the low memory and computational requirements of the BiCG method, which are due to the use of the Bi-Lanczos algorithm, with the property of the smooth residual behavior of the GMRES method, which is based on the minimization of the residual. For this purpose, the look-ahead variant of the Bi-Lanczos method, the look-ahead Lanczos algorithm, was used to compute a basis $\{v_1, \ldots, v_m\}$ of K_m. The matrix $V_m = (v_1 \ldots v_m)$, consisting of the basis vectors, is not necessarily orthogonal due to the use of the Bi-Lanczos algorithm, which is why, in contrast to the GMRES method, the QMR method only performs a quasi-minimization of the function given by (4.3.60). Naturally, the BiCG algorithm implies a necessary access to the matrix A^T in the QMR method. The TFQMR method (Transpose-Free QMR) described below was developed by Freund [29] based on the CGS method instead of the BiCG method, thereby eliminating matrix-vector multiplications with A^T. Before we proceed to the direct derivation of the method, we will use the following two lemmas to define a basis of the Krylov subspace K_m and prove a property of the basis vectors. For this, let the mapping $\lfloor\ \rfloor : \mathbb{R} \to \mathbb{Z}$ be defined by

$$x \xrightarrow{\lfloor\ \rfloor} \lfloor x \rfloor \in \,]x - 1, x]. \tag{4.3.133}$$

Lemma 4.102 *Suppose the CGS method does not break down before the computation of the vectors* $q_{\lfloor\frac{m-1}{2}\rfloor}$ *and* $u_{\lfloor\frac{m-1}{2}\rfloor}$, *then the vectors* $y_1, \ldots, y_m$ *defined by*

$$y_k := \begin{cases} u_{\lfloor\frac{k-1}{2}\rfloor} & , \text{ if } k \text{ is odd,} \\[2mm] q_{\lfloor\frac{k-1}{2}\rfloor} & , \text{ if } k \text{ is even} \end{cases} \tag{4.3.134}$$

form a basis of K_m.

Proof:
Let $n \in \mathbb{N}_0$, then for $k = 2n + 1$ it follows that

$$\begin{aligned}
\boldsymbol{y}_k &= \boldsymbol{u}_n = \varphi_n\left(\boldsymbol{A}\right)\psi_n\left(\boldsymbol{A}\right)\boldsymbol{r}_0 \\
&= \alpha_0^2\cdot\ldots\cdot\alpha_{n-1}^2\boldsymbol{A}^{k-1}\boldsymbol{r}_0 + \pi\left(\boldsymbol{A}\right)\boldsymbol{r}_0 \ \text{ with } \ \pi\in\mathcal{P}_{k-2}
\end{aligned}$$

and for $k = 2n > 0$

$$\begin{aligned}
\boldsymbol{y}_k &= \boldsymbol{q}_{n-1} = \varphi_n\left(\boldsymbol{A}\right)\psi_{n-1}\left(\boldsymbol{A}\right)\boldsymbol{r}_0 \\
&= -\alpha_0^2\cdot\ldots\cdot\alpha_{n-2}^2\alpha_{n-1}\boldsymbol{A}^{k-1}\boldsymbol{r}_0 + \pi\left(\boldsymbol{A}\right)\boldsymbol{r}_0 \ \text{ with } \ \pi\in\mathcal{P}_{k-2}.
\end{aligned}$$

By assumption, $\alpha_i \neq 0$ for $i = 0,\ldots,n-1$ and thus

$$\operatorname{span}\{\boldsymbol{y}_1,\ldots,\boldsymbol{y}_m\} = K_m.$$

$$\square$$

Lemma 4.103 *Let*

$$\boldsymbol{w}_k := \begin{cases} \varphi_{\lfloor\frac{k-1}{2}\rfloor}^2\left(\boldsymbol{A}\right)\boldsymbol{r}_0 & , \ \text{if } k \text{ is odd,} \\ \varphi_{\lfloor\frac{k-1}{2}\rfloor}\left(\boldsymbol{A}\right)\varphi_{\lfloor\frac{k}{2}\rfloor}\left(\boldsymbol{A}\right)\boldsymbol{r}_0 & , \ \text{if } k \text{ is even,} \end{cases} \tag{4.3.135}$$

then

$$\boldsymbol{w}_{k+1} = \boldsymbol{w}_k - \alpha_{\lfloor\frac{k-1}{2}\rfloor}\boldsymbol{A}\boldsymbol{y}_k \ \text{ for } \ k = 1,\ldots,m$$

with the vectors $\boldsymbol{y}_1,\ldots,\boldsymbol{y}_m$ *given by* (4.3.134).

Proof:
From (4.3.102) we obtain

$$\lambda\psi_j(\lambda) = \frac{1}{\alpha_j}\left(\varphi_j(\lambda) - \varphi_{j+1}(\lambda)\right). \tag{4.3.136}$$

Let $n \in \mathbb{N}_0$, then for $k = 2n+1$ it follows that

$$\begin{aligned}
\boldsymbol{A}\boldsymbol{y}_k &= \boldsymbol{A}\boldsymbol{u}_n \\[4pt]
&\overset{\substack{(4.3.111)\\(4.3.136)}}{=} \frac{1}{\alpha_n}\left(\varphi_n^2\left(\boldsymbol{A}\right)\boldsymbol{r}_0 - \varphi_n\left(\boldsymbol{A}\right)\varphi_{n+1}\left(\boldsymbol{A}\right)\boldsymbol{r}_0\right) \\[4pt]
&= \frac{1}{\alpha_{\lfloor\frac{k-1}{2}\rfloor}}\left(\varphi_{\lfloor\frac{k-1}{2}\rfloor}^2\left(\boldsymbol{A}\right)\boldsymbol{r}_0 - \varphi_{\lfloor\frac{(k+1)-1}{2}\rfloor}\left(\boldsymbol{A}\right)\varphi_{\lfloor\frac{k+1}{2}\rfloor}\left(\boldsymbol{A}\right)\boldsymbol{r}_0\right) \\[4pt]
&= \frac{1}{\alpha_{\lfloor\frac{k-1}{2}\rfloor}}\left(\boldsymbol{w}_k - \boldsymbol{w}_{k+1}\right)
\end{aligned}$$

and for $k = 2n > 0$

$$\boldsymbol{Ay}_k \quad = \quad \boldsymbol{Aq}_{n-1}$$

$$\overset{(4.3.110)}{\underset{(4.3.136)}{=}} \quad \frac{1}{\alpha_{n-1}} \left(\varphi_{n-1}\left(\boldsymbol{A}\right) \varphi_n\left(\boldsymbol{A}\right) \boldsymbol{r}_0 - \varphi_n^2\left(\boldsymbol{A}\right) \boldsymbol{r}_0 \right)$$

$$= \quad \frac{1}{\alpha_{\lfloor \frac{k-1}{2}\rfloor}} \left(\varphi_{\lfloor \frac{k-1}{2}\rfloor}\left(\boldsymbol{A}\right) \varphi_{\lfloor \frac{k}{2}\rfloor}\left(\boldsymbol{A}\right) \boldsymbol{r}_0 - \varphi_{\lfloor \frac{(k+1)-1}{2}\rfloor}^2\left(\boldsymbol{A}\right) \boldsymbol{r}_0 \right)$$

$$= \quad \frac{1}{\alpha_{\lfloor \frac{k-1}{2}\rfloor}} \left(\boldsymbol{w}_k - \boldsymbol{w}_{k+1} \right).$$

$\square$

Due to the definition of the vector $\boldsymbol{w}_k$, it always holds that $\boldsymbol{w}_{2k+1} = \boldsymbol{r}_k$. With

$$\boldsymbol{Y}_m = \left(\boldsymbol{y}_1 \ldots \boldsymbol{y}_m\right), \tag{4.3.137}$$

$$\boldsymbol{W}_{m+1} = \left(\boldsymbol{w}_1 \ldots \boldsymbol{w}_{m+1}\right) \tag{4.3.138}$$

and

$$\overline{\boldsymbol{B}}_m = \begin{pmatrix} \alpha_{\lfloor \frac{1-1}{2}\rfloor}^{-1} & & & \\ -\alpha_{\lfloor \frac{1-1}{2}\rfloor}^{-1} & \alpha_{\lfloor \frac{2-1}{2}\rfloor}^{-1} & & \\ & \ddots & \ddots & \\ & & -\alpha_{\lfloor \frac{m-2}{2}\rfloor}^{-1} & \alpha_{\lfloor \frac{m-1}{2}\rfloor}^{-1} \\ & & & -\alpha_{\lfloor \frac{m-1}{2}\rfloor}^{-1} \end{pmatrix} \in \mathbb{R}^{(m+1)\times m}$$

Lemma 4.103 can be written in the form

$$\boldsymbol{A}\boldsymbol{Y}_m = \boldsymbol{W}_{m+1}\overline{\boldsymbol{B}}_m. \tag{4.3.139}$$

Moreover, due to Lemma 4.102, for every vector $\boldsymbol{x}_m \in \boldsymbol{x}_0 + K_m$ there exists the representation

$$\boldsymbol{x}_m = \boldsymbol{x}_0 + \boldsymbol{Y}_m \boldsymbol{\alpha}_m$$

with $\boldsymbol{\alpha}_m \in \mathbb{R}^m$. From (4.3.135) it follows that $\boldsymbol{w}_1 = \boldsymbol{r}_0$ and therefore, for the residual vector corresponding to $\boldsymbol{x}_m$, using equation (4.3.139), we obtain

$$\boldsymbol{r}_m \quad = \quad \boldsymbol{b} - \boldsymbol{A}\boldsymbol{x}_m$$

$$= \quad \boldsymbol{r}_0 - \boldsymbol{A}\boldsymbol{Y}_m \boldsymbol{\alpha}_m$$

$$= \quad \boldsymbol{W}_{m+1} \left(\boldsymbol{e}_1 - \overline{\boldsymbol{B}}_m \boldsymbol{\alpha}_m \right) \tag{4.3.140}$$

with $\boldsymbol{e}_1 = (1,0,\ldots,0)^T \in \mathbb{R}^{m+1}$. The matrix $\boldsymbol{W}_{m+1}$ is generally not norm-preserving, which is why in quasi-minimization, a complete minimization of the residual is not considered. With the additional use of the scaling matrix

$$\boldsymbol{S}_{m+1} := \begin{pmatrix} \|\boldsymbol{w}_1\|_2 & & \\ & \ddots & \\ & & \|\boldsymbol{w}_{m+1}\|_2 \end{pmatrix} \in \mathbb{R}^{(m+1)\times(m+1)}$$

one can conclude

$$\|\boldsymbol{r}_m\|_2 \le \left\|\boldsymbol{W}_{m+1}\boldsymbol{S}_{m+1}^{-1}\right\|_2 \underbrace{\left\|\|\boldsymbol{w}_1\|_2\boldsymbol{e}_1 - \boldsymbol{S}_{m+1}\overline{\boldsymbol{B}}_m\boldsymbol{\alpha}_m\right\|_2}_{=:\,\tau_m}. \tag{4.3.141}$$

The scaling in the above formulation serves as a preconditioner, since the condition number of the matrix $\boldsymbol{W}_{m+1}\boldsymbol{S}_{m+1}^{-1}$ is a criterion for the quality of the estimate of the residual by the quantity τ_m. We obtain the following estimate, which is proved in the next lemma:

Lemma 4.104 *Let* $\tau_m := \left\|\|\boldsymbol{w}_1\|_2\boldsymbol{e}_1 - \boldsymbol{S}_{m+1}\overline{\boldsymbol{B}}_m\boldsymbol{\alpha}_m\right\|_2$, *then it follows that*

$$\|\boldsymbol{r}_m\|_2 \le \sqrt{m+1}\,\tau_m$$

for the residual vector $\boldsymbol{r}_m$ *given by* (4.3.140).

Proof:
Since the Euclidean norm of any matrix $\boldsymbol{A} = (a_{ij}) \in \mathbb{R}^{m\times n}$ satisfies the estimate

$$\|\boldsymbol{A}\|_2 \le \left(\sum_{i=1}^m \sum_{j=1}^n |a_{ij}|^2\right)^{1/2} = \|\boldsymbol{A}\|_F$$

the claim follows by inserting the inequality

$$\left\|\boldsymbol{W}_{m+1}\boldsymbol{S}_{m+1}^{-1}\right\|_2 \le \left(\sum_{i=1}^{m+1}\left\|\frac{\boldsymbol{w}_i}{\|\boldsymbol{w}_i\|_2}\right\|_2^2\right)^{1/2} = \sqrt{m+1}$$

into equation (4.3.141). $\qquad\Box$

Analogous to the mapping introduced in the GMRES method by (4.3.65), we use, due to the estimate (4.3.141), the function $J_m : \mathbb{R}^m \to \mathbb{R}$ with

$$J_m(\boldsymbol{\alpha}) := \left\|\|\boldsymbol{w}_1\|_2\boldsymbol{e}_1 - \boldsymbol{S}_{m+1}\overline{\boldsymbol{B}}_m\boldsymbol{\alpha}\right\|_2 \tag{4.3.142}$$

and describe, with the following two theorems, an iterative method for solving the minimization problem

$$\boldsymbol{\alpha}_m = \arg\min_{\boldsymbol{\alpha}\in\mathbb{R}^m} J_m(\boldsymbol{\alpha}),$$

which yields the TFQMR iterate via

$$\boldsymbol{x}_m = \boldsymbol{x}_0 + \boldsymbol{Y}_m\boldsymbol{\alpha}_m. \tag{4.3.143}$$

Theorem 4.105 *Let* $m \ge 1$ *and suppose that*

$$\overline{\boldsymbol{T}}_m = \begin{pmatrix} \boldsymbol{T}_m \\ h_{m+1,1}\dots h_{m+1,m} \end{pmatrix} := \boldsymbol{S}_{m+1}\overline{\boldsymbol{B}}_m \in \mathbb{R}^{(m+1)\times m}$$

with a non-singular matrix $\boldsymbol{T}_m$. *For* $k = m-1, m$ *let further*

$$\boldsymbol{\alpha}_k = \arg \min_{\boldsymbol{\alpha} \in \mathbb{R}^k} J_k(\boldsymbol{\alpha}),$$

and

$$\tau_k := J_k(\boldsymbol{\alpha}_k)$$

with the function J_k given by (4.3.142). Then

$$\boldsymbol{\alpha}_m = \left(1 - c_m^2\right) \begin{pmatrix} \boldsymbol{\alpha}_{m-1} \\ 0 \end{pmatrix} + c_m^2 \tilde{\boldsymbol{\alpha}}_m \tag{4.3.144}$$

with

$$\tilde{\boldsymbol{\alpha}}_m := \left(\alpha_{\lfloor \frac{j-1}{2} \rfloor}\right)^T_{j=1,\dots,m} = \left(\alpha_0, \alpha_0, \alpha_1, \dots, \alpha_{\lfloor \frac{m-1}{2} \rfloor}\right)^T \in \mathbb{R}^m,$$

as well as

$$\tau_m = \tau_{m-1} \vartheta_m c_m$$

with

$$\vartheta_m = \frac{\|\boldsymbol{w}_{m+1}\|_2}{\tau_{m-1}}$$

and

$$c_m = \left(1 + \vartheta_m^2\right)^{-1/2}. \tag{4.3.145}$$

Proof:
A straightforward calculation yields

$$\overline{\boldsymbol{T}}_m = \begin{pmatrix} \frac{\|\boldsymbol{w}_1\|_2}{\alpha_0} & & & \\ -\frac{\|\boldsymbol{w}_2\|_2}{\alpha_0} & \frac{\|\boldsymbol{w}_2\|_2}{\alpha_0} & & \\ & \ddots & \ddots & \\ & & -\frac{\|\boldsymbol{w}_m\|_2}{\alpha_{\lfloor \frac{m-2}{2} \rfloor}} & \frac{\|\boldsymbol{w}_m\|_2}{\alpha_{\lfloor \frac{m-1}{2} \rfloor}} \\ & & & -\frac{\|\boldsymbol{w}_{m+1}\|_2}{\alpha_{\lfloor \frac{m-1}{2} \rfloor}} \end{pmatrix},$$

which leads to

$$\tilde{\boldsymbol{\alpha}}_m = \|\boldsymbol{w}_1\|_2 \boldsymbol{T}_m^{-1} \boldsymbol{e}_1 \tag{4.3.146}$$

and thus

$$\|\boldsymbol{w}_{m+1}\|_2 = \left\| \|\boldsymbol{w}_1\|_2 \boldsymbol{e}_1 - \overline{\boldsymbol{T}}_m \tilde{\boldsymbol{\alpha}}_m \right\|_2 \tag{4.3.147}$$

holds. Analogous to Lemma 4.91, we perform an orthogonal transformation of the matrix $\overline{\boldsymbol{T}}_m$ using Givens rotations and obtain the representation

$$\boldsymbol{Q}_m \overline{\boldsymbol{T}}_m = \overline{\boldsymbol{R}}_m = \begin{pmatrix} \boldsymbol{R}_m \\ \boldsymbol{0}^T \end{pmatrix}. \tag{4.3.148}$$

Furthermore, we define

$$\overline{\boldsymbol{g}}_m := \|\boldsymbol{w}_1\|_2 \boldsymbol{Q}_m \boldsymbol{e}_1.$$

Due to the special structure of the matrix $\overline{\boldsymbol{T}}_m$, the mapping $\boldsymbol{R}_m$ is an upper right triangular matrix with bandwidth two. For the vector $\overline{\boldsymbol{g}}_m$, we obtain, using

$$\overline{\boldsymbol{g}}_m = \begin{pmatrix} \boldsymbol{g}_m \\ \tilde{\gamma}_{m+1} \end{pmatrix} = \begin{pmatrix} \gamma_1 \\ \vdots \\ \gamma_m \\ \tilde{\gamma}_{m+1} \end{pmatrix} \quad \text{and} \quad \overline{\boldsymbol{g}}_{m-1} = \begin{pmatrix} \boldsymbol{g}_{m-1} \\ \tilde{\gamma}_m \end{pmatrix}$$

and the transformation given by (4.3.148), the form

$$\begin{aligned}
\overline{\boldsymbol{g}}_m &= \begin{pmatrix} \boldsymbol{I} & & \\ & c_m & s_m \\ & -s_m & c_m \end{pmatrix} \begin{pmatrix} \boldsymbol{Q}_{m-1} & \boldsymbol{0} \\ \boldsymbol{0}^T & 1 \end{pmatrix} \begin{pmatrix} \|\boldsymbol{w}_1\|_2 \\ \boldsymbol{0} \end{pmatrix} \\[2mm]
&= \begin{pmatrix} \boldsymbol{I} & & \\ & c_m & s_m \\ & -s_m & c_m \end{pmatrix} \begin{pmatrix} \boldsymbol{g}_{m-1} \\ \tilde{\gamma}_m \\ 0 \end{pmatrix} = \begin{pmatrix} \boldsymbol{g}_{m-1} \\ c_m \tilde{\gamma}_m \\ -s_m \tilde{\gamma}_m \end{pmatrix},
\end{aligned} \qquad (4.3.149)$$

where, without loss of generality, $c_m \geq 0$ is assumed. Analogous to the GMRES method, $\boldsymbol{R}_m$ is a non-singular matrix, and we obtain

$$\boldsymbol{\alpha}_m = \arg \min_{\boldsymbol{\alpha} \in \mathbb{R}^m} J_m(\boldsymbol{\alpha}) = \boldsymbol{R}_m^{-1} \boldsymbol{g}_m.$$

Furthermore, it holds that

$$\begin{pmatrix} \boldsymbol{R}_m \\ \boldsymbol{0}^T \end{pmatrix} = \overline{\boldsymbol{R}}_m = \begin{pmatrix} \boldsymbol{I} & & \\ & c_m & s_m \\ & -s_m & c_m \end{pmatrix} \begin{pmatrix} \boldsymbol{Q}_{m-1} \boldsymbol{T}_m \\ 0 \ldots 0 \, h_{m+1,m} \end{pmatrix},$$

which, with $c_m^2 + s_m^2 = 1$, yields the equation

$$\boldsymbol{Q}_{m-1} \boldsymbol{T}_m = \begin{pmatrix} \boldsymbol{I} & \boldsymbol{0} \\ \boldsymbol{0}^T & c_m \end{pmatrix} \boldsymbol{R}_m.$$

With (4.3.146), we therefore obtain, taking into account the assumed invertibility of the matrix $\boldsymbol{T}_m$, the property $c_m \neq 0$ and consequently the representation

$$\begin{aligned}
\tilde{\boldsymbol{\alpha}}_m &= \|\boldsymbol{w}_1\|_2 \boldsymbol{T}_m^{-1} \boldsymbol{e}_1 \\[2mm]
&= \left(\boldsymbol{Q}_{m-1} \boldsymbol{T}_m \right)^{-1} \overline{\boldsymbol{g}}_{m-1} \\[2mm]
&= \boldsymbol{R}_m^{-1} \begin{pmatrix} \boldsymbol{I} & \boldsymbol{0} \\ \boldsymbol{0}^T & c_m^{-1} \end{pmatrix} \begin{pmatrix} \boldsymbol{g}_{m-1} \\ \tilde{\gamma}_m \end{pmatrix} \\[2mm]
&= \boldsymbol{R}_m^{-1} \begin{pmatrix} \boldsymbol{R}_{m-1} \boldsymbol{\alpha}_{m-1} \\ \tilde{\gamma}_m c_m^{-1} \end{pmatrix}.
\end{aligned} \qquad (4.3.150)$$

It is easily seen that the matrix $\boldsymbol{R}_m$ takes the form

$$R_m = \begin{pmatrix} R_{m-1} & \begin{matrix} 0 \\ r_{m-1,m} \end{matrix} \\ 0^T & r_{m,m} \end{pmatrix}$$

and utilizing (4.3.149) one can write

$$g_m = \begin{pmatrix} g_{m-1} \\ c_m \tilde{\gamma}_m \end{pmatrix}.$$

Based on $g_{m-1} = R_{m-1}\alpha_{m-1}$ and taking account the expression of the upper triangular matrix R_m derived above, we obtain

$$\begin{aligned} \alpha_m &= R_m^{-1} g_m = R_m^{-1} \begin{pmatrix} R_{m-1}\alpha_{m-1} \\ \tilde{\gamma}_m c_m \end{pmatrix} \\[2mm] &= R_m^{-1} \left((1 - c_m^2) \begin{pmatrix} R_{m-1}\alpha_{m-1} \\ 0 \end{pmatrix} + c_m^2 \begin{pmatrix} R_{m-1}\alpha_{m-1} \\ \tilde{\gamma}_m c_m^{-1} \end{pmatrix} \right) \\[2mm] &= (1 - c_m^2) \begin{pmatrix} \alpha_{m-1} \\ 0 \end{pmatrix} + c_m^2 \tilde{\alpha}_m \end{aligned} \tag{4.3.151}$$

such that the claimed form (4.3.144) is proven. Applying the function J_m to $\tilde{\alpha}_m$, we obtain, using the identity $c_m^{-2} - 1 = c_m^{-2} s_m^2$, the equation

$$\begin{aligned} \|w_{m+1}\|_2 \overset{(4.3.147)}{=} \; & J_m(\tilde{\alpha}_m) = \left\| Q_m \left[\begin{pmatrix} \|w_1\|_2 \\ 0 \end{pmatrix} - T_m \tilde{\alpha}_m \right] \right\|_2 \\[2mm] \overset{\substack{(4.3.148)\\(4.3.149)}}{=} \; & \left\| \begin{pmatrix} g_{m-1} \\ c_m \tilde{\gamma}_m \\ -s_m \tilde{\gamma}_m \end{pmatrix} - \begin{pmatrix} R_m \tilde{\alpha}_m \\ 0 \end{pmatrix} \right\|_2 \\[2mm] \overset{(4.3.150)}{=} \; & \left\| \begin{pmatrix} g_{m-1} \\ c_m \tilde{\gamma}_m \\ -s_m \tilde{\gamma}_m \end{pmatrix} - \begin{pmatrix} g_{m-1} \\ \tilde{\gamma}_m c_m^{-1} \\ 0 \end{pmatrix} \right\|_2 \\[2mm] = \; & \sqrt{\tilde{\gamma}_m^2 (c_m^2 - 2 + c_m^{-2} + s_m^2)} = |\tilde{\gamma}_m| |s_m| c_m^{-1}, \end{aligned}$$

which, together with (4.3.149) and (4.3.151), yields

$$\begin{aligned} \tau_m &= \left\| \begin{pmatrix} g_{m-1} \\ c_m \tilde{\gamma}_m \\ -s_m \tilde{\gamma}_m \end{pmatrix} - \begin{pmatrix} R_m \alpha_m \\ 0 \end{pmatrix} \right\|_2 = |\tilde{\gamma}_m s_m| \\[2mm] &= \tau_{m-1} \frac{\|w_{m+1}\|_2}{\tau_{m-1}} c_m = \tau_{m-1} \vartheta_m c_m \end{aligned}$$

with $\vartheta_m = \|w_{m+1}\|_2 \tau_{m-1}^{-1}$, and furthermore, since $s_m^2 + c_m^2 = 1$ and taking into account $\tau_{m-1} = |\tilde{\gamma}_m|$, we have

$$c_m = \sqrt{\frac{1}{1 + \dfrac{s_m^2}{c_m^2}}} = \sqrt{\frac{1}{1 + \vartheta_m^2}}.$$

$\square$

Theorem 4.106 *Let* $c_m, \vartheta_m, \boldsymbol{y}_m$ *be given by Lemma 4.102 and Theorem 4.105, and* $m \geq 1$. *Then the TFQMR iterate can be written in the recursive form*

$$\boldsymbol{x}_m = \boldsymbol{x}_{m-1} + \eta_m \boldsymbol{d}_m$$

using

$$\eta_m := \alpha_{\lfloor \frac{m-1}{2} \rfloor} c_m^2 \qquad (4.3.152)$$

and

$$\boldsymbol{d}_m := \boldsymbol{y}_m + \frac{\vartheta_{m-1}^2 \eta_{m-1}}{\alpha_{\lfloor \frac{m-1}{2} \rfloor}} \boldsymbol{d}_{m-1}.$$

Proof:

Let us define

$$\tilde{\boldsymbol{x}}_m = \boldsymbol{x}_0 + \boldsymbol{Y}_m \tilde{\boldsymbol{\alpha}}_m$$

with the vector $\tilde{\boldsymbol{\alpha}}_m$ given by Theorem 4.105 and the matrix $\boldsymbol{Y}_m$ according to (4.3.137), then it follows that

$$\tilde{\boldsymbol{x}}_m = \boldsymbol{x}_0 + \boldsymbol{Y}_{m-1} \tilde{\boldsymbol{\alpha}}_{m-1} + \alpha_{\lfloor \frac{m-1}{2} \rfloor} \boldsymbol{y}_m = \tilde{\boldsymbol{x}}_{m-1} + \alpha_{\lfloor \frac{m-1}{2} \rfloor} \boldsymbol{y}_m. \qquad (4.3.153)$$

For the TFQMR iterate $\boldsymbol{x}_m$, using (4.3.143) and (4.3.144), the recursion formula is obtained as

$$
\begin{aligned}
\boldsymbol{x}_m &= \left(1 - c_m^2\right)\left(\boldsymbol{x}_0 + \boldsymbol{Y}_m \begin{pmatrix} \boldsymbol{\alpha}_{m-1} \\ 0 \end{pmatrix}\right) + c_m^2 \left(\boldsymbol{x}_0 + \boldsymbol{Y}_m \tilde{\boldsymbol{\alpha}}_m\right) \\[2mm]
&= \left(1 - c_m^2\right) \boldsymbol{x}_{m-1} + c_m^2 \tilde{\boldsymbol{x}}_m \\[2mm]
&\overset{(4.3.152)}{=} \boldsymbol{x}_{m-1} + \frac{\eta_m}{\alpha_{\lfloor \frac{m-1}{2} \rfloor}} \left(\tilde{\boldsymbol{x}}_m - \boldsymbol{x}_{m-1}\right).
\end{aligned}
\qquad (4.3.154)
$$

If we define the vector $\tilde{\boldsymbol{d}}_m$ as

$$\tilde{\boldsymbol{d}}_m := \frac{\tilde{\boldsymbol{x}}_m - \boldsymbol{x}_{m-1}}{\alpha_{\lfloor \frac{m-1}{2} \rfloor}} \qquad (4.3.155)$$

then the equation

$$
\begin{aligned}
\tilde{\boldsymbol{d}}_m &\overset{\substack{(4.3.153)\\(4.3.154)}}{=} \boldsymbol{y}_m + \frac{1}{\alpha_{\lfloor \frac{m-1}{2} \rfloor}} \left(\tilde{\boldsymbol{x}}_{m-1} - \boldsymbol{x}_{m-2} - \eta_{m-1} \tilde{\boldsymbol{d}}_{m-1}\right) \\[3mm]
&\overset{\substack{(4.3.152)\\(4.3.155)}}{=} \boldsymbol{y}_m + \frac{1}{\alpha_{\lfloor \frac{m-1}{2} \rfloor}} \left(\frac{1}{c_{m-1}^2} \eta_{m-1} \tilde{\boldsymbol{d}}_{m-1} - \eta_{m-1} \tilde{\boldsymbol{d}}_{m-1}\right) \\[3mm]
&\overset{(4.3.145)}{=} \boldsymbol{y}_m + \frac{\vartheta_{m-1}^2 \eta_{m-1}}{\alpha_{\lfloor \frac{m-1}{2} \rfloor}} \tilde{\boldsymbol{d}}_{m-1}
\end{aligned}
$$

follows, which allows us to identify $\boldsymbol{d}_m$ with $\tilde{\boldsymbol{d}}_m$. Substituting (4.3.155) into (4.3.154) thus yields the claimed representation of the TFQMR iterates. $\qquad\square$

Let us consider the vectors $\boldsymbol{q}_j, \boldsymbol{u}_{j+1}$, and $\boldsymbol{p}_{j+1}$ that appear in the CGS algorithm. Using Lemmas 4.102 and 4.103 as well as the auxiliary variables

$$\boldsymbol{v}_j := \boldsymbol{A}\boldsymbol{p}_j$$

we obtain the recursion formulas

$$\boldsymbol{y}_{2j} \;=\; \boldsymbol{q}_{j-1} = \boldsymbol{u}_{j-1} - \alpha_{j-1}\boldsymbol{v}_{j-1} = \boldsymbol{y}_{2j-1} - \alpha_{j-1}\boldsymbol{v}_{j-1},$$

$$\boldsymbol{y}_{2j+1} \;=\; \boldsymbol{u}_j = \boldsymbol{r}_j + \beta_{j-1}\boldsymbol{q}_{j-1} = \boldsymbol{w}_{2j+1} + \beta_{j-1}\boldsymbol{y}_{2j}.$$

Furthermore, we have

$$\boldsymbol{p}_j = \boldsymbol{u}_j + \beta_{j-1}\left(\boldsymbol{q}_{j-1} + \beta_{j-1}\boldsymbol{p}_{j-1}\right) = \boldsymbol{y}_{2j+1} + \beta_{j-1}\left(\boldsymbol{y}_{2j} + \beta_{j-1}\boldsymbol{p}_{j-1}\right),$$

so that for the auxiliary variable one can conclude

$$\boldsymbol{v}_j = \boldsymbol{A}\boldsymbol{y}_{2j+1} + \beta_{j-1}\left(\boldsymbol{A}\boldsymbol{y}_{2j} + \beta_{j-1}\boldsymbol{v}_{j-1}\right).$$

With lemmas 4.103 and 4.104 as well as the theorems 4.105 and 4.106, we obtain, in summary, the TFQMR algorithm in the following formulation, whose implementation in MATLAB is listed in Appendix A.

TFQMR Algorithm —

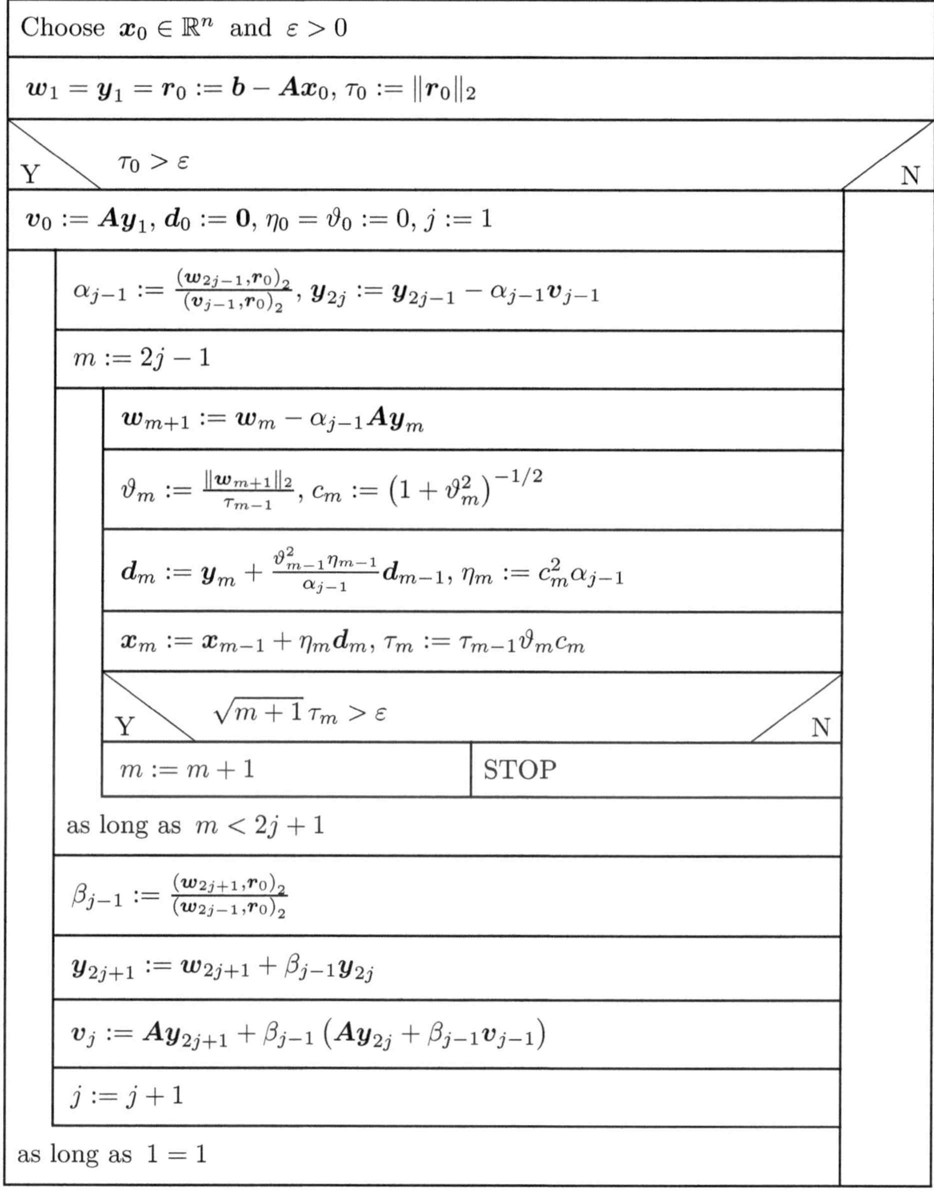

Choose $\boldsymbol{x}_0 \in \mathbb{R}^n$ and $\varepsilon > 0$

$\boldsymbol{w}_1 = \boldsymbol{y}_1 = \boldsymbol{r}_0 := \boldsymbol{b} - \boldsymbol{A}\boldsymbol{x}_0,\ \tau_0 := \|\boldsymbol{r}_0\|_2$

Y $\tau_0 > \varepsilon$ N

$\boldsymbol{v}_0 := \boldsymbol{A}\boldsymbol{y}_1,\ \boldsymbol{d}_0 := \boldsymbol{0},\ \eta_0 = \vartheta_0 := 0,\ j := 1$

$\alpha_{j-1} := \dfrac{(\boldsymbol{w}_{2j-1}, \boldsymbol{r}_0)_2}{(\boldsymbol{v}_{j-1}, \boldsymbol{r}_0)_2},\ \boldsymbol{y}_{2j} := \boldsymbol{y}_{2j-1} - \alpha_{j-1}\boldsymbol{v}_{j-1}$

$m := 2j - 1$

$\boldsymbol{w}_{m+1} := \boldsymbol{w}_m - \alpha_{j-1}\boldsymbol{A}\boldsymbol{y}_m$

$\vartheta_m := \dfrac{\|\boldsymbol{w}_{m+1}\|_2}{\tau_{m-1}},\ c_m := \left(1 + \vartheta_m^2\right)^{-1/2}$

$\boldsymbol{d}_m := \boldsymbol{y}_m + \dfrac{\vartheta_{m-1}^2 \eta_{m-1}}{\alpha_{j-1}}\boldsymbol{d}_{m-1},\ \eta_m := c_m^2 \alpha_{j-1}$

$\boldsymbol{x}_m := \boldsymbol{x}_{m-1} + \eta_m \boldsymbol{d}_m,\ \tau_m := \tau_{m-1}\vartheta_m c_m$

Y $\sqrt{m+1}\,\tau_m > \varepsilon$ N

$m := m + 1$ | STOP

as long as $m < 2j + 1$

$\beta_{j-1} := \dfrac{(\boldsymbol{w}_{2j+1}, \boldsymbol{r}_0)_2}{(\boldsymbol{w}_{2j-1}, \boldsymbol{r}_0)_2}$

$\boldsymbol{y}_{2j+1} := \boldsymbol{w}_{2j+1} + \beta_{j-1}\boldsymbol{y}_{2j}$

$\boldsymbol{v}_j := \boldsymbol{A}\boldsymbol{y}_{2j+1} + \beta_{j-1}\left(\boldsymbol{A}\boldsymbol{y}_{2j} + \beta_{j-1}\boldsymbol{v}_{j-1}\right)$

$j := j + 1$

as long as $1 = 1$

4.3.2.9   The QMRCGSTAB Method

Motivated by the development of the TFQMR method from the CGS algorithm, Chan et al. published the QMRCGSTAB method [18], which can be derived from the BiCGSTAB method using the principle of quasi-minimization. To this end, we define, with the following lemma, a basis of the Krylov subspace K_m using the mapping given by (4.3.133).

Lemma 4.107 *Assuming that the BiCGSTAB method does not break down before the computation of the vectors $\boldsymbol{p}_{\lfloor \frac{m-1}{2} \rfloor}$ and $\boldsymbol{s}_{\lfloor \frac{m-1}{2} \rfloor}$, then the vectors $\boldsymbol{y}_1,\dots,\boldsymbol{y}_m$ defined by*

$$\boldsymbol{y}_k := \begin{cases} \boldsymbol{p}_{\lfloor \frac{k-1}{2} \rfloor} & , \quad \text{if } k \text{ is odd}, \\ \boldsymbol{s}_{\lfloor \frac{k-1}{2} \rfloor} & , \quad \text{if } k \text{ is even} \end{cases} \tag{4.3.156}$$

form a basis of K_m.

Proof:
The proof follows analogously to the proof of Lemma 4.102 using equations (4.3.124) and (4.3.125). $\qquad\square$

Analogous to Lemma 4.103, we obtain the following lemma.

Lemma 4.108 *Let*

$$\boldsymbol{w}_k := \begin{cases} \boldsymbol{r}_{\lfloor \frac{k-1}{2} \rfloor} & , \quad \text{if } k \text{ is odd}, \\ \boldsymbol{s}_{\lfloor \frac{k-1}{2} \rfloor} & , \quad \text{if } k \text{ is even} \end{cases} \tag{4.3.157}$$

and

$$\delta_k := \begin{cases} \alpha_{\lfloor \frac{k-1}{2} \rfloor} & , \quad \text{if } k \text{ is odd}, \\ \omega_{\lfloor \frac{k-1}{2} \rfloor} & , \quad \text{if } k \text{ is even}, \end{cases}$$

then

$$\boldsymbol{w}_{k+1} = \boldsymbol{w}_k - \delta_k \boldsymbol{A}\boldsymbol{y}_k \quad \text{for } k = 1,\dots,m$$

holds, with the vectors $\boldsymbol{y}_1,\dots,\boldsymbol{y}_m$ given by (4.3.156).

Proof:
Using equations (4.3.121), (4.3.123), and (4.3.125), the assertion can be proven by straightforward calculation. $\qquad\square$

If we define the matrices $\boldsymbol{Y}_m$ and $\boldsymbol{W}_{m+1}$ according to (4.3.137) and (4.3.138) with the vectors given by (4.3.156) and (4.3.157), the residual vector can be written as

$$\boldsymbol{r}_m = \boldsymbol{W}_{m+1} \boldsymbol{S}_{m+1}^{-1} \boldsymbol{S}_{m+1} \left(\boldsymbol{e}_1 - \overline{\boldsymbol{B}}_m \boldsymbol{\alpha}_m \right)$$

with $\boldsymbol{S}_{m+1} = \operatorname{diag}\{\|\boldsymbol{w}_1\|_2,\dots,\|\boldsymbol{w}_{m+1}\|_2\}$,

$$\overline{\boldsymbol{B}}_m = \begin{pmatrix} \delta_1^{-1} & & & \\ -\delta_1^{-1} & \delta_2^{-1} & & \\ & \ddots & \ddots & \\ & & -\delta_{m-1}^{-1} & \delta_m^{-1} \\ & & & -\delta_m^{-1} \end{pmatrix} \in \mathbb{R}^{(m+1)\times m}$$

and $e_1 = (1,0,\ldots,0)^T \in \mathbb{R}^{m+1}$. The QMRCGSTAB algorithm differs from the TFQMR method only in the matrix $\overline{B}_m$, which has the same structure in both methods but different entries. Therefore, in the QMRCGSTAB method, the mapping J_m is also introduced according to (4.3.142), and the QMRCGSTAB iterate is defined by (4.3.143), so that the estimate for the norm of the residual vector given by Lemma 4.104 also holds. Moreover, Theorems 4.105 and 4.106 can be directly rewritten for the present QMR-CGSTAB method, and with the following two theorems, we obtain the mathematical foundation of the method.

Theorem 4.109 *Let* $m \geq 1$ *and furthermore for* $k = m-1,m$

$$\boldsymbol{\alpha}_k = \arg \min_{\boldsymbol{\alpha} \in \mathbb{R}^k} J_k(\boldsymbol{\alpha}) \text{ and } \tau_k := J_k(\boldsymbol{\alpha}_k)$$

with the function J_k *given by (4.3.142). Then*

$$\boldsymbol{\alpha}_m = \left(1 - c_m^2\right) \left(\begin{array}{c} \boldsymbol{\alpha}_{m-1} \\ \end{array} \right) + c_m^2 \tilde{\boldsymbol{\alpha}}_m$$

with $\tilde{\boldsymbol{\alpha}}_m := (\delta_1,\delta_2,\ldots,\delta_m)^T \in \mathbb{R}^m$ *as well as* $\tau_m = \tau_{m-1}\vartheta_m c_m$ *with* $\vartheta_m = \frac{\|\boldsymbol{w}_{m+1}\|_2}{\tau_{m-1}}$
and $c_m = \left(1 + \vartheta_m^2\right)^{-1/2}$.

Proof:
The proof follows analogously to the proof of Theorem 4.105. $\qquad\square$

Theorem 4.110 *Let* $c_m,\vartheta_m,\boldsymbol{y}_m$ *be given by Lemma 4.107 and Theorem 4.109, and* $m \geq 1$. *Then the QMRCGSTAB iterate can be written as*

$$\eta_m := \delta_m c_m^2$$

and

$$\boldsymbol{d}_m := \boldsymbol{y}_m + \frac{\vartheta_{m-1}^2 \eta_{m-1}}{\delta_m} \boldsymbol{d}_{m-1}$$

in the recursive form

$$\boldsymbol{x}_m = \boldsymbol{x}_{m-1} + \eta_m \boldsymbol{d}_m.$$

Proof:
The assertion can be derived analogously to the proof of Theorem 4.106. $\qquad\square$

If we take into account the representation of the vectors $\boldsymbol{y}_1,\ldots,\boldsymbol{y}_m$ and $\boldsymbol{w}_1,\ldots,\boldsymbol{w}_m$, the QMRCGSTAB algorithm can be written in the following form. An implementation in MATLAB can be found in Appendix A.

QMRCGSTAB Algorithm —

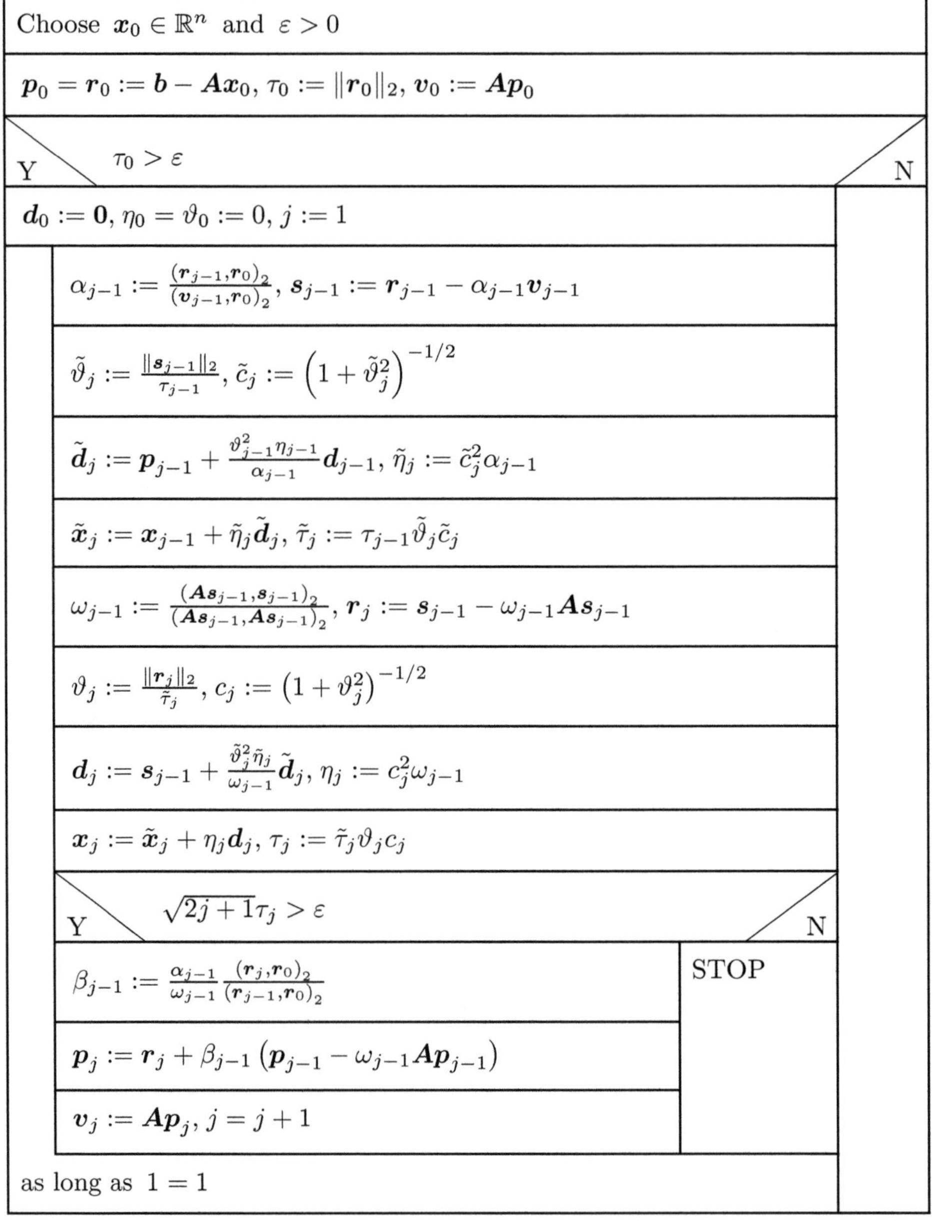

4.3.2.10   *Convergence Analyses*

In this section, we investigate the convergence behavior of the methods applicable to non-singular matrices: GMRES, CGS, BiCGSTAB, TFQMR, and QMRCGSTAB. For the analysis, we use the nonsymmetric matrix derived in Example 1.4 (1.0.9), where the advection parameter is set to $a = 1$ and the diffusion parameter to $\varepsilon = 0.1$, and with $N = 100$, there are always 10,000 unknowns.

All presented figures show the base-10 logarithm of the residual plotted against the iteration number. Due to the structure of the system of equations, a matrix-vector multiplication always requires $5N^2 - 4N$ operations, which is therefore always on the order of a scalar multiplication with N^2 operations. The CGS, BiCGSTAB, TFQMR, and QMRCGSTAB methods each require two matrix-vector multiplications per iteration. In contrast, the GMRES method only requires one such operation per iteration. However, the Arnoldi algorithm included in the GMRES method for computing the orthonormal basis results in a number of inner products that depends on the dimension of the Krylov subspace. In our case, the GMRES method was used in a restarted version with a maximum Krylov subspace dimension of $m = 30$. A smaller upper limit led to an increase in the number of iterations and was therefore avoided. Within one restart, 465 inner products must be computed as part of the Arnoldi algorithm, which corresponds to about 3 matrix-vector multiplications per iteration. All other methods do not involve any iteration-dependent number of inner products. Per iteration, 3 (CGS), 4 (BiCGSTAB, TFQMR), or 6 (QMRCGSTAB) inner products are computed, so these methods have a comparable computational effort per loop cycle. In contrast, the GMRES method on average requires significantly more computational effort.

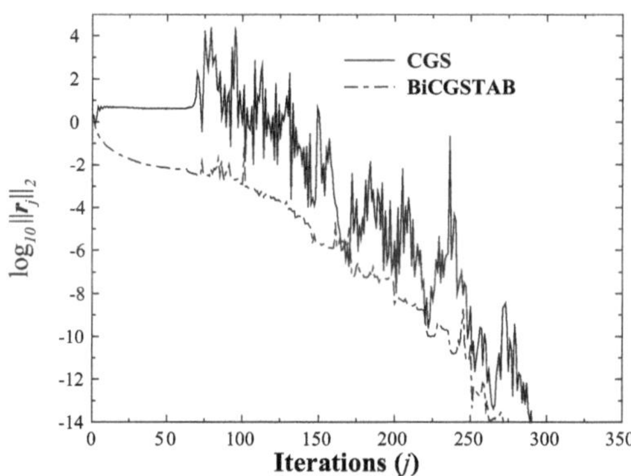

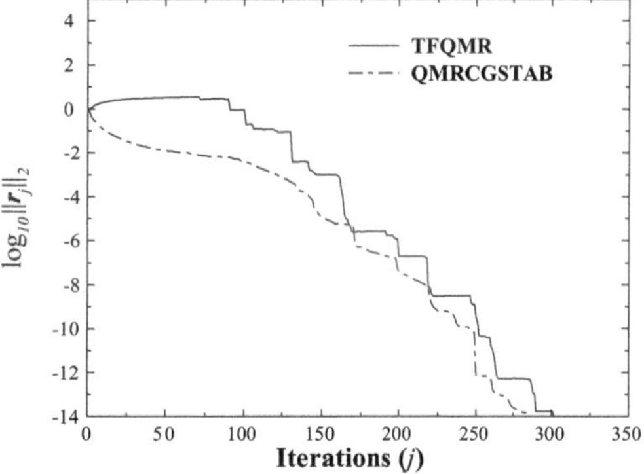

Figure 4.39 Convergence histories of the CGS and BiCGSTAB methods

Figure 4.40 Convergence histories of the TFQMR and QMRCGSTAB methods

The convergence histories shown display the expected typical behavior of the linear system solvers. While the CGS method exhibits very strong oscillations, the BiCGSTAB method, due to the introduced one-dimensional residual minimization, shows a much smoother decrease (see Figure 4.39). The use of quasi-minimization within the TFQMR and QMRCGSTAB methods, as seen in Figure 4.40, leads to an almost monotonic and oscillation-free convergence behavior. A consistently monotonic decay of the residual is theoretically expected for the GMRES method and its GMRES(m) variant. This property is also reflected in Figure 4.41.

From the determined convergence histories, it is evident that all considered linear system solvers achieved a reduction of the residual by 14 orders of magnitude in less than 1000 loop cycles. Here, the methods BiCGSTAB (272 iterations), CGS (291 iterations),

TFQMR (302 iterations), and QMRCGSTAB (286 iterations) required a comparable computational effort. Only the GMRES method required a significantly higher number of iterations, with 838 loop cycles.

In the context of the Euler and Navier-Stokes equations, [50] reports similar behavior of the linear system solvers, where the GM-RES method, without the use of suitable preconditioning, even proved to be impractical in some cases.

The methods CGS, BiCGSTAB, TFQMR, and QMRCGSTAB, which are based on the Bi-Lanczos algorithm, have proven to be robust and stable in many practically relevant problems. However, it should be mentioned that these four methods inherit the possible premature breakdowns of the Bi-Lanczos algorithm, and therefore convergence cannot be guaranteed even when possible rounding errors are neglected.

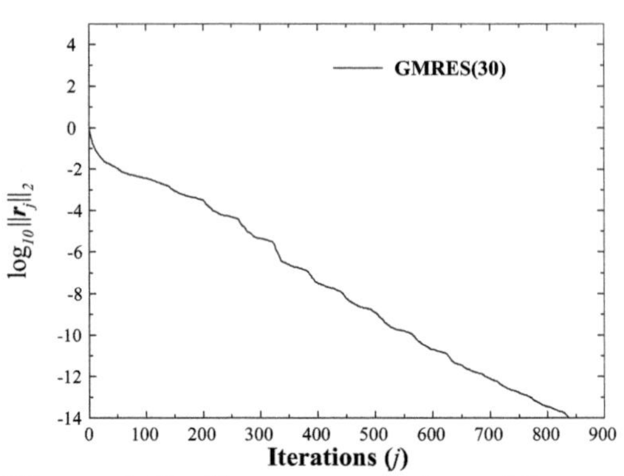

Figure 4.41   Convergence history of the GMRES(30) method

4.4   Exercises

Problem 1:
Show that the Jacobi method and the Gauss-Seidel method converge for a strictly diagonally dominant matrix $\boldsymbol{A} \in \mathbb{R}^{n \times n}$.
Hint: A matrix $\boldsymbol{A} = (a_{ij})_{i,j=1,\ldots,n} \in \mathbb{R}^{n \times n}$ is called strictly diagonally dominant if

$$|a_{kk}| > \sum_{\substack{i=1 \\ i \neq k}}^{n} |a_{ki}|, \quad k = 1,\ldots,n$$

holds.

Problem 2:
Let $\boldsymbol{x}, \boldsymbol{y} \in \mathbb{R}^n$ with $\boldsymbol{y} \neq \boldsymbol{0}$. Let $\psi : \mathbb{R} \to \mathbb{R}$ be defined by

$$\psi(a) := \|\boldsymbol{x} - a\boldsymbol{y}\|_2.$$

Show that:

$$\frac{\boldsymbol{x}^T \boldsymbol{y}}{\boldsymbol{y}^T \boldsymbol{y}} = \arg \min_{a \in \mathbb{R}} \psi(a).$$

Problem 3:
Let $A \in \mathbb{C}^{n \times n}$ be Hermitian and positive definite. Show that

$$\|\cdot\|_A : \mathbb{C}^n \;\rightarrow\; \mathbb{R}$$
$$x \;\mapsto\; \|x\|_A := \|A^{1/2}x\|_2$$

defines a norm on $\mathbb{C}^n$.

Problem 4:
Show that the matrix

$$A = \begin{pmatrix} 2 & -1 & & & \\ -1 & \ddots & \ddots & & \\ & \ddots & \ddots & \ddots & \\ & & \ddots & \ddots & -1 \\ & & & -1 & 2 \end{pmatrix} \in \mathbb{R}^{n \times n}$$

is irreducible, without using the previously introduced graph-theoretical approach.

Problem 5:
Given the matrix
$$A = \begin{pmatrix} 1 - \frac{7}{8}\cos\frac{\pi}{4} & \frac{7}{8}\sin\frac{\pi}{4} \\ -\frac{7}{8}\sin\frac{\pi}{4} & 1 - \frac{7}{8}\cos\frac{\pi}{4} \end{pmatrix}$$
and the vector
$$b = \begin{pmatrix} 0 \\ 0 \end{pmatrix}.$$

Show that the linear iteration method

$$x_{n+1} = (I - A)x_n + b, \quad n = 0,1,2,\ldots$$

converges to the uniquely determined solution $x = (0,0)^T$ for any initial vector $x_0 \in \mathbb{R}^2$, even though there exists an induced matrix norm with

$$\|I - A\| > 1.$$

Illustrate both facts graphically as well.

Problem 6:
Give an example where the linear iteration methods ψ and ϕ do not converge, but the product iteration $\psi \circ \phi$ does converge.

Problem 7:
The single-step method for solving the system $Ax = y$ is convergent for symmetric, real, positive definite $n \times n$ matrices A. Prove this statement for $n = 2$ without referring to the general proof (Theorem 4.18).

Problem 8:
A matrix $\boldsymbol{A} \in \mathbb{R}^{n \times n}$ is called an M-matrix if

- $a_{ii} > 0$, $i = 1, \ldots, n$,

- $a_{ij} \leq 0$, $i, j = 1, \ldots, n$, $i \neq j$, and

- $\boldsymbol{A}$ is non-singular with $\boldsymbol{A}^{-1} \geq \boldsymbol{0}$

hold.

(a) Give an example of an M-matrix that is not a diagonal matrix.

(b) Show for an M-matrix $\boldsymbol{A} \in \mathbb{R}^{n \times n}$:

(1) Let $\boldsymbol{A}\boldsymbol{x} = \boldsymbol{b}$ and $\boldsymbol{A}\boldsymbol{x}' = \boldsymbol{b}'$, then $\boldsymbol{b} \leq \boldsymbol{b}'$ implies $\boldsymbol{x} \leq \boldsymbol{x}'$.

(2) For the Jacobi iteration matrix associated with $\boldsymbol{A}$,

$$\boldsymbol{M}_J \geq \boldsymbol{0}$$

and

$$\rho(\boldsymbol{M}_J) < 1.$$

Hint: The relations $\boldsymbol{A} \geq \boldsymbol{0}$ and $\boldsymbol{x} \geq \boldsymbol{0}$ are always to be understood componentwise. Use the statement that for $\boldsymbol{A} \geq \boldsymbol{0}$, there always exists an eigenvector $\boldsymbol{x} \geq \boldsymbol{0}$ corresponding to $\lambda = \rho(\boldsymbol{A})$.

Problem 9:
We consider matrices $\boldsymbol{A} = (a_{ij})_{i,j=1,\ldots,n} \in \mathbb{R}^{n \times n}$ $(n \geq 2)$ with $a_{ii} = 1$, $i = 1, \ldots, n$ and $a_{ij} = a$ for $i \neq j$.

(a) What does the Jacobi iteration matrix for $\boldsymbol{A}$ look like? Compute its eigenvalues and the eigenvalues of $\boldsymbol{A}$.

(b) For which a does the Jacobi method converge? For which a is $\boldsymbol{A}$ positive definite?

(c) Are there positive definite matrices $\boldsymbol{A} \in \mathbb{R}^{n \times n}$ for which the Jacobi method does not converge?

Problem 10:
Let $U := \mathrm{span}\{\boldsymbol{u}_1, \ldots, \boldsymbol{u}_m\}$ be a subspace of $\mathbb{R}^n$

(a) Show: If for $\boldsymbol{x}, \boldsymbol{y} \in \mathbb{R}^n$, $\boldsymbol{x} - \boldsymbol{y} \in U$, then it follows that

$$\mathrm{span}\{U, \boldsymbol{x}\} = \mathrm{span}\{U, \boldsymbol{y}\}.$$

(b) Let $\boldsymbol{A} \in \mathbb{R}^{n \times n}$, then

$$\boldsymbol{A}U := \{\boldsymbol{A}\boldsymbol{x} \mid \boldsymbol{x} \in U\}.$$

Show that:

$$\boldsymbol{A}U = \mathrm{span}\{\boldsymbol{A}\boldsymbol{u}_1, \ldots, \boldsymbol{A}\boldsymbol{u}_m\}.$$

Problem 11:
Prove:

(a) If A has at least one positive and one negative real eigenvalue, then the Richardson method diverges for any choice of $\Theta \in \mathbb{C}$.

(b) If A has, among others, two complex eigenvalues λ_1, λ_2 with opposite sign $(\lambda_1/|\lambda_1| = -\lambda_2/|\lambda_2|)$, then the Richardson method diverges for any choice of $\Theta \in \mathbb{C}$.

(c) If the spectrum of A lies in a closed disk around $\mu \in \mathbb{C}\backslash\{0\}$ with radius $r < |\mu|$, then the choice $\Theta = 1/\mu$ leads to convergence of the Richardson method with

$$\rho(I - \Theta A) \leq r/|\mu| < 1.$$

Problem 12:
For the system

$$\begin{pmatrix} 4 & 0 & 2 \\ 0 & 5 & 2 \\ 5 & 4 & 10 \end{pmatrix} \begin{pmatrix} x_1 \\ x_2 \\ x_3 \end{pmatrix} = \begin{pmatrix} 4 \\ -3 \\ 2 \end{pmatrix}$$

determine the spectral radii of the iteration matrices for the simultaneous and single-step methods and write both methods in components. Show that the matrix is consistently ordered and determine the optimal relaxation parameter for the single-step method.

Problem 13:
Given the iterative scheme

$$x_{k+1} = x_k + a_k(b - Ax_k), \quad a_k \in \mathbb{R} \tag{4.4.1}$$

for solving the equation $Ax = b$ with $A \in \mathbb{R}^{n \times n}$ and $b \in \mathbb{R}^n$.

(a) Write the residual vector $r_{k+1} = b - Ax_{k+1}$ as a function of A, r_k and a_k, that is,

$$r_{k+1} = r_{k+1}(A, r_k, a_k).$$

(b) Compute $a_k \in \mathbb{R}$ such that r_{k+1} is minimal with respect to the Euclidean norm.

(c) Let $\mathcal{F}(A) := \left\{ \dfrac{y^T A y}{y^T y} : y \in \mathbb{R}^n\backslash\{0\} \right\}$.

Show: For the iterative scheme (4.4.1) with $a_k = \dfrac{(r_k, Ar_k)}{(Ar_k, Ar_k)}$ it holds for any $r_k \neq 0$

$$\|r_{k+1}\|_2 < \|r_k\|_2$$

if and only if $0 \notin \mathcal{F}(A^T)$.

Problem 14:
Given the system of equations $Ax = b$ with

$$A = \begin{pmatrix} 0.78 & 0.563 \\ 0.913 & 0.659 \end{pmatrix}, \qquad b = \begin{pmatrix} 0.217 \\ 0.254 \end{pmatrix}.$$

(a) What is the exact solution?

(b) For the two approximate solutions $u = (0.999, -1.001)^T$ and $v = (0.341, -0.087)^T$ compute the residuals $r(u)$ and $r(v)$. Does the *more accurate* solution u have the smaller residual? Explain the discrepancy by considering the residual function r.

 Note: The residual $r(z)$ of an approximate solution z of the equation $Ax = b$ is defined by $r(z) = \|b - Az\|_2$.

Problem 15:
The system of equations

$$\begin{pmatrix} 3 & -1 \\ -1 & 3 \end{pmatrix} \begin{pmatrix} x_1 \\ x_2 \end{pmatrix} = \begin{pmatrix} 1 \\ -1 \end{pmatrix}$$

is to be solved using the Jacobi and Gauss-Seidel methods. Approximately how many iterations are required in each case to reduce the error $\|x_n - x\|_2$ by a factor of 10^{-6}?

Problem 16:
Show that the method of steepest descent is consistent but not linear.

Problem 17:
Consider the method of conjugate directions and construct a matrix $A \in \mathbb{R}^{4 \times 4}$ as well as four A-orthogonal search directions $p_0, \ldots, p_3$ such that for the given right-hand side $b = (10,9,8,7)^T$ and given initial vector $x_0 = (1,2,3,4)^T$, the error vector determined by the method

$$e_m = x_m - A^{-1}b$$

satisfies the condition

$$\|e_m\|_2 \geq 2.749 \quad \text{for} \quad m = 0,1,2,3.$$

Problem 18:
Show: Let $A \in \mathbb{R}^{n \times n}$ be symmetric and positive definite and suppose $A^j = A$ for some $j \in \mathbb{N}$ with $n > j > 1$, then the CG method yields the exact solution of the equation $Ax = b$ for any $b \in \mathbb{R}^n$ at the latest with x_j.

Problem 19:
Show: Let $A \in \mathbb{R}^{n \times n}$ and $\xi \in \mathbb{R}^n$ be given. Furthermore, let ϕ_j, $\psi_j \in P_j$, then

$$\left(\phi_j(A^T)\eta, \psi_j(A)\xi \right)_2 = 0 \quad \forall \eta \in \mathbb{R}^n,$$

if $\xi \in \mathrm{Ker}(\phi_j(A)\psi_j(A))$.

Problem 20:
Given is the linear system

$$\begin{pmatrix} a & 14 \\ 7 & 50 \end{pmatrix} x = \begin{pmatrix} 1 \\ 1 \end{pmatrix} \quad \text{with } a \in \mathbb{R}^+ \setminus \left\{ \frac{49}{25} \right\}.$$

(a) For which values of a does the Jacobi method, respectively the Gauss-Seidel method, converge?

(b) Can the convergence rate be increased by relaxation of the Gauss-Seidel method?

5   Preconditioners

We have already studied in detail the influence of the condition number on the properties of the system of equations and the relationship between the error and residual vectors in Section 2.3. The results obtained (Theorems 2.41 and 2.42), as well as the convergence statements for the method of steepest descent (Theorem 4.68), the CG method (Theorem 4.80), and the GMRES method (Corollary 4.94), emphatically illustrate the advantage of a small condition number of the matrix $A \in \mathbb{R}^{n \times n}$ of the linear system $Ax = b$.

Motivated by these facts, it is natural to use an equivalent reformulation of the system of equations with the aim of reducing the condition number of the matrix. Such techniques are referred to as preconditioning the system and have proven to be an efficient means of accelerating and stabilizing Krylov subspace methods in the context of model problems [20, 34, 71, 76] as well as in practical applications [2, 23, 41, 50, 51, 52, 58]. In addition to the use of a suitable linear solver, the choice of preconditioner is of crucial importance for the resulting overall method. For this reason, in this section we will discuss and examine possible preconditioning techniques in detail. A current overview can also be found in the work of Benzi [9]. After an introductory definition, we will present different variants of preconditioning. We then present, in addition to the preconditioned CG method, a special formulation of a preconditioned BiCGSTAB method that allows a comparison of left and right preconditioning. For the convergence analysis, we use the Poisson equation and the convection-diffusion equation.

Definition 5.1 Let P_L and $P_R \in \mathbb{R}^{n \times n}$ be non-singular. Then

$$P_L A P_R x^P = P_L b \tag{5.0.1}$$

$$x = P_R x^P \tag{5.0.2}$$

is called the preconditioned system associated with the system $Ax = b$. If $P_L \neq I$, then P_L is called a left preconditioner and the system is left preconditioned. If $P_R \neq I$, then P_R is called a right preconditioner and the system is right preconditioned. A system that is both left and right preconditioned is called two-sided preconditioned.

Due to the invertibility of the matrix A, for example, A^T is a possible preconditioning matrix. With $P_L = A^T$ and $P_R = I$, we obtain the normal equations

$$A^T A x = A^T b. \tag{5.0.3}$$

Supplementary Information The online version contains supplementary material available at https://doi.org/10.1007/978-3-658-50260-7_5.

A. Meister, *Numerical Methods for Linear Systems of Equations*,
Mathematics Study Resources 26, https://doi.org/10.1007/978-3-658-50260-7_5

The matrix $\boldsymbol{B} = \boldsymbol{A}^T\boldsymbol{A}$ is symmetric and positive definite, which makes the use of the CG method possible. This approach is referred to as CGNR (CG Normal equations Residual minimizing). Analogously, $\boldsymbol{P}_L = \boldsymbol{I}$ and $\boldsymbol{P}_R = \boldsymbol{A}^T$ leads to the CGNE method (CG Normal equations Error minimizing). Both methods can thus be interpreted as preconditioned CG methods. However, these approaches often prove to be inefficient in the case of poorly conditioned matrices, since $\mathrm{cond}_2(\boldsymbol{A}\boldsymbol{A}^T) \gg \mathrm{cond}_2(\boldsymbol{A})$. If, on the other hand, $\boldsymbol{A}$ has a small condition number, both methods are good alternatives to the methods discussed in Section 4.3.2, since the positive properties of the CG algorithm carry over to these methods. Especially in the special case of orthogonal matrices, CGNR and CGNE yield the exact solution of the system in the first iteration step, since $\boldsymbol{A}^T\boldsymbol{A} = \boldsymbol{I}$.

With $\boldsymbol{P}_L\boldsymbol{A}\boldsymbol{P}_R = \boldsymbol{I}$, the matrices $\boldsymbol{P}_L$ and $\boldsymbol{P}_R$ are chosen optimally in terms of convergence. However, such a requirement is equivalent to inverting the matrix $\boldsymbol{A}$ and is therefore usually not feasible for reasons of storage and computational cost. The goal of preconditioning is to define matrices $\boldsymbol{P}_L$ and $\boldsymbol{P}_R$ that are easy to compute, require little storage, and for which $\boldsymbol{P}_L\boldsymbol{A}\boldsymbol{P}_R$ is a good approximation of the identity matrix, so that hopefully $\mathrm{cond}(\boldsymbol{P}_L\boldsymbol{A}\boldsymbol{P}_R) \ll \mathrm{cond}(\boldsymbol{A})$. The two-sided variant, as mentioned in Section 4.1.5, is used in methods for positive definite and symmetric matrices and the choice $\boldsymbol{P}_L^T = \boldsymbol{P}_R$ is used so that $\boldsymbol{P}_L\boldsymbol{A}\boldsymbol{P}_R$ again results in a positive definite and symmetric matrix.

In practice, a distinction is often made between explicit and implicit preconditioners. This terminology is used to emphasize whether the preconditioning matrix is available explicitly or whether the preconditioner is only known implicitly through its effect during matrix-vector multiplications. An example of the implicit variant is the incomplete LU factorization, in which the inverse of the preconditioner is available in factored form as the product of two triangular matrices. The product of the matrices $\boldsymbol{LU}$ can, even for sparse matrices $\boldsymbol{L}$ and $\boldsymbol{U}$, result in a dense matrix. The explicit computation of the product matrix could therefore potentially exceed the available memory resources. Furthermore, to determine the preconditioning matrix, the resulting matrix would have to be inverted at great expense, whereas the inversion of the product $\boldsymbol{LU}$ is implicitly given by simple backward and forward elimination.

5.1   Scalings

Scalings represent the simplest form of preconditioning. The advantages of low memory requirements and easy computation are offset by the disadvantage that scaling usually only provides a very rough approximation of the inverse of the matrix $\boldsymbol{A}$, and consequently, only a slight acceleration is typically achieved.

Definition 5.2 A non-singular diagonal matrix $\boldsymbol{D} = \mathrm{diag}\{d_{11}, \ldots, d_{nn}\} \in \mathbb{R}^{n \times n}$ is called scaling.

With the following list, we will present some common forms of scaling based on a non-singular matrix $\boldsymbol{A} \in \mathbb{R}^{n \times n}$:

(a) Scaling with the diagonal element:
Assuming $a_{ii} \neq 0$ for $i = 1, \ldots, n$, one chooses

$$d_{ii} := \frac{1}{a_{ii}} \text{ for } i = 1, \ldots, n. \tag{5.1.1}$$

(b) Row/column scaling with respect to the 1-norm (sum of absolute values):

$$d_{ii} := \frac{1}{\sum\limits_{j=1}^{n} |a_{ij}|} \text{ or } d_{jj} := \frac{1}{\sum\limits_{i=1}^{n} |a_{ij}|} \tag{5.1.2}$$

for $i = 1, \ldots, n$ or $j = 1, \ldots, n$, respectively.

(c) Row/column scaling with respect to the Euclidean norm:

$$d_{ii} := \frac{1}{\left(\sum\limits_{j=1}^{n} |a_{ij}|^2 \right)^{\frac{1}{2}}} \text{ or } d_{jj} := \frac{1}{\left(\sum\limits_{i=1}^{n} |a_{ij}|^2 \right)^{\frac{1}{2}}} \tag{5.1.3}$$

for $i = 1, \ldots, n$ or $j = 1, \ldots, n$, respectively.

(d) Row/column scaling with respect to the maximum norm:

$$d_{ii} := \frac{1}{\max\limits_{j=1,\ldots,n} |a_{ij}|} \text{ or } d_{jj} := \frac{1}{\max\limits_{i=1,\ldots,n} |a_{ij}|} \tag{5.1.4}$$

for $i = 1, \ldots, n$ or $j = 1, \ldots, n$, respectively.

As expected, the quality of a scaling also depends crucially on the chosen norm. In the 1-norm and the maximum norm, optimal scalings are even possible.

Theorem 5.3 *Let $A \in \mathbb{R}^{n \times n}$ be invertible and $\widetilde{D} \in \mathbb{R}^{n \times n}$ be given as row scaling with respect to the 1-norm according to (5.1.2). Then*

$$\mathrm{cond}_\infty(\widetilde{D}A) \leq \mathrm{cond}_\infty(DA) \tag{5.1.5}$$

holds for any scaling $D \in \mathbb{R}^{n \times n}$.

Proof:
Due to the assumed invertibility of the matrix A, $\widetilde{D}$ is well-defined, and with $C = \widetilde{D}A$, we have a matrix whose rows have length 1 with respect to the 1-norm. Let an arbitrary scaling $D \in \mathbb{R}^{n \times n}$ be given, then we define $\overline{D} = \mathrm{diag}\{\bar{d}_{11}, \ldots, \bar{d}_{11}\} = D\widetilde{D}^{-1}$. From the row equilibrium of the matrix C it follows that

$$\|DA\|_\infty = \|\overline{D}C\|_\infty = \max_{i=1,\ldots,n} |\bar{d}_{ii}| \, \|C\|_\infty = \max_{i=1,\ldots,n} |\bar{d}_{ii}| \, \|\widetilde{D}A\|_\infty,$$

whereas always

$$\|(DA)^{-1}\|_\infty = \|(\overline{D}C)^{-1}\|_\infty = \|C^{-1}\overline{D}^{-1}\|_\infty \geq \frac{\|C^{-1}\|_\infty}{\|\overline{D}\|_\infty} = \frac{1}{\max\limits_{i=1,\ldots,n} |\bar{d}_{ii}|} \|(\widetilde{D}A)^{-1}\|_\infty$$

holds. In summary, this yields the assertion. $\qquad\qquad\square$

Taking into account that $\|A\|_1 = \|A^T\|_\infty$, we obtain the following corollary.

Corollary 5.4 *Let* $A \in \mathbb{R}^{n \times n}$ *be invertible and* $\widetilde{D} \in \mathbb{R}^{n \times n}$ *be given as column scaling with respect to the 1-norm according to (5.1.2). Then*

$$\mathrm{cond}_1(A\widetilde{D}) \leq \mathrm{cond}_1(AD) \tag{5.1.6}$$

for any scaling $D \in \mathbb{R}^{n \times n}$.

Remark:
The condition number of a given matrix is invariant under multiplication by $\lambda \in \mathbb{R} \setminus \{0\}$. Thus, all optimal scalings are always only uniquely determined up to a multiplicative factor.

For the condition number in the Euclidean norm, the following estimate holds:

Theorem 5.5 *Let* $A \in \mathbb{R}^{n \times n}$ *be non-singular and let the scaling* $\widetilde{D} \in \mathbb{R}^{n \times n}$ *be chosen such that all columns of* $A\widetilde{D}$ *have the same length with respect to the Euclidean norm. Then,*

$$\mathrm{cond}_2(A\widetilde{D}) \leq \sqrt{n} \inf_{D} \mathrm{cond}_2(A D), \tag{5.1.7}$$

where the infimum is taken over all non-singular diagonal matrices $D \in \mathbb{R}^{n \times n}$.

Proof:
Let $B \in \mathbb{R}^{n \times n}$ be an arbitrary matrix and let b_j denote the j -th column of B , then the following chain of inequalities holds:

$$\max_{j=1,\ldots,n} \|b_j\|_2 \leq \|B\|_2 \leq \left(\sum_{i,j=1}^{n} |b_{ij}|^2 \right)^{\frac{1}{2}} \leq \sqrt{n} \max_{j=1,\ldots,n} \|b_j\|_2. \tag{5.1.8}$$

We define $C = A\widetilde{D}$. Since for any scaling $D \in \mathbb{R}^{n \times n}$ there exists a scaling $\overline{D} \in \mathbb{R}^{n \times n}$ with $D = \widetilde{D}^{-1}\overline{D}$, the statement can be formulated in the equivalent form

$$\mathrm{cond}_2(C) \leq \sqrt{n} \inf_{\overline{D}} \mathrm{cond}_2(C\overline{D}),$$

where the infimum is again taken over all invertible diagonal matrices $\overline{D} \in \mathbb{R}^{n \times n}$. Based on the above remark, we can also assume without loss of generality for the columns c_j of the matrix C the property

$$\|c_j\|_2 = 1. \tag{5.1.9}$$

Now, let an arbitrary scaling $\overline{D} \in \mathbb{R}^{n \times n}$ be given, and define $\hat{D} = \mathrm{diag}\{\hat{d}, \ldots, \hat{d}\} \in \mathbb{R}^{n \times n}$ with $\hat{d} = \max_{j=1,\ldots,n} |\overline{d}_{jj}|$. This choice implies

$$\|\hat{D}^{-1}\|_2 = \frac{1}{\|\overline{D}\|_2},$$

from which, using the submultiplicativity according to Theorem 2.17, the inequality

$$\|C^{-1}\|_2 = \hat{d}\|(C\hat{D})^{-1}\|_2 = \hat{d}\|(\hat{D}^{-1}\overline{D}\,\overline{D}^{-1}C^{-1}\|_2 \leq \hat{d}\|(C\overline{D})^{-1}\|_2 \tag{5.1.10}$$

follows. From (5.1.8) we obtain

$$\frac{\hat{d}}{\sqrt{n}}\|\boldsymbol{C}\|_2 = \frac{1}{\sqrt{n}}\|\boldsymbol{C}\hat{\boldsymbol{D}}\|_2 \leq \max_{j=1,\ldots,n} \|(\boldsymbol{C}\hat{\boldsymbol{D}})_j\|_2 \overset{(5.1.9)}{=} \max_{j=1,\ldots,n} \|(\boldsymbol{C}\overline{\boldsymbol{D}})_j\|_2 \leq \|\boldsymbol{C}\overline{\boldsymbol{D}}\|_2.$$

$$(5.1.11)$$

A combination of the inequalities (5.1.10) and (5.1.11) completes the proof. $\qquad\square$

Of course, this statement is unsatisfactory for practical purposes, since in practice the dimension of the system of equations is generally so large that even the factor $\sqrt{n}$ is unacceptably high. However, it is possible to prove sharpenings of this theorem, as can be found in the works [67] and [68].

It should be noted that scaling does not necessarily lead to an improvement in the condition number of the matrix. Theorem 5.3 states that in the case of an invertible matrix $\boldsymbol{A}$ whose rows already have the same length with respect to the maximum absolute row sum norm, no improvement in the condition number with respect to the row sum norm can be achieved by left preconditioning in the form of scaling. An analogous statement holds for right preconditioning with Corollary 5.4 for non-singular matrices whose columns have the same length with respect to the maximum column sum norm. Let us consider the matrix

$$\boldsymbol{A} = \begin{pmatrix} 2 & 100 \\ 100 & 100 \end{pmatrix},$$

then a left preconditioning of the system of equations by scaling according to (5.1.1) yields the matrix

$$\boldsymbol{B} = \begin{pmatrix} 1 & 50 \\ 1 & 1 \end{pmatrix}$$

with $\mathrm{cond}_2(\boldsymbol{B}) = 51.062$, so that the scaling, due to $\mathrm{cond}_2(\boldsymbol{A}) = 2.6899$, has led to a significant increase in the condition number.

5.2   Polynomial Preconditioners

The fundamental idea of polynomial preconditioners is based on representing the inverse of a matrix in the form of a Neumann series.

Theorem 5.6 *Let* $\rho\,(\boldsymbol{I} - \boldsymbol{A}) < 1$*, then the matrix* $\boldsymbol{A} \in \mathbb{R}^{n\times n}$ *is non-singular, and the inverse* $\boldsymbol{A}^{-1}$ *can be represented in the form of a Neumann series*

$$\boldsymbol{A}^{-1} = \sum_{k=0}^{\infty} (\boldsymbol{I} - \boldsymbol{A})^k. \tag{5.2.1}$$

Proof:
With $\rho\,(\boldsymbol{I} - \boldsymbol{A}) < 1$, according to Theorem 2.37, there exists an induced matrix norm $\|.\|$ with
$\|\boldsymbol{I} - \boldsymbol{A}\| < 1$. From this, it follows by the geometric series

$$\|\sum_{k=0}^{\infty}(\boldsymbol{I} - \boldsymbol{A})^k\| \leq \sum_{k=0}^{\infty} \|(\boldsymbol{I} - \boldsymbol{A})^k\| \leq \sum_{k=0}^{\infty} \|\boldsymbol{I} - \boldsymbol{A}\|^k = \frac{1}{1 - \|\boldsymbol{I} - \boldsymbol{A}\|},$$

so that with $\boldsymbol{B} = \sum_{k=0}^{\infty}(\boldsymbol{I} - \boldsymbol{A})^k$ we have a bounded linear operator, which, since $\|(\boldsymbol{I} - \boldsymbol{A})^{m+1}\| \leq \|\boldsymbol{I} - \boldsymbol{A}\|^{m+1} \to 0$ for $m \to \infty$, satisfies the property

$$\boldsymbol{BA} = \lim_{m\to\infty} \left[\sum_{k=0}^{m}(\boldsymbol{I} - \boldsymbol{A})^k\right] \boldsymbol{A} = \lim_{m\to\infty} \left(\boldsymbol{I} - (\boldsymbol{I} - \boldsymbol{A})^{m+1}\right) = \boldsymbol{I}$$

and thus represents the inverse of the matrix $\boldsymbol{A}$. $\qquad\qquad\square$

The condition on the spectral radius is, of course, quite restrictive. As the following theorem shows, however, this condition can always be satisfied for certain classes of matrices by means of a suitable scaling.

Theorem 5.7 *Let $\boldsymbol{A} = (a_{ij})_{i,j=1,\ldots,n} \in \mathbb{R}^{n\times n}$ be invertible and strictly diagonally dominant, and let the scaling $\boldsymbol{D} = \mathrm{diag}\{d_{11},\ldots,d_{nn}\}$ be given by*

$$d_{ii} := \frac{1}{a_{ii}} \text{ for } i = 1,\ldots,n. \tag{5.2.2}$$

Then

$$\rho\left(\boldsymbol{I} - \boldsymbol{DA}\right) < 1, \tag{5.2.3}$$

and it holds that

$$\boldsymbol{A}^{-1} = \left[\sum_{k=0}^{\infty}\left(\boldsymbol{I} - \boldsymbol{DA}\right)^k\right] \boldsymbol{D}. \tag{5.2.4}$$

Proof:
The existence and invertibility of the matrix $\boldsymbol{D}$ follows from the strict diagonal dominance of $\boldsymbol{A}$. Furthermore, $\|\boldsymbol{I} - \boldsymbol{DA}\|_\infty < 1$, so that (5.2.3) is satisfied, and the representation of the inverse $\boldsymbol{A}^{-1}$ follows from Theorem 5.6. $\qquad\qquad\square$

Of course, when using the matrix $\boldsymbol{D}$ as a suitable approximation of the inverse $\boldsymbol{A}^{-1}$, the condition (5.2.3) can in principle be ensured. However, this matrix $\boldsymbol{D}$ does not necessarily have to be a diagonal matrix, and we are thus back to the original problem of preconditioning.

The use of a truncated Neumann series as a preconditioner arises quite naturally from Theorem 5.6 and was first proposed in [22].

Definition 5.8 Let $\rho\left(\boldsymbol{I} - \boldsymbol{A}\right) < 1$ and $m \in \mathbb{N}$ arbitrary, then

$$\boldsymbol{P}^{(m)} := \sum_{k=0}^{m}\left(\boldsymbol{I} - \boldsymbol{A}\right)^k \tag{5.2.5}$$

is called the truncated Neumann series of order m for the matrix $\boldsymbol{A}$.

Remark:
The matrices $\boldsymbol{P}^{(m)}$ constructed in this way can be used for both left and right preconditioning. In implementation, it should be noted that $\boldsymbol{P}^{(m)}$ is only needed for matrix-vector multiplications, which suggests the use of Horner's scheme.

Implementation of a matrix-vector multiplication $P^{(m)}z$ —

$$b_0 := z$$

for $i = 1,\ldots,m$

$$b_i := z + (I - A)\, b_{i-1}$$

$$P^{(m)}z = b_m$$

The polynomial approach (5.2.5) suggests considering other polynomials besides the truncated Neumann series. For example, by using the ansatz

$$P := \sum_{k=0}^{m} \theta_k \, (I - A)^k \tag{5.2.6}$$

one obtains a parameter-dependent polynomial, where the weights θ_k can be used to specifically control the convergence rate of the iterative method.

If the matrix $A \in \mathbb{C}^{n\times n}$ is Hermitian, then from the space $\mathcal{P}_m$ of all polynomials of degree at most m, one can determine the best approximation p^*, which in a given norm $\|.\|$ satisfies the condition

$$\|I - A\, p^*(A)\| \leq \|I - A\, p(A)\| \quad \forall p \in \mathcal{P}_m. \tag{5.2.7}$$

We will demonstrate this approach using the Euclidean norm. According to Theorem 2.38, A is unitarily diagonalizable, and we obtain the condition equivalent to (5.2.7)

$$\|I - D\, p^*(D)\|_2 \leq \|I - D\, p(D)\|_2 \quad \forall p \in \mathcal{P}_m,$$

where $D = \operatorname{diag}\{\lambda_1,\ldots,\lambda_n\} \in \mathbb{R}^{n\times n}$ is the diagonal matrix representing the eigenvalues of A. Consequently, among all polynomials $p \in \mathcal{P}_m$, p^* minimizes the expression

$$\max_{i=1,\ldots,n} |1 - \lambda_i\, p(\lambda_i)|.$$

For the unknown eigenvalues, the Gerschgorin theorem [70] can be used to determine an interval $I = [a,b]$ with $a < 0 < b$ that contains all eigenvalues. Then, p^* is determined such that the expression

$$\max_{x\in I} |1 - x\, p(x)|$$

is minimized over all $p \in \mathcal{P}_m$. At this point, the natural question arises as to the existence of such a polynomial. First, with $\|q\|_\infty := \max_{x\in I} |q(x)|$, we have a norm on the linear space $\mathcal{P}_{m+1}(I) := \{q : I \mapsto \mathbb{R} \mid q \in \mathcal{P}_{m+1}\}$. Furthermore, $\mathcal{P}^0_{m+1}(I) := \{q \in \mathcal{P}_{m+1}(I) \mid q(0) = 0\}$ is a finite-dimensional subspace of $\mathcal{P}_{m+1}(I)$. If we choose a fixed but arbitrary polynomial $\tilde{q} \in \mathcal{P}^0_{m+1}(I)$, then

$$\mathcal{M} := \left\{ q \in \mathcal{P}^0_{m+1}(I) \mid \|1 - q\|_\infty \leq \|1 - \tilde{q}\|_\infty \right\}$$

is a closed and bounded subset of $\mathcal{P}^0_{m+1}(I)$, and is therefore compact. Thus, $\mathcal{M}$ is also a compact subset of $\mathcal{P}_{m+1}(I)$, so that for every $\hat{q} \in \mathcal{P}_{m+1}(I)$ there exists at least one $q \in \mathcal{M}$ with

$$\|\hat{q} - q\|_\infty \leq \|\hat{q} - \overline{q}\|_\infty \ \forall \overline{q} \in \mathcal{M}.$$

The choice $\hat{q}(x) = 1$ yields

$$\|1 - q\|_\infty \leq \|1 - \overline{q}\|_\infty \leq \|1 - \breve{q}\|_\infty$$

for all $\breve{q} \in \mathcal{P}^0_{m+1}(I)$. With

$$p(x) = q(x)/x \text{ for } x \neq 0$$

and $p(0) = \lim_{x \to 0} q(x)/x$ we obtain the desired polynomial. For details on the practical computation, see [5].

The determination of polynomial preconditioners as best approximations in a weighted L_2-norm can be found, for example, in [42] and [61].

5.3 Splitting-Associated Preconditioners

In Section 4.1, we discussed in detail the derivation of various splitting methods and the discussion of their properties. These methods are based on a decomposition of the matrix A in the form $A = B + (A - B)$. Here, B is intended to be either an easily invertible approximation of the matrix A or at least a matrix that enables an easy calculation of $B^{-1}x$, so that the resulting iteration matrix $M = B^{-1}(B - A)$ has a spectral radius as small as possible. This requirement is very similar to the desired properties of preconditioners, which suggests using the matrix $N = B^{-1}$ as a preconditioner.

Definition 5.9 Let a splitting method for solving $Ax = b$ with an invertible matrix N be given by $x_{j+1} = Mx_j + Nb$, then $P = N$ is called the preconditioner associated with the splitting method.

The preconditioning matrices introduced by the above definition can be used for both left and right preconditioning. The possible choices for the matrix N listed below are based exclusively on the iterative methods presented in Section 4.1, which also ensures the non-singularity of the diagonal matrix $D = \text{diag}\{a_{11}, \ldots, a_{nn}\}$ necessary for the invertibility of the matrix B. Furthermore, the matrices L and R according to (4.1.12) and (4.1.13) represent the strict lower left and strict upper right triangular matrices with respect to A, respectively. The following table provides a selection of possible splitting-associated preconditioners.

Splitting Method	Associated Preconditioner
Jacobi Method	$P_{Jac} = D^{-1}$
Gauss-Seidel Method	$P_{GS} = (D + L)^{-1}$
SOR Method	$P_{GS}(\omega) = \omega (D + \omega L)^{-1}$
Symm. Gauss-Seidel Method	$P_{SGS} = (D + R)^{-1} D (D + L)^{-1}$
SSOR Method	$P_{SGS}(\omega) = \omega(2 - \omega)(D + \omega R)^{-1} D (D + \omega L)^{-1}$

In implementation, one takes advantage of the fact that for the preconditioners presented, the matrices to be inverted are always diagonal or triangular in form, and thus the corresponding elimination technique can be used for the matrix-vector multiplication. The Jacobi-associated preconditioner shown is equivalent to scaling with the diagonal elements (5.1.1). The properties of the preconditioners $\boldsymbol{P}_{GS}(\omega)$ and $\boldsymbol{P}_{SGS}(\omega)$, which are related to a relaxation method, depend crucially on the choice of the relaxation parameter ω. In the case of a symmetric, positive definite matrix $\boldsymbol{A}$, [6] provides estimates for the spectrum of the preconditioned system and an ω that is optimal in the sense of cond_2.

5.4 The Incomplete LU Factorization

If a large sparse matrix is given, the computation of an LU decomposition proves to be inefficient and often even impractical in terms of computation time and memory requirements, even when rounding errors are neglected.

The idea of the incomplete LU factorization is to perform a decomposition of the form $\boldsymbol{A} = \boldsymbol{LU} + \boldsymbol{F}$ with a lower triangular matrix $\boldsymbol{L}$ and an upper triangular matrix $\boldsymbol{U}$, where the total memory requirement of the matrices $\boldsymbol{L}$ and $\boldsymbol{U}$ should be identical to that of the matrix $\boldsymbol{A}$. Neglecting the remainder matrix $\boldsymbol{F}$ yields an easily invertible matrix $\widetilde{\boldsymbol{A}} = \boldsymbol{LU}$, whose inverse can be used as an approximation of $\boldsymbol{A}^{-1}$ for preconditioning the system of equations $\boldsymbol{Ax} = \boldsymbol{b}$. To ensure the properties mentioned above, $n^2 + n$ conditions must be imposed on the incomplete LU decomposition.

Before we derive an easily programmable algorithm, we introduce the helpful concepts of matrix, column, and row patterns, as well as sparsity structure.

Definition 5.10 Any set

$$\mathcal{M} \subset \{(i,j) \mid i,j \in \{1,\ldots,n\}\}$$

is called a matrix pattern in the space $\mathbb{R}^{n \times n}$. Given a matrix pattern $\mathcal{M}$, the

$$\mathcal{M}_\mathcal{C}(j) := \{i \mid (i,j) \in \mathcal{M}\}$$

is called the j-th column pattern associated with $\mathcal{M}$, and

$$\mathcal{M}_\mathcal{R}(j) := \{i \mid (j,i) \in \mathcal{M}\}$$

is called the j-th row pattern associated with $\mathcal{M}$. For a given matrix $\boldsymbol{A} \in \mathbb{R}^{n \times n}$,

$$\mathcal{M}^{\boldsymbol{A}} := \{(i,j) \mid a_{ij} \neq 0\}$$

denotes the sparsity structure of $\boldsymbol{A}$.

With the help of the sparsity structure, we now specify the conditions for computing the incomplete LU factorization using the following definition.

Definition 5.11 Let $\boldsymbol{A} \in \mathbb{R}^{n \times n}$. If the decomposition

$$\boldsymbol{A} = \boldsymbol{LU} + \boldsymbol{F} \tag{5.4.1}$$

exists under the conditions

1) $u_{ii} = 1$ for $i = 1,\ldots,n$,

2) $\ell_{ij} = u_{ij} = 0$, if $(i,j) \notin \mathcal{M}^{\boldsymbol{A}}$,

3) $(\boldsymbol{LU})_{ij} = a_{ij}$, if $(i,j) \in \mathcal{M}^{\boldsymbol{A}}$,

and let $\boldsymbol{L} = (\ell_{ij})_{i,j=1,\ldots,n}$ and $\boldsymbol{U} = (u_{ij})_{i,j=1,\ldots,n}$ be a non-singular lower and upper triangular matrix, respectively, then (5.4.1) is called incomplete LU factorization (incomplete LU, ILU) of the matrix $\boldsymbol{A}$ with respect to the pattern $\mathcal{M}^{\boldsymbol{A}}$.

The existence results for the LR decomposition obtained in Section 3.1 cannot be transferred to the incomplete formulation. For example, due to condition 2, for

$$\boldsymbol{A} = \begin{pmatrix} 1 & -1 \\ 1 & 0 \end{pmatrix}$$

there is no incomplete LU factorization, although with $\det \boldsymbol{A}[k] \neq 0$ ($k = 1,2$), the existence and a form of uniqueness of the LR decomposition is guaranteed by Theorems 3.7 and 3.8. Nevertheless, the incomplete LU factorization is a very efficient form of preconditioning, which is often successfully applied in simulations of engineering problems, since the systems of equations that arise in this context usually do not contradict conditions 1 and 2 (see Examples 1.1 and 1.4). Existence results for the case of an M-matrix can be found in the work of Meijerink and van der Vorst [49] and for H-matrices in the publication by Manteuffel [48].

To derive the decomposition, we successively develop for $i = 1,\ldots,n$ first the i-th column of $\boldsymbol{L}$ and then the i-th row of $\boldsymbol{U}$. From conditions 1 and 3, we obtain, since $u_{mi} = 0$ ($m > i$), the equation

$$a_{ki} = \sum_{m=1}^{n} \ell_{km} u_{mi} = \sum_{m=1}^{i} \ell_{km} u_{mi} = \sum_{m=1}^{i-1} \ell_{km} u_{mi} + \ell_{ki},$$

from which the i-th column of $\boldsymbol{L}$ immediately follows in the form

$$\ell_{ki} = a_{ki} - \sum_{m=1}^{i-1} \ell_{km} u_{mi} \quad \text{for} \quad k = i,\ldots,n. \tag{5.4.2}$$

With $\ell_{im} = 0$ ($m > i$), it follows for $k = i+1,\ldots,n$ the equation

$$a_{ik} = \sum_{m=1}^{n} \ell_{im} u_{mk} = \sum_{m=1}^{i} \ell_{im} u_{mk},$$

and we obtain the representation of the i-th row of $\boldsymbol{U}$ according to

$$u_{ik} = \frac{1}{\ell_{ii}} \left(a_{ik} - \sum_{m=1}^{i-1} \ell_{im} u_{mk} \right) \quad \text{for} \quad k = i+1,\ldots,n. \tag{5.4.3}$$

Taking condition 2 into account finally yields the incomplete LU factorization in the following notation:

The incomplete LU factorization —

For $i = 1, \ldots, n$

> For $k = i, \ldots, n, \ k \in \mathcal{M}_{\mathcal{C}}^{\boldsymbol{A}}(i)$
>
>> $$\ell_{ki} = a_{ki} - \sum_{\substack{m=1 \\ m \in \mathcal{M}_{\mathcal{R}}^{\boldsymbol{A}}(k) \cap \mathcal{M}_{\mathcal{C}}^{\boldsymbol{A}}(i)}}^{i-1} \ell_{km} u_{mi}$$
>
> For $k = i+1, \ldots, n, \ k \in \mathcal{M}_{\mathcal{R}}^{\boldsymbol{A}}(i)$
>
>> $$u_{ik} = \frac{1}{\ell_{ii}} \left(a_{ik} - \sum_{\substack{m=1 \\ m \in \mathcal{M}_{\mathcal{R}}^{\boldsymbol{A}}(i) \cap \mathcal{M}_{\mathcal{C}}^{\boldsymbol{A}}(k)}}^{i-1} \ell_{im} u_{mk} \right)$$

Definition 5.12 Let

$$\boldsymbol{A} = \boldsymbol{L}\boldsymbol{U} + \boldsymbol{F}$$

be an incomplete LU factorization of the matrix $\boldsymbol{A}$, then

$$\boldsymbol{P} := \boldsymbol{U}^{-1}\boldsymbol{L}^{-1} \tag{5.4.4}$$

is the corresponding ILU preconditioner.

We have already noted that the matrix $\boldsymbol{P}$ is not computed explicitly, but is only used implicitly in its effect on the basis of a forward elimination and a subsequent backward elimination. Moreover, there is no need to store the diagonal elements of the matrix $\boldsymbol{U}$, so that the two matrices $\boldsymbol{L}$ and $\boldsymbol{U}$ together require the same amount of storage as $\boldsymbol{A}$.

For matrices with a relatively small bandwidth, the incomplete LU decomposition with *fill-in* often proves to be effective. Here, it is exploited that the choice of the sparsity pattern $\mathcal{M}^{\boldsymbol{A}}$ within conditions 2 and 3 is not mandatory. If, in these requirements, the sparsity pattern $\mathcal{M}^{\boldsymbol{A}}$ is replaced by a pattern $\mathcal{M} \supset \mathcal{M}^{\boldsymbol{A}}$, then the error matrix $\boldsymbol{F}$ has fewer nonzero entries, so that $\boldsymbol{L}\boldsymbol{U}$ usually provides a better approximation of the matrix $\boldsymbol{A}$. This extension is described by ILU(p), where the natural number p represents a degree for the number of *fill-in* elements. For a detailed description of the generalization mentioned, see Saad [62].

5.5 The Incomplete Cholesky Factorization

As we already observed in Section 3.2, in the case of a symmetric and positive definite matrix $\boldsymbol{A}$, the effort required for the LU decomposition can be reduced by specifically exploiting the properties of the matrix. The Cholesky factorization derived from this will now also be considered for preconditioning in an incomplete formulation.

Definition 5.13 Let $A \in \mathbb{R}^{n \times n}$ be symmetric and positive definite. Let the decomposition

$$A = LL^T + F \tag{5.5.1}$$

exists under the conditions

1) $\ell_{ij} = 0$ if $(i,j) \notin \mathcal{M}^A$,

2) $(LL^T)_{ij} = a_{ij}$ if $(i,j) \in \mathcal{M}^A$,

and let $L = (\ell_{ij})_{i,j=1,\dots,n}$ be a non-singular lower left triangular matrix, then (5.5.1) is called incomplete Cholesky factorization (IC) of the matrix A with respect to the pattern $\mathcal{M}^A$.

The main advantages of such a factorization lie in the computational and storage requirements, and in the property that, in the case of a positive definite, symmetric matrix A, due to

$$\left(LAL^T\right)^T = LAL^T$$

and

$$\left(LAL^T x, x\right)_2 = \left(AL^T x, L^T x\right)_2 > 0, \ \forall x \in \mathbb{R}^n \setminus \{0\}$$

the symmetry and positive definiteness are also transferred to the product LAL^T, which allows methods for symmetric, positive definite matrices to also be applied to the preconditioned system. From equations (3.2.2) and (3.2.3), taking into account the conditions imposed on the incomplete factorization, the computation of the incomplete Cholesky factorization is obtained in the following form:

The incomplete Cholesky factorization —

For $k = 1,\dots,n$

$$\ell_{kk} = \sqrt{a_{kk} - \sum_{\substack{j=1 \\ j \in \mathcal{M}_{\mathcal{R}}^A(k)}}^{k-1} \ell_{kj}^2}$$

For $i = k+1,\dots,n, \ i \in \mathcal{M}_{\mathcal{R}}^A(k)$

$$\ell_{ik} = \frac{1}{\ell_{kk}} \left(a_{ik} - \sum_{\substack{j=1 \\ j \in \mathcal{M}_{\mathcal{R}}^A(i) \cap \mathcal{M}_{\mathcal{R}}^A(k)}}^{k-1} \ell_{ij}\ell_{kj} \right)$$

In contrast to ILU, IC does not require an additional condition on the diagonal elements (see condition 1 in Definition 5.11), since, due to positive definiteness, $(k,k) \in \mathcal{M}^A$ always holds for $i = 1,\dots,n$.

5.6 The Incomplete QR Factorization

In the construction of the incomplete LU factorization, the main idea was to restrict the memory required for the preconditioner a priori by demanding a sparsity pattern equivalent to that of A. This idea can also be transferred to other preconditioners.

Definition 5.14 If, by means of a modified variant of any algorithm for computing a QR factorization, a representation of the matrix A in the form

$$A = QR + F \tag{5.6.1}$$

is computed, where the upper triangular matrix R and the matrix Q satisfy the conditions

1) $r_{ij} = 0$, if $(i,j) \notin \mathcal{M}^{A}$

2) $q_{ij} = 0$, if $(i,j) \notin \mathcal{M}^{A}$

and if, when replacing the sparsity pattern $\mathcal{M}^{A}$ by $\{(i,j) \mid i,j = 1,\dots,n\}$, Q is an orthogonal matrix and F is a zero matrix, then (5.6.1) is called incomplete QR factorization (IQR) of the matrix A with respect to the pattern $\mathcal{M}^{A}$.

With the Gram-Schmidt scheme, we have already encountered a method for determining a QR decomposition in Section 3.3.1. Taking into account the conditions 1 and 2 stated above, the Gram-Schmidt process can be used in the following form to compute an incomplete QR factorization.

The incomplete QR factorization —

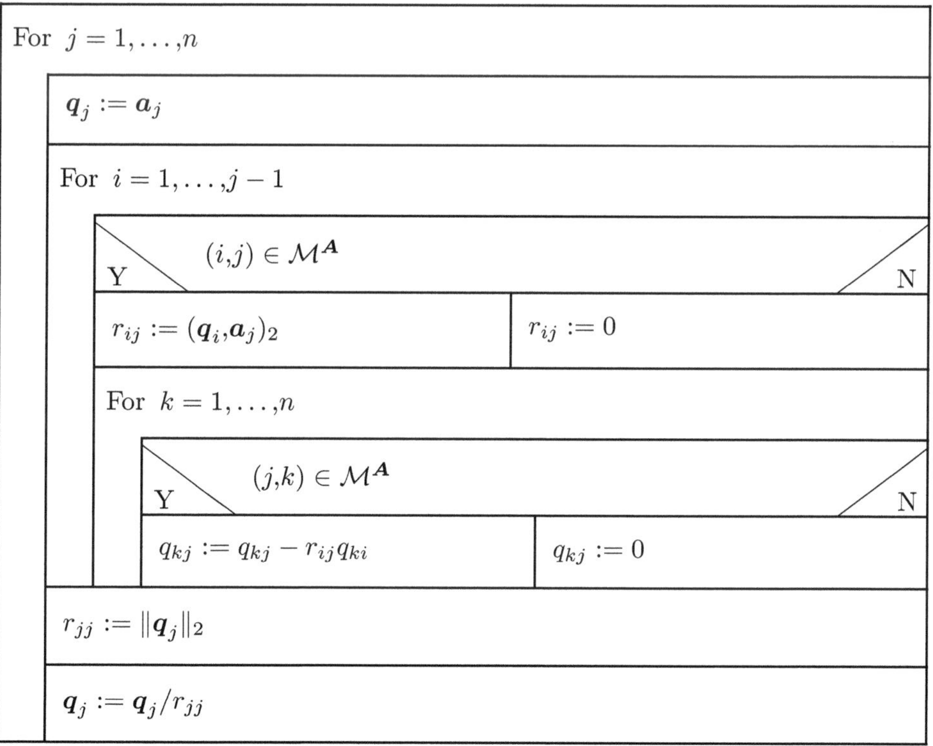

Definition 5.15 Let

$$A = QR + F$$

be an incomplete QR factorization of the matrix A. If Q and R are invertible, then

$$P := R^{-1}Q^T \tag{5.6.2}$$

is the corresponding IQR preconditioner.

The memory requirement for the matrices Q and R together is slightly more than 1.5 times that of the matrix A. Here, Q does not necessarily represent an orthogonal matrix, so that Q^T only provides an approximation to Q^{-1}.

5.7 The Incomplete Frobenius Inverse

In this section, we introduce a preconditioning matrix called the incomplete Frobenius inverse, which, for a given pattern, is optimal with respect to the Frobenius norm. Here, we will always consider the sparsity pattern of the matrix A as the pattern. However, there are also strategies for adaptively choosing the pattern [10].

Definition 5.16 A pattern $\mathcal{M}$ is called regular in $\mathbb{R}^{n \times n}$ if

$$\mathcal{P}_\mathcal{M} := \left\{ P \in \mathbb{R}^{n \times n} \mid P \text{ is non-singular and } \mathcal{M}^P = \mathcal{M} \right\} \neq \emptyset$$

holds. Provided that for a given regular pattern $\mathcal{M}$ the following minimum exists, then

$$P_\mathcal{M}^R \in \arg \min_{P \in \mathcal{P}_\mathcal{M}} \|AP - I\|_F^2$$

is called the incomplete Frobenius inverse (Frobenius right inverse) of the matrix A with respect to the pattern $\mathcal{M}$.

As the following lemma shows, the incomplete Frobenius inverse can be determined column by column. Since each column of $P_\mathcal{M}^R$ can be computed independently of the others, this construction is very well suited for parallel implementation; see, for example, [35].

Lemma 5.17 *Let $\mathcal{M}$ be a non-singular pattern in $\mathbb{R}^{n \times n}$ and*

$$\mathcal{V}_{\mathcal{M}_\mathcal{S}(j)} := \left\{ v \in \mathbb{R}^n \mid v_i = 0 \text{ for } i \notin \mathcal{M}_\mathcal{S}(j) \right\},$$

then $P_\mathcal{M}^R = (p_1, \dots, p_n)$ is an incomplete Frobenius inverse with respect to the pattern $\mathcal{M}$ if and only if each column vector p_j satisfies the condition

$$p_j \in \arg \min_{p \in \mathcal{V}_{\mathcal{M}_\mathcal{S}(j)} \setminus \{0\}} \|Ap - e_j\|_2^2 \tag{5.7.1}$$

where e_j denotes the j-th unit vector in $\mathbb{R}^n$.

Proof:

The claim follows directly from $\|\boldsymbol{A}\boldsymbol{P}_{\mathcal{M}}^R - \boldsymbol{I}\|_F^2 = \sum_{j=1}^n \|\boldsymbol{A}\boldsymbol{p}_j - \boldsymbol{e}_j\|_2^2.$ $\qquad\square$

To compute the incomplete Frobenius inverse, Lemma 5.17 requires solving a restricted minimization problem (5.7.1) for each column vector $\boldsymbol{p}_j$. In the following theorem, we will derive an equivalent linear least squares problem for this purpose.

Theorem 5.18 *Let $\boldsymbol{A} \in \mathbb{R}^{n \times n}$ be non-singular and $m_j = |\mathcal{M}_{\mathcal{S}}^{\boldsymbol{A}}(j)|$ for $j = 1,\ldots,n$. Furthermore, let $\mathcal{M}_{\mathcal{S}}^{\boldsymbol{A}}(j) = \left\{ i_1^j,\ldots,i_{m_j}^j \right\}$ and for $j = 1,\ldots,n$ let the matrix $\boldsymbol{Z}_j \in \mathbb{R}^{m_j \times n}$ be defined by*

$$(z_j)_{m\ell} := \delta_{\ell,i_m^j} \text{ for } m = 1,\ldots,m_j \text{ and } \ell = 1,\ldots,n$$

using the Kronecker symbol

$$\delta_{i,j} = \left\{ \begin{array}{ll} 0, & i \neq j, \\ 1, & i = j. \end{array} \right.$$

Then $\boldsymbol{P}_{\mathcal{M}^{\boldsymbol{A}}}^R = (\boldsymbol{p}_1,\ldots,\boldsymbol{p}_n)$ is a Frobenius inverse (Frobenius right inverse) of the matrix $\boldsymbol{A}$ with respect to the pattern $\mathcal{M}^{\boldsymbol{A}}$ if and only if for each column vector $\boldsymbol{p}_j$ the corresponding projected vector $\widetilde{\boldsymbol{p}}_j = \boldsymbol{Z}_j\boldsymbol{p}_j$ satisfies the condition

$$\widetilde{\boldsymbol{p}}_j \in \arg \min_{\boldsymbol{p} \in \mathbb{R}^{m_j}\setminus\{\boldsymbol{0}\}} \|\boldsymbol{A}\boldsymbol{Z}_j^T\boldsymbol{p} - \boldsymbol{e}_j\|_2^2. \tag{5.7.2}$$

Moreover, there exists at most one Frobenius inverse of the matrix $\boldsymbol{A}$ with respect to the pattern $\mathcal{M}^{\boldsymbol{A}}$.

Proof:

The matrix $\boldsymbol{Z}_j \in \mathbb{R}^{m_j \times n}$ defines an isomorphism between $\mathbb{R}^{m_j}\setminus\{\boldsymbol{0}\}$ and $\mathcal{V}_{\mathcal{M}_{\mathcal{S}}^{\boldsymbol{A}}(j)}\setminus\{\boldsymbol{0}\}$. Furthermore, $\boldsymbol{Z}_j^T\boldsymbol{Z}_j\boldsymbol{p} = \boldsymbol{p}$ holds for all $\boldsymbol{p} \in \mathcal{V}_{\mathcal{M}_{\mathcal{S}}^{\boldsymbol{A}}(j)}\setminus\{\boldsymbol{0}\}$ and $\boldsymbol{Z}_j\boldsymbol{Z}_j^T = \boldsymbol{I} \in \mathbb{R}^{m_j \times m_j}$. Thus, it follows that

$$\boldsymbol{Z}_j \left(\arg \min_{\boldsymbol{p} \in \mathcal{V}_{\mathcal{M}_{\mathcal{S}}^{\boldsymbol{A}}(j)}\setminus\{\boldsymbol{0}\}} \|\boldsymbol{A}\boldsymbol{p} - \boldsymbol{e}_j\|_2^2 \right) = \boldsymbol{Z}_j\boldsymbol{Z}_j^T \left(\arg \min_{\boldsymbol{p} \in \mathbb{R}^{m_j}\setminus\{\boldsymbol{0}\}} \|\boldsymbol{A}\boldsymbol{Z}_j^T\boldsymbol{p} - \boldsymbol{e}_j\|_2^2 \right)$$

$$= \arg \min_{\boldsymbol{p} \in \mathbb{R}^{m_j}\setminus\{\boldsymbol{0}\}} \|\boldsymbol{A}\boldsymbol{Z}_j^T\boldsymbol{p} - \boldsymbol{e}_j\|_2^2$$

and

$$\boldsymbol{Z}_j^T \left(\arg \min_{\boldsymbol{p} \in \mathbb{R}^{m_j}\setminus\{\boldsymbol{0}\}} \|\boldsymbol{A}\boldsymbol{Z}_j^T\boldsymbol{p} - \boldsymbol{e}_j\|_2^2 \right) = \boldsymbol{Z}_j^T\boldsymbol{Z}_j \left(\arg \min_{\boldsymbol{p} \in \mathcal{V}_{\mathcal{M}_{\mathcal{S}}^{\boldsymbol{A}}(j)}\setminus\{\boldsymbol{0}\}} \|\boldsymbol{A}\boldsymbol{Z}_j^T\boldsymbol{Z}_j\boldsymbol{p} - \boldsymbol{e}_j\|_2^2 \right)$$

$$= \arg \min_{\boldsymbol{p} \in \mathcal{V}_{\mathcal{M}_{\mathcal{S}}^{\boldsymbol{A}}(j)}\setminus\{\boldsymbol{0}\}} \|\boldsymbol{A}\boldsymbol{p} - \boldsymbol{e}_j\|_2^2,$$

whereby $\boldsymbol{Z}_j$ and $\boldsymbol{Z}_j^T$ represent isomorphisms on the sets listed in (5.7.1) and (5.7.2), and thus the first part of the assertion follows.

We now turn to the uniqueness of the Frobenius inverse. The matrix $A_j := AZ_j^T \in \mathbb{R}^{n \times m_j}$ consists of the m_j linearly independent columns a_i, $i \in \mathcal{M}_{\mathcal{S}}^A(j)$ of the matrix A. Thus, by Theorem 3.14, there exists a decomposition of the matrix A_j in the form

$$A_j = Q \begin{pmatrix} R \\ 0 \end{pmatrix}$$

with an orthogonal matrix $Q \in \mathbb{R}^{n \times n}$ and an invertible upper triangular matrix $R \in \mathbb{R}^{m_j \times m_j}$. With

$$f_j = \begin{pmatrix} f_{j1} \\ f_{j2} \end{pmatrix} = Q^T e_j$$

we obtain

$$\|AZ_j^T p - e_j\|_2^2 = \|Q^T(A_j p - e_j)\|_2^2 = \left\| \begin{pmatrix} R \\ 0 \end{pmatrix} p - \begin{pmatrix} f_{j1} \\ f_{j2} \end{pmatrix} \right\|_2^2 .$$

Due to the required existence of the minimum $\min\limits_{P \in \mathcal{P}_{\mathcal{M}}} \|AP - I\|_F^2$ it holds that $f_{j1} \neq 0$. Taking into account

$$\|AZ_j^T p - e_j\|_2^2 = \|Rp - f_{j1}\|_2^2 + \|f_{j2}\|_2^2$$

we obtain the uniquely determined j-th column of the Frobenius inverse by

$$p_j = Z_j^T \left(\arg \min\limits_{p \in \mathbb{R}^{m_j} \setminus \{0\}} \|Rp - f_{j1}\|_2^2 \right) = Z_j^T R^{-1} f_{j1} \neq 0.$$

□

Thus, the Frobenius inverse can be determined on the basis of the procedures for QR decomposition described in Section 3.3, whereby of course other algorithms (for example, the Householder transformation) can also be used.

The requirement for the existence of the minimum within Definition 5.16 is essential, since the set $\mathcal{P}_{\mathcal{M}}$ is not closed with respect to the norm. Let us consider the example $A = (e_3, e_1, e_2) \in \mathbb{R}^{3 \times 3}$, then it follows that $\mathcal{M}_{\mathcal{S}}^A(1) = 3$ and we obtain for $p \in \mathcal{V}_{\mathcal{M}_{\mathcal{S}}^A(1)} \setminus \{0\} = \{(0,0,\lambda)^T | \lambda \in \mathbb{R} \setminus \{0\}\}$ the equation

$$\|Ap - e_1\|_2^2 = \left\| \begin{pmatrix} 0 \\ \lambda \\ 0 \end{pmatrix} - \begin{pmatrix} 1 \\ 0 \\ 0 \end{pmatrix} \right\|_2^2 = 1 + \lambda^2 .$$

Thus, $\min\limits_{p \in \mathcal{V}_{\mathcal{M}_{\mathcal{S}}^A(1)} \setminus \{0\}} \|Ap - e_1\|_2^2$ does not exist, and we only obtain

$$\arg \inf\limits_{p \in \mathcal{V}_{\mathcal{M}_{\mathcal{S}}^A(1)} \setminus \{0\}} \|Ap - e_1\|_2^2 = 0 \notin \mathcal{V}_{\mathcal{M}_{\mathcal{S}}^A(1)} \setminus \{0\}.$$

5.8   The Preconditioned CG Method

In this section, we will study the use of the preconditioners introduced above using the example of the CG method. Let $A \in \mathbb{R}^{n \times n}$ be a symmetric and positive definite matrix of the system of equations

$$Ax = b. \tag{5.8.1}$$

Given this, we transform the system (5.8.1) using a non-singular matrix P_L into the equivalent form

$$A^P x^P = b^P \tag{5.8.2}$$

with $A^P = P_L A P_L^T$, $x^P = P_L^{-T} x$, and $b^P = P_L b$. The matrix P_L can be determined, for example, by the incomplete Cholesky factorization or a symmetric splitting method (for example, the Jacobi method or the symmetric Gauss-Seidel method) where the Jacobi method only leads to a change in the condition number for matrices with different diagonal elements and thus, in the case of the two-dimensional Poisson equation, has no effect on the condition number.

If, as already indicated above, we denote all preconditioned quantities with the superscript P, then the preconditioned CG method (PCG) can be written in the form of the original conjugate gradient method by inserting the superscript, with a final computation of the approximate solution $x_{m+1} = P_L^T x_{m+1}^P$ needing to be added.

We will now derive a special representation of the PCG method that is formulated in the original variables. A formal use of the original variables yields, using the definitions

$$\hat{p}_m := P_L^T p_m^P \quad \text{and} \quad \hat{v}_m := P_L^{-1} v_m^P$$

the preconditioned CG method in the preliminary form:

Preliminary PCG Method —

<table>
<tr><td colspan="2">Choose $P_L^{-T} x_0 \in \mathbb{R}^n$</td></tr>
<tr><td>$P_L r_0 := P_L b - P_L A P_L^T P_L^{-T} x_0$</td><td>(5.8.3)</td></tr>
<tr><td colspan="2">$P_L^{-T} \hat{p}_0 := P_L r_0, \quad \alpha_0^P := r_0^T P_L^T P_L r_0$</td></tr>
<tr><td colspan="2">For $m = 0, \ldots, n-1$</td></tr>
<tr><td colspan="2">

Y $\alpha_m^P \neq 0$ N   STOP

$$P_L \hat{v}_m := P_L A P_L^T P_L p_m = P_L A \hat{p}_m \tag{5.8.4}$$

$$\lambda_m^P := \frac{\alpha_m^P}{(P_L \hat{v}_m, P_L p_m)_2} = \frac{\alpha_m^P}{(\hat{v}_m, \hat{p}_m)_2}$$

$$P_L^{-T} x_{m+1} := P_L^{-T} x_m + \lambda_m^P P_L p_m = P_L^{-T} x_m + \lambda_m^P P_L^{-T} \hat{p}_m \tag{5.8.5}$$

$$P_L r_{m+1} := P_L r_m - \lambda_m^P P_L \hat{v}_m \tag{5.8.6}$$

$$\alpha_{m+1}^P := r_{m+1}^T P_L^T P_L r_{m+1}$$

$$\underbrace{P_L p_{m+1}}_{=P_L^{-T} \hat{p}_{m+1}} := P_L r_{m+1} + \frac{\alpha_{m+1}^P}{\alpha_m^P} \underbrace{P_L p_m}_{=P_L^{-T} \hat{p}_m} \tag{5.8.7}$$

</td></tr>
</table>

Since $\boldsymbol{P}_L^T \in \mathbb{R}^{n \times n}$ is invertible and $\boldsymbol{x}_0 \in \mathbb{R}^n$ can be chosen arbitrarily, the specification $\boldsymbol{P}_L^{-T} \boldsymbol{x}_0 \in \mathbb{R}^n$ that appears in the algorithm can be replaced by simply choosing $\boldsymbol{x}_0 \in \mathbb{R}^n$.

Left-multiplying equations (5.8.3), (5.8.4), and (5.8.6) by $\boldsymbol{P}_L^{-1}$ and equations (5.8.5) and (5.8.7) by $\boldsymbol{P}_L^T$, and using $\boldsymbol{P} = \boldsymbol{P}_L^T \boldsymbol{P}_L$, yields the preconditioned CG method:

Algorithm PCG Method —

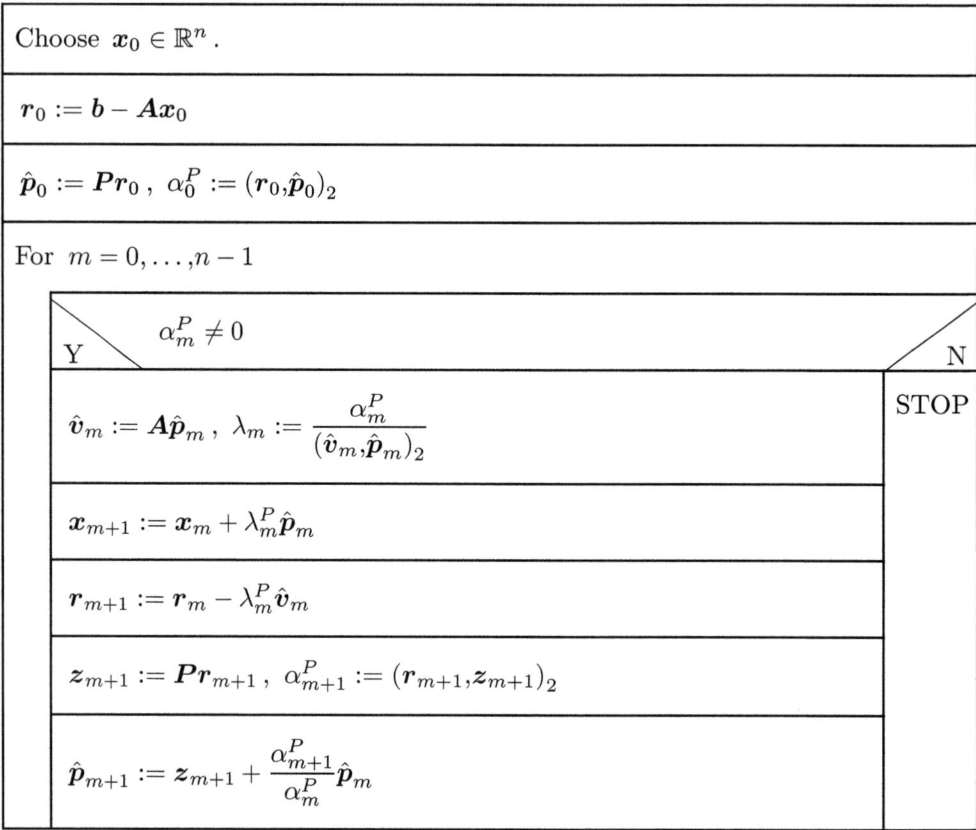

The notations ^{P} and $\hat{}$ highlight the remaining difference from the original CG method. The PCG method is also generally used as an iterative method. In the CG algorithm, the scalar quantity α_m allows direct control of the residual, so that a stopping criterion of the form $\sqrt{\alpha_m} \leq \varepsilon$ can be used instead of $\alpha_m = 0$. An analogous approach to control the residual is not possible in the PCG method, since $\alpha_m^P = \|\boldsymbol{P}_L \boldsymbol{r}_m\|_2^2$ does not provide a direct evaluation of the residual. Therefore, in this case, $\|\boldsymbol{r}_m\|_2 \leq \varepsilon$ should always be implemented as the stopping condition.

Example 5.19 We use the system of equations already considered in Example 4.79, which results from the discretization of the two-dimensional Poisson equation, to compare the CG method with the PCG algorithm. Again, it is a system of equations with 40,000 unknowns. For preconditioning, the symmetric Gauss-Seidel method is used with

$$\boldsymbol{P} = \boldsymbol{P}_{SGS} = (\boldsymbol{D} + \boldsymbol{R})^{-1} \boldsymbol{D} (\boldsymbol{D} + \boldsymbol{L})^{-1}$$

The results presented in the following table show the expected acceleration of the CG method.

	CG Method	PCG Method
Iterations	Residual Progression	
0	140.348	140.348
50	491.151	8.58174
100	150.025	0.0105147
150	1.83245	4.23371e-05
200	0.148948	5.42568e-08
250	0.00307128	1.69676e-11
300	2.40822e-05	6.69697e-15
336	5.07545e-07	9.04322e-17

5.9 The Preconditioned BiCGSTAB Method

In this section, we will study some aspects of preconditioning using the example of the BiCGSTAB method. In principle, a preconditioned version can be obtained from any iterative method for solving $\boldsymbol{Ax} = \boldsymbol{b}$ by applying the method to equation (5.0.1). In the algorithm, the matrix $\boldsymbol{A}$ is then replaced by the product $\boldsymbol{P}_L\boldsymbol{AP}_R$ and the vector $\boldsymbol{b}$ by $\boldsymbol{P}_L\boldsymbol{b}$, and after the iteration is terminated, the approximate solution $\boldsymbol{x}_m$ to the original problem is obtained according to (5.0.2) by $\boldsymbol{x}_m = \boldsymbol{P}_R\boldsymbol{x}_m^P$ from the approximation $\boldsymbol{x}_m^P$. Within the iterative method, this results in different values compared to the base algorithm, which we denote with the superscript P. Naturally, it is desirable that, for a given initial vector $\boldsymbol{x}_0$ and a specified accuracy $\varepsilon > 0$, after termination of both the base method and the preconditioned variant, approximate solutions $\boldsymbol{x}_m$ are obtained that satisfy the condition $\|\boldsymbol{b} - \boldsymbol{Ax}_m\| \leq \varepsilon$. If one considers the preconditioned method in the above sense, using $\boldsymbol{P}_L\boldsymbol{AP}_R$ and $\boldsymbol{P}_L\boldsymbol{b}$, the algorithm is always terminated based on the norm of the residual vector $\boldsymbol{r}_m^P = \boldsymbol{P}_L\boldsymbol{b} - \boldsymbol{P}_L\boldsymbol{AP}_R\boldsymbol{x}_m^P$, which for $\boldsymbol{x}_m = \boldsymbol{P}_R\boldsymbol{x}_m^P$ yields the residual $\|\boldsymbol{P}_L(\boldsymbol{b} - \boldsymbol{Ax}_m)\|$. Consequently, when using left preconditioning within the method, there is no control over the relevant residual. Therefore, even upon termination of the method, no direct statement can be made about the quality of the approximation. In contrast, right preconditioning yields $\boldsymbol{r}_m^P = \boldsymbol{r}_m$, so that a reliable comparison of right and left preconditioners in this form is not possible.

In [51], a special formulation of a preconditioned BiCGSTAB method is presented, which allows control of the relevant residual even in the case of left preconditioning. We will describe this variant in the following. The approach is general and can therefore be applied analogously to other methods. Fundamental to the derivation is the fact that the explicit computation of the matrix $\boldsymbol{P}_L\boldsymbol{AP}_R$ is not advisable for two reasons. On the one hand, the preconditioners are often only given implicitly (for example, ILU and IQR), and on the other hand, $\boldsymbol{P}_L\boldsymbol{AP}_R$ may be a dense matrix, even though $\boldsymbol{A}$ is sparse. Therefore, the matrix-vector multiplications are carried out successively, and the computation

$$\boldsymbol{v}^P = \boldsymbol{P}_L\boldsymbol{AP}_R\boldsymbol{p}^P$$

can be performed in two steps according to

$$\begin{aligned} v &= AP_R p^P \\ v^P &= P_L v \end{aligned} \tag{5.9.1}$$

without additional computational effort.

Right preconditioning has no effect on the control of the residual, so for the sake of clarity, we will restrict our further considerations to the interesting case of left preconditioning and subsequently include right preconditioning.

With the initial vector x_0, the residual vectors are given by $r_0 = b - Ax_0$ and $r_0^P = P_L(b - Ax_0) = P_L r_0$. Since $p_0^P = r_0^P$, the vectors r_j, r_j^P, and p_j^P can be assumed to be known in the $(j+1)$-th iteration step. Inserting the quantity

$$v_j = Ap_j^P \tag{5.9.2}$$

yields $v_j^P = P_L v_j$. Thus, with

$$s_j = r_j - \alpha_j^P v_j \tag{5.9.3}$$

we obtain the equation $s_j^P = r_j^P - \alpha_j^P v_j^P = P_L s_j$, and, using

$$t_j = As_j^P \tag{5.9.4}$$

we obtain the relations

$$t_j^P = P_L t_j \quad \text{and} \quad r_{j+1}^P = s_j^P - \omega_j^P t_j^P = P_L \left(s_j - \omega_j^P t_j \right).$$

Due to the invertibility of the matrix P_L, it follows directly from $r_{j+1}^P = P_L r_{j+1}$ that $r_{j+1} = s_j - \omega_j^P t_j$. By introducing the auxiliary vectors v_j, s_j, and t_j according to (5.9.2), (5.9.3), and (5.9.4), we are able to monitor the residual $\|r_j\|$ instead of $\|r_j^P\|$ without any additional computational effort. Thus, the preconditioned BiCGSTAB algorithm results in the form shown in the following diagram.

Based on the representation

$$\begin{aligned} x_{j+1} &= P_R x_{j+1}^P = P_R x_j^P + \alpha_j^P P_R p_j^P + \omega_j^P P_R s_j^P \\ &= x_j + P_R \left(\alpha_j^P p_j^P + \omega_j^P s_j^P \right) \end{aligned} \tag{5.9.5}$$

equation (5.9.6) can be replaced by (5.9.5), which eliminates equation (5.9.7), and the algorithm is formulated directly in the desired approximate solution x_j. While monitoring the residual $\|r_j\|$ does not cause any additional computational effort, using equation (5.9.5) per iteration requires a matrix-vector multiplication with P_R, which, in the form shown in the diagram, only needs to be performed once outside the iteration.

The preconditioned BiCGSTAB method —

$$\text{Choose } \boldsymbol{x}_0 \in \mathbb{R}^n \text{ and } \varepsilon > 0$$

$$\boldsymbol{r}_0 = \boldsymbol{p}_0 := \boldsymbol{b} - \boldsymbol{A}\boldsymbol{x}_0, \ \boldsymbol{r}_0^P = \boldsymbol{p}_0^P := \boldsymbol{P}_L \boldsymbol{r}_0$$

$$\rho_0^P := \left(\boldsymbol{r}_0^P, \boldsymbol{r}_0^P\right)_2, \ j := 0$$

$$\text{While } \|\boldsymbol{r}_j\|_2 > \varepsilon$$

$$\boldsymbol{v}_j := \boldsymbol{A}\,\boldsymbol{P}_R\,\boldsymbol{p}_j^P, \ \boldsymbol{v}_j^P := \boldsymbol{P}_L\,\boldsymbol{v}_j$$

$$\alpha_j^P := \frac{\rho_j^P}{\left(\boldsymbol{v}_j^P, \boldsymbol{r}_0^P\right)_2}, \ \boldsymbol{s}_j := \boldsymbol{r}_j - \alpha_j^P \boldsymbol{v}_j, \ \boldsymbol{s}_j^P := \boldsymbol{P}_L \boldsymbol{s}_j$$

$$\boldsymbol{t}_j := \boldsymbol{A}\,\boldsymbol{P}_R\,\boldsymbol{s}_j^P, \ \boldsymbol{t}_j^P := \boldsymbol{P}_L\,\boldsymbol{t}_j$$

$$\omega_j^P := \frac{\left(\boldsymbol{t}_j^P, \boldsymbol{s}_j^P\right)_2}{\left(\boldsymbol{t}_j^P, \boldsymbol{t}_j^P\right)_2}$$

$$\boldsymbol{x}_{j+1}^P := \boldsymbol{x}_j^P + \alpha_j^P \boldsymbol{p}_j^P + \omega_j^P \boldsymbol{s}_j^P \tag{5.9.6}$$

$$\boldsymbol{r}_{j+1} := \boldsymbol{s}_j - \omega_j^P \boldsymbol{t}_j, \ \boldsymbol{r}_{j+1}^P := \boldsymbol{s}_j^P - \omega_j^P \boldsymbol{t}_j^P$$

$$\rho_{j+1}^P := \left(\boldsymbol{r}_{j+1}^P, \boldsymbol{r}_0^P\right)_2, \ \beta_j^P := \frac{\alpha_j^P}{\omega_j^P} \frac{\rho_{j+1}^P}{\rho_j^P}$$

$$\boldsymbol{p}_{j+1}^P := \boldsymbol{r}_{j+1}^P + \beta_j^P \left(\boldsymbol{p}_j^P - \omega_j^P \boldsymbol{v}_j^P\right), \ j := j + 1$$

$$\boldsymbol{x}_j := \boldsymbol{P}_R \boldsymbol{x}_j^P \tag{5.9.7}$$

5.10 Comparison of Preconditioners

For the analysis of preconditioning techniques, we again consider the model problem of the convection-diffusion equation with $N^2 = 10{,}000$ points. Figures 5.1 to 5.5 show the convergence histories of the individual projection methods for a diffusion parameter $\varepsilon = 0.1$, as already discussed in Section 4.3.2.10. The second curve in each case shows the acceleration of the basic algorithm achieved by right preconditioning using an incomplete LU factorization . Here, the required number of iterations was reduced to about 30% of the respective base method. Although the use of an incomplete LU factorization always leads to a doubling of the matrix-vector multiplications in addition to its computation, all equation solvers show an overall reduction in computation time compared to the unpreconditioned method.

In [50], a comparison of computation times for various preconditioned Krylov subspace methods in the simulation of inviscid and viscous flow fields is presented. For the wide range of different problem settings considered, the BiCGSTAB method in combination with an incomplete LU factorization proved to be very efficient, with the use of preconditioning leading to a tremendous acceleration of the overall method.

Figures 5.6 to 5.10 correspond to the preconditioned variant of the BiCGSTAB method presented in Section 5.9, allowing a comparison of left and right preconditioning. In contrast to the previous figures, the convection-diffusion equation was used with a diffusion parameter $\varepsilon = 0.01$, so that the nonsymmetric part of the matrix (1.0.9) has a greater influence compared to $\varepsilon = 0.1$. To distinguish the preconditioners, we introduce the abbreviations listed in the following table.

Abbreviation	Preconditioner
ILU	Incomplete LU factorization
Jacobi	Jacobi preconditioner
GS	Gauss-Seidel preconditioner
SGS	Symmetric Gauss-Seidel preconditioner
POLY(3)	Polynomial preconditioner according to (5.2.5) with $m = 3$
Z_1	Row scaling according to the sum norm
Z_2	Row scaling according to the Euclidean norm
S_1	Column scaling according to the sum norm
S_2	Column scaling according to the Euclidean norm

In addition, the suffix L or R indicates the use of the respective preconditioner in a left or right formulation, respectively.

From the perspective of efficiency, the preconditioners can be divided into three groups. First, we combine the incomplete LU factorization shown in Figure 5.6 and the symmetric Gauss-Seidel preconditioner also included in Figure 5.6 into the first group. The second group includes all preconditioners shown in Figures 5.7, 5.9, and 5.10. Compared to the algorithms of the first group, these methods require about three times as many iterations. The last group of methods, which are inefficient for this test case, includes the polynomial preconditioner shown in Figure 5.8. This method resulted in a significantly higher number of iterations and, in the form used ($m = 3$), requires four times as many matrix-vector multiplications as the base algorithm. Furthermore, the incomplete QR factorization and the incomplete Frobenius inverse can also be assigned to this group classify. Both preconditioners already proved to be unacceptably expensive to compute and were therefore not included in the figures. However, a reduction in computation time for the calculation of the Frobenius inverse can be achieved due to the good parallelizability of this method.

When comparing left and right preconditioning no significant differences were found, although the right preconditioning usually resulted in a slightly lower number of iterations.

In conclusion, we can state that right preconditioning based on an incomplete LU factorization or the symmetric Gauss-Seidel method has proven to be a very efficient approach. These results are in excellent agreement with the study of different preconditioners in the context of the Euler equations presented in [51]. The preconditioner associated with the symmetric Gauss-Seidel method, in contrast to the incomplete LU factorization, requires no storage space and is directly given by the matrix of the system of equations, which eliminates the need to compute the preconditioner. In terms of minimizing memory requirements, the symmetric Gauss-Seidel method thus represents an efficient alternative to the well-established incomplete LU factorization.

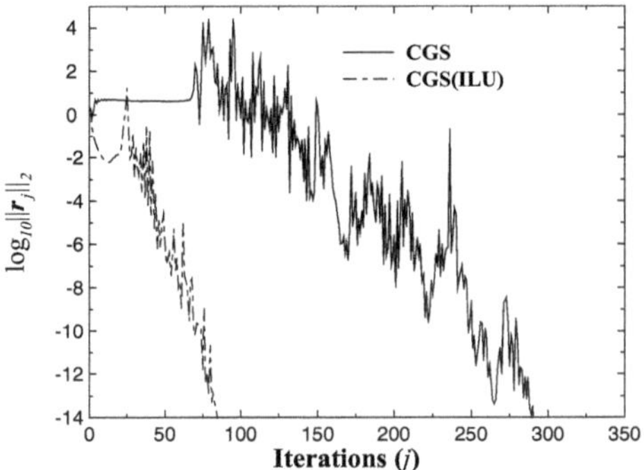

Figure 5.1   Convergence history of the CGS method

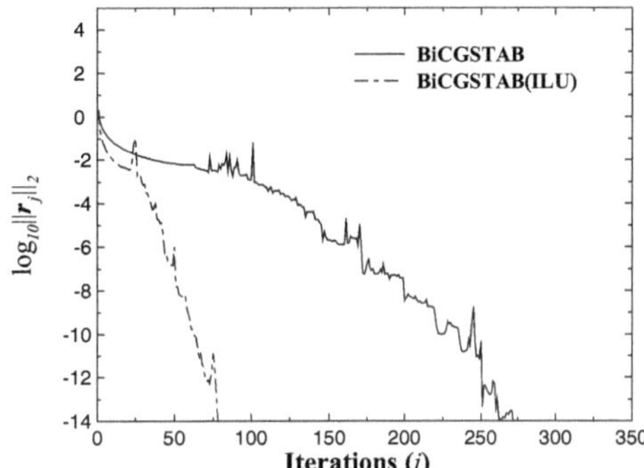

Figure 5.2   Convergence history of the BiCGSTAB method

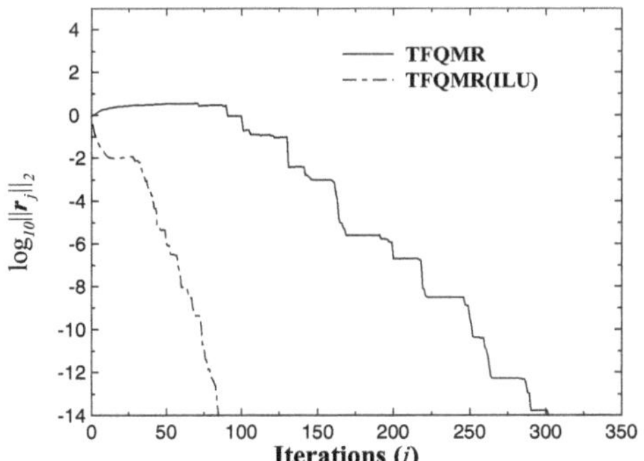

Figure 5.3   Convergence history of the TFQMR method

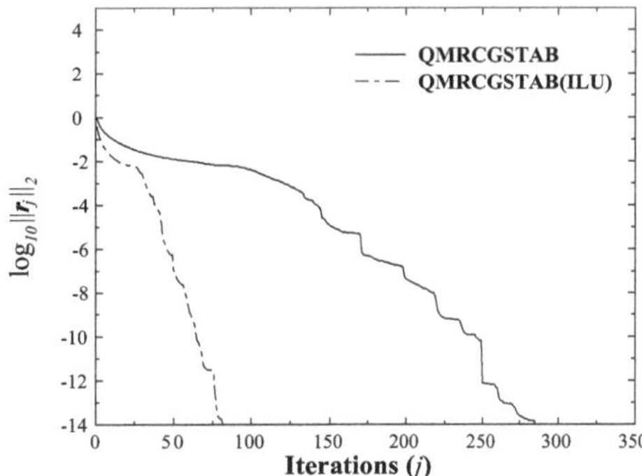

Figure 5.4   Convergence history of the QMRCGSTAB method

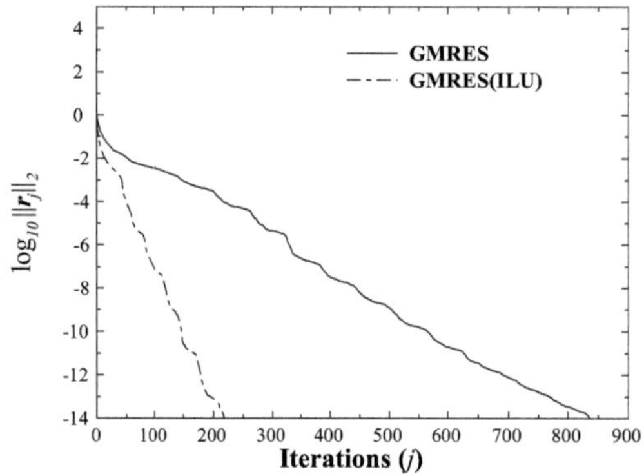

Figure 5.5   Convergence history of the GMRES method

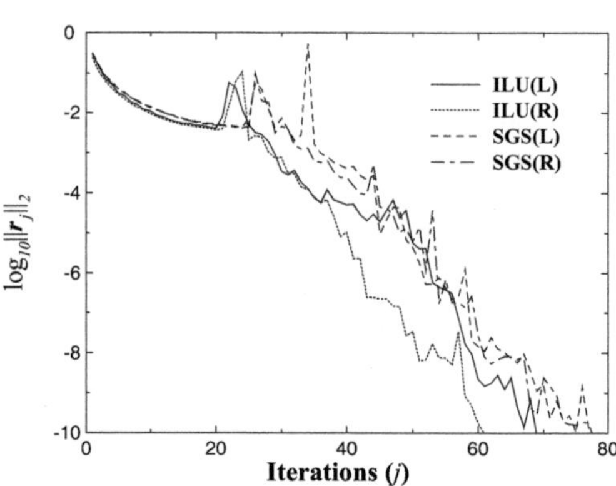

Figure 5.6   Convergence histories of the preconditioned BiCGSTAB method

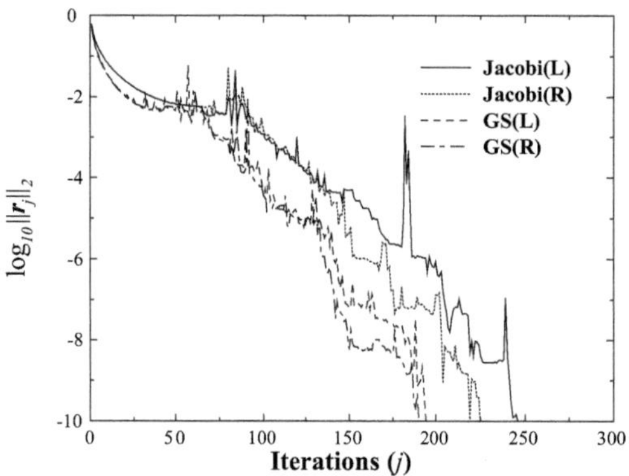

Figure 5.7   Convergence histories of the preconditioned BiCGSTAB method

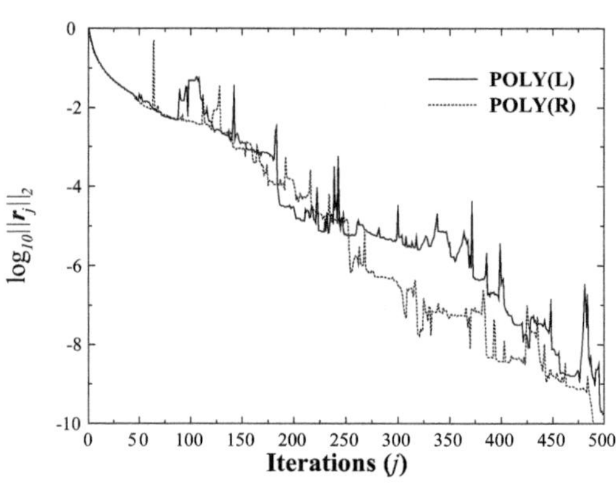

Figure 5.8   Convergence histories of the preconditioned BiCGSTAB method

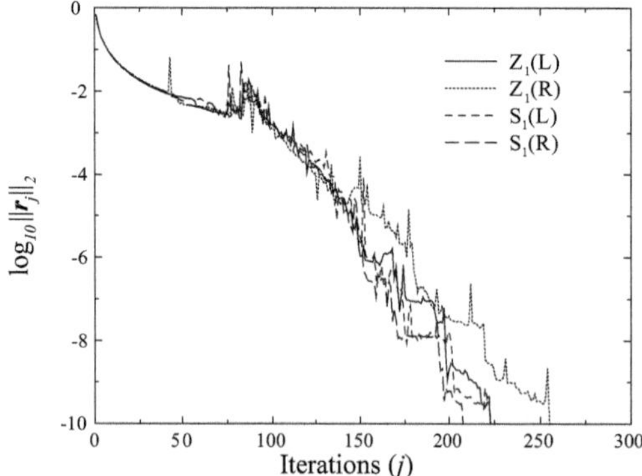

Figure 5.9   Convergence histories of the preconditioned BiCGSTAB method

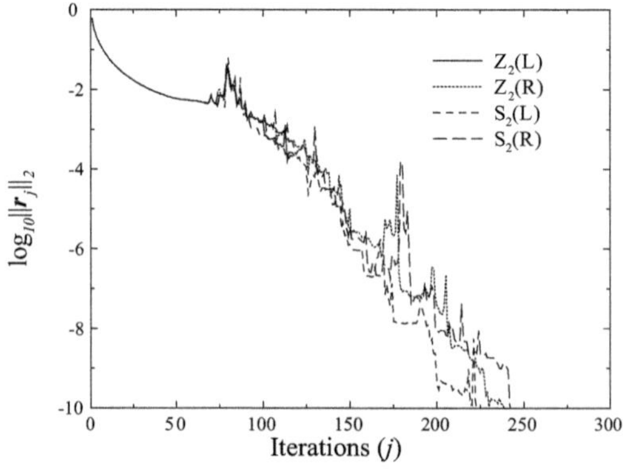

Figure 5.10   Convergence histories of the preconditioned BiCGSTAB method

5.11 Exercises

Problem 1:
Show: Let $P_L = P_R^T \in \mathbb{R}^{n \times n}$ be non-singular and $A \in \mathbb{R}^{n \times n}$ symmetric and positive definite, then the matrix

$$A^P := P_L \, A \, P_R$$

is also symmetric and positive definite.

Problem 2:
Compute an incomplete LU factorization of the matrix

$$A = \begin{pmatrix} 3 & 13 & 1 & 0 & 0 \\ 0 & 1 & 0 & 0 & 2 \\ 0 & 0 & 1 & 1 & 0 \\ 1 & 0 & 1 & 3 & 1 \\ 1 & 4 & 0 & 1 & 4 \end{pmatrix}$$

and compare it with a complete LU decomposition, i.e., the LR decomposition.

Problem 3:
Determine the preconditioners P_{Jac}, P_{GS}, and P_{SGS} for the matrix

$$A = \begin{pmatrix} 10 & 7 & 1 \\ 4 & 6 & 2 \\ 8 & 5 & 12 \end{pmatrix}$$

and compute the condition numbers

$$\mathrm{cond}_\infty(A), \ \mathrm{cond}_\infty(P_{Jac}A), \ \mathrm{cond}_\infty(P_{GS}A) \text{ and } \mathrm{cond}_\infty(P_{SGS}A).$$

Problem 4:
Show that for a symmetric, positive definite matrix $A \in \mathbb{R}^{n \times n}$ the two-sided preconditioning

$$A^P = P_L A P_R$$

with

$$P_L = P_R^T = (D + L)^{-1} D^{1/2}$$

exists. Compute A^P for

$$A = \begin{pmatrix} 10 & 4 & 1 \\ 4 & 12 & 2 \\ 1 & 2 & 6 \end{pmatrix}$$

and compare the condition numbers $\mathrm{cond}_\infty(A)$ and $\mathrm{cond}_\infty(A^P)$.

A   Implementations in MATLAB

This appendix presents possible implementations of the frequently used methods in MAT-LAB [1]. The algorithms were developed by C. Vömel and are provided as supplementary materials online on SpringerLink.

With the exception of the Arnoldi algorithm, all listed methods can be used to solve the problem $\boldsymbol{Ax} = \boldsymbol{b}$ with $\boldsymbol{A} \in \mathbb{R}^{n \times n}$, $\boldsymbol{b} \in \mathbb{R}^n$ using the following notation:

```
x =   Name(A,b);
x =   Name(A,b,tol);
x =   Name(A,b,tol,maxit);
x =   Name(A,b,tol,maxit,x0);
```

where `Name` must be replaced by the desired method name. The optional arguments have the default settings specified in the following table.

Input variable	Description	Default
tol	Accuracy requirement (tolerance)	10^{-6}
maxint	Maximum number of iterations	$\min\{n,30\}$
x0	Initial vector	Zero vector

For each of these methods, a check of the input parameters and for the presence of a trivial solution is performed using the programs Arguments.m and Trivial.m, respectively.

Program SteepestDescent.m: Method of steepest descent

```
function x = SteepestDescent(A,b,tol,maxit,x0)
%
% CHECK THE INPUT ARGUMENTS
   Arguments

% CHECK FOR TRIVIAL SOLUTION
   Trivial

% MAIN ALGORITHM
%
   tolb = tol * normb;
   r = b - A * x;
% iterate
   for i = 1:maxit
     normr = norm(r);
     if (normr <= tolb)
```

```matlab
%      converged
         break
      end
      q = A*r;
      if (q == 0)
         error('Matrix is singular.');
      end
      rtq = r'*q;
      if(rtq <=0)
         error('Matrix is not positive definite.');
      end
      lambda = (r'*r)/rtq;
      x = x + lambda * r;
      r = r - lambda * q;
   end
   normr = norm(b-A*x);
   if (normr > tolb)
      warning('Method has not converged.')
   end
   return
```

Program ConjugGrad.m: CG method

```matlab
function x = ConjugGrad(A,b,tol,maxit,x0)
%
% CHECK THE INPUT ARGUMENTS
   Arguments

% CHECK FOR TRIVIAL SOLUTION
   Trivial

% MAIN ALGORITHM
%
   tolb = tol * normb;
   r = b - A * x;
   p = r;
   normr = norm(r);
   alpha = normr^2;
   % iterate
   for i = 1:maxit
      if (normr <= tolb)
%      converged
         break
      end
      v = A*p;
      if (v == 0)
         error('Matrix is singular.');
      end
```

```
    vr = (v'*r);
    if(vr <=0)
      error('Matrix is not positive definite.');
    end
    lambda = alpha/vr;
    x = x + lambda * p;
    r = r - lambda * v;
    alpha2 = alpha;
    normr = norm(r);
    alpha = normr^2;
    p = r + alpha/alpha2 * p;
  end
  normr = norm(b-A*x);
  if (normr > tolb)
    warning('Method has not converged.')
  end
  return
```

```
Program Arnoldi.m: Arnoldi algorithm
```

```
function [V,H] =  Arnoldi(A,r0,m);
%
% OUTPUT VARIABLEN:
% ----------------
% V -- n*m orthogonal matrix
% H -- m*m upper Hessenberg matrix
%
% INPUT VARIABLEN:
% ----------------
% A: quadratic n*n Matrix.
% r0: Column vector
% m: Number of columns of Q and H

% CHECK THE INPUT ARGUMENTS
  if (nargin < 3)
    error('Funktion requires more parameters.');
  else
    [mm,n] = size(A);
    if (mm ~= n)
      error('Matrix is not quadratic.');
    end
    if ~isequal(size(r0),[mm,1])
      error('r0 has not the right dimension.');
    end
    if ((m > n)||(m<=0))
      error('Wrong column number.');
    end
  end
```

```
%
% MAIN ALGORITHM
%
  V = zeros(n,m);
  H = zeros(m,m);
  w = r0;
  normw = norm(w);
  if (normw == 0)
    error('Arnoldi break-down.');
  end
  V(:,1) = w/normw;
  for j = 1:m
    w = A*V(:,j);
    H(1:j,j) = V(:,1:j)'*w;
    w = w - V(:,1:j)*H(1:j,j);
    if(j == m), break; end
    normw = norm(w);
    if (normw == 0)
      error('Arnoldi break-down.');
    end
    V(:,j+1) = w/normw;
    H(j+1,j) = normw;
  end
  return
```

Program RestartedFOM.m: Restarted FOM

```
function x = RestartedFOM(A,b,tol,maxit,x0)
%
% CHECK THE INPUT ARGUMENTS
  Arguments

% CHECK FOR TRIVIAL SOLUTION
  Trivial

% MAIN ALGORITHM
%
  tolb = tol * normb;
  r = b - A * x;
  normr = norm(r);
% iterate
  for i = 1:maxit
    if (normr <= tolb)
%     converged
      break
    end
    [V,H] = Arnoldi(A,r,i);
    e = normr*eye(i,1);
```

```
  alpha = H\e;
  x = x + V * alpha;
  r = b - A * x;
  normr = norm(r);
end
if (normr > tolb)
  warning('Method has not converged.')
end
return
```

```
Program Gmres.m: GMRES algorithm
```

```
function x = Gmres(A,b,tol,maxit,x0)
%
% CHECK THE INPUT ARGUMENTS
  Arguments

% CHECK FOR TRIVIAL SOLUTION
  Trivial

% MAIN ALGORITHM
%
  tolb = tol * normb;
  r = b - A * x;
  normr = norm(r);
  Gamma=normr*eye(maxit+1,1);
% Hessenberg matrix
  H = zeros(maxit+1);
% orthogonal basis
  V = zeros(n,maxit+1);
% Givens parameter
  c=zeros(maxit+1,1);
  s=zeros(maxit+1,1);
% Dimension of the Krylov subspace
  kdim = 0;

  V(:,1) = r/normr;
% iterate
  for j = 1:maxit
    if (normr <= tolb)
%     converged
      break
    else
      kdim = kdim+1;
    end
    w = A*V(:,j);
%   Modified Gram-Schmidt
    for i =1:j
```

```
      H(i,j) = w'*V(:,i);
      w = w - H(i,j)*V(:,i);
    end
    H(j+1,j) = norm(w);
    for i = 1:j-1
      H(i:i+1,j) = [c(i),s(i);-s(i),c(i)] * H(i:i+1,j);
    end
    beta=norm(H(j:j+1,j));
    if(beta ~= 0)
        V(:,j+1)=w/H(j+1,j);
    else
%       'happy breakdown'
    end
    if beta~=0
%        Calculation of a Givens rotation
        s(j)=H(j+1,j)/beta;
        c(j)=H(j,j)/beta;
%        Applictaion to  H(j:j+1,j)
        H(j,j) = beta;
        H(j+1,j)=0;
%        Application to  b(j:j+1,j)
        Gamma(j+1) = -s(j)*Gamma(j);
        Gamma(j) = c(j)*Gamma(j);
    end
    normr = abs(Gamma(j+1));
  end
% Calculate the coefficients, solve H*Alpha = Gamma
  Alpha = H(1:kdim,1:kdim)\Gamma(1:kdim);
  x = x + V(:,1:kdim) * Alpha;
  r = b - A*x;
  normr = norm(r);
  if (normr > tolb)
    warning('Method has not converged.')
  end
  return
```

Program Bicg.m: BiCG algorithm

```
function x = Bicg(A,b,tol,maxit,x0)
%
% CHECK THE INPUT ARGUMENTS
  Arguments

% CHECK FOR TRIVIAL SOLUTION
  Trivial

% MAIN ALGORITHM
%
  tolb = tol * normb;
  r = b - A * x;
  p = r;
  rstar = r;
  pstar = r;
  w = r'*rstar;
% iterate
  for i = 1:maxit
    normr = norm(r);
    if (normr <= tolb)
%     converged
      break
    end
    v = A*p;
    vp = (v'*pstar);
    if(vp ==0)
      error('Bicg break-down.');
    end
    if(w ==0)
      error('Bicg solution stagnates.');
    end
    alpha = w/vp;
    x = x + alpha * p;
    r = r - alpha * v;
    rstar = rstar - alpha * A' * pstar;
    w1 = r'*rstar;
    beta = w1/w;
    p = r + beta * p;
    pstar = rstar + beta * pstar;
    w = w1;
  end
  normr = norm(b-A*x);
  if (normr > tolb)
     warning('Method has not converged.')
  end
  return
```

Program Cgs.m: CGS algorithm

```
function x = Cgs(A,b,tol,maxit,x0)
%
% CHECK THE INPUT ARGUMENTS
  Arguments
% CHECK FOR TRIVIAL SOLUTION
  Trivial

% MAIN ALGORITHM
%
  tolb = tol * normb;
  r0 = b - A * x;
  r = r0;
  rr0 = r'*r0;
  p = r;
  u = r;
% iterate
  for i = 1:maxit
    normr = norm(r);
    if (normr <= tolb)
%       konvergiert
        break
    end
    v = A*p;
    vr0 = v'*r0;
    if(vr0 ==0)
      error('Cgs break-down.');
    end
    if(rr0 ==0)
      error('Cgs solution stagnates.');
    end
    alpha = rr0/vr0;
    q = u - alpha * v;
    t = u + q;
    x = x + alpha * t;
    r = r - alpha * A * t;
    r1r0 = r'*r0;
    beta = r1r0/rr0;
    u = r + beta * q;
    p = u + beta * ( q + beta * p );
    rr0 = r1r0;
  end
  normr = norm(b-A*x);
  if ( normr > tolb)
      warning('Method has not converged.')
  end
  return
```

Program Bicgstab.m: BiCGSTAB algorithm

```matlab
function x = Bicgstab(A,b,tol,maxit,x0)
%
% CHECK THE INPUT ARGUMENTS
  Arguments

% CHECK FOR TRIVIAL SOLUTION
  Trivial

% MAIN ALGORITHM
%
  tolb = tol * normb;
  r0 = b - A * x;
  r = r0;
  rr0 = r'*r0;
  p = r;
% iterate
  for i = 1:maxit
    normr = norm(r);
    if (normr <= tolb)
%     converged
      break
    end
    v = A*p;
    vr0 = v'*r0;
    if(vr0 ==0)
      error('Bicgstab break-down.');
    end
    if(rr0 ==0)
      error('Bicgstab solution stagnates.');
    end
    alpha = rr0/vr0;
    s = r - alpha * v;
    t = A * s;
    ts = s'*t;
    tt = t'*t;
    if((tt ==0)||(ts==0))
      error('Bicgstab break-down.');
    end
    omega = ts/tt;
    x = x + alpha * p + omega * s;
    r = s - omega * t;
    r1r0 = r'*r0;
    beta = (alpha*r1r0)/(omega*rr0);
    p = r + beta * (p - omega * v);
    rr0 = r1r0;
  end
```

```
  normr = norm(b-A*x);
  if (normr > tolb)
     warning('Method has not converged.')
  end
  return
```

Program Tfqmr.m: TFQMR algorithm

```
function x = Tfqmr(A,b,tol,maxit,x0)
%
% CHECK THE INPUT ARGUMENTS
  Arguments

% CHECK FOR TRIVIAL SOLUTION
  Trivial

% MAIN ALGORITHM
%
  tolb = tol * normb;
  r0 = b - A * x;
  y1 = r0;
  w = r0;
  tau = norm(r0);

  if(tau > tolb)
    v = A * y1;
    d = zeros(n,1);
    eta = 0;
    theta = 0;
    rho1 = tau^2;

    for j = 1:maxit
      rho2 = v'*r0;
      if(rho2 ==0)
        error('Tfqmr break-down.');
      end
      if(rho1 ==0)
        error('Tfqmr solution stagnates.');
      end
      alpha = rho1/rho2;
      y2 = y1 - alpha * v;
      for m=2*j-1:2*j
        w = w - alpha * A * y1;
        d = y1 + theta^2*eta/alpha * d;
        theta = norm(w)/tau;
        c = 1/sqrt(1+theta^2);
        eta = c^2*alpha;
```

```matlab
        x = x + eta * d;
        tau = tau * theta * c;
        if (sqrt(m+1)*tau <= tolb)
%          convergence
           break
        end
         y1 = y2;
      end
      rho2 = w'*r0;
      beta = rho2/rho1;
      y1= w + beta * y2;
      v = A*y1 + beta *(A*y2+beta*v);
      rho1 = rho2;
    end
  end
  normr = norm(b-A*x);
  if (normr > tolb)
     warning('Method has not converged.')
  end
  return
```

Program Qmrcgstab.m: QMRCGSTAB algorithm

```matlab
function x = Qmrcgstab(A,b,tol,maxit,x0)
%
% CHECK THE INPUT ARGUMENTS
   Arguments

% CHECK FOR TRIVIAL SOLUTION
   Trivial

% MAIN ALGORITHM
%
   tolb = tol * normb;
   r0 = b - A * x;
   p = r0;
   tau = norm(r0);
   v = A * p;
   r = r0;
   if(tau > tolb)
     d = zeros(n,1);
     eta = 0;
     theta = 0;
     rho1 = tau^2;
     for j = 1:maxit
       rho2 = v'*r0;
       if(rho2 ==0)
         error('Qmrcgstab break-down.');
```

```
           end
           if(rho1 ==0)
             error('Qmrcgstab solution stagnates.');
           end
           alpha = rho1/rho2;
           s = r - alpha * v ;
%          First quasi minimization step
           theta2 = norm(s)/tau;
           c = 1/sqrt(1+(theta2)^2);
           tau2 = tau * theta2 * c;
           eta2 = c^2 * alpha;
           d2 = p + ((theta^2)*eta/alpha) * d;
           x2 = x + eta2 * d2;
%          Calculate t, omega, and update the residual
           t = A * s;
           uu = s'*t;
           vv = t'*t;
           if( v==0)
             error('Matrix is singular.');
           end
           omega = uu/vv;
           if( omega==0 )
             error('Qmrcgstab solution stagnates.');
           end
           r = s - omega * t;
%          Second quasi minimization step
           theta = norm(r)/tau2;
           c = 1/sqrt(1+theta^2);
           tau = tau2 * theta * c;
           eta = c^2 * omega;
           d = s + (((theta2)^2)*eta2/omega) * d2;
           x = x2 + eta * d;
           if (sqrt(2*j+1)*abs(tau) <= tolb)
%             convergence
             break
           end
           rho2 = r'*r0;
           if(rho2 ==0)
             error('Qmrcgstab solution stagnates.');
           end
           beta = (alpha*rho2)/(omega*rho1);
           p = r + beta * (p - omega * v);
           v = A * p;
           rho1 = rho2;
        end
     end
     normr = norm(b-A*x);
     if (normr > tolb)
```

```
    warning('Method has not converged.')
end
return
```

Program Arguments.m: Check of input parameters

```
if (nargin < 2)
  error('Funktion requires more parameters.');
else
  [m,n] = size(A);
  if (m ~= n)
    error('Matrix is not quadratic.');
  end
  if ~isequal(size(b),[m,1])
    error('Right hand side has a wrong dimension. ');
  end
end
if (nargin < 3) | isempty(tol)
  tol = 1.0e-6;
end
if (nargin < 4) | isempty(maxit)
  maxit = min(n,30);
end
if (nargin < 5) | isempty(x0)
  x = zeros(n,1);
else
  if ~isequal(size(x0),[n,1])
    error('Initial guess x0 has a wrong dimension.');
  end
  x = x0;
end
```

Program Trivial.m: Check for trivial solution

```
normb = norm(b);
if (normb == 0)
  x = zeros(n,1);
  return
end
```

Bibliography

[1] *MATLAB*. http://www.mathworks.com.

[2] K. AJMANI, W.-F. NG, M. LIOU. Preconditioned Conjugate Gradient Methods for the Navier-Stokes Equations. *J. Comput. Phys.*, 110: 68–81, 1994.

[3] E. ANDERSON, Z. BAI, C. BISCHOF, S. BLACKFORD, J. DEMMEL, J. DONGARRA, J. DU CROZ, A. GREENBAUM, S. HAMMARLING, A. MCKENNEY, D. SORENSEN. *LAPACK Users' Guide*. SIAM, Philadelphia, 3. edition , 1999.

[4] A. ARNONE, M.-S. LIOU, L. A. POVINELLI. Integration of Navier-Stokes Equations Using Dual Time Stepping and a Multigrid. *AIAA Journal*, 33 (6): 985–990, 1995.

[5] S. F. ASHBY, T. A. MANTEUFFEL, P. E. SAYLOR. Adaptive Polynomial Preconditioning for Hermitian Indefinite Linear Systems. *BIT*, 29: 583–609, 1989.

[6] O. AXELSSON. A survey of preconditioned iterative methods for linear systems of algebraic equations. *BIT*, 25: 166–187, 1985.

[7] O. AXELSSON. *Iterative Solution Methods*. Cambridge University Press, Cambridge, 1996.

[8] R. BARRETT, M. BERRY, T. CHAN, J. DEMMEL, J. DONATO, J. DONGARRA, V. EIJKHOUT, R. POZO, C. ROMINE, H. VAN DER VORST. *Templates for the Solution of Linear Systems*. SIAM, Philadelphia, 1994.

[9] M. BENZI. Preconditioning Techniques for Large Linear Systems: A Survey. *J. Comput. Phys.*, 182: 418–477, 2002.

[10] M. BENZI, M. TUMA. A comparative study of Sparse Approximate Inverse Preconditioners. *Los Alamos National Laboratory Technical Report*, LA-UR-98-24, 1998.

[11] A. BRANDT. Guide to Multigrid Development. In *Multigrid Methods*, (Eds.) W. HACKBUSCH, U. TROTTENBERG, number 960 in Lecture Notes in Mathematics, pp. 220–312, Berlin, Heidelberg, New York, 1981. Springer.

[12] C. BREZINSKI, M. REDIVO-ZAGLIA. Look-ahead in BiCGSTAB and other product methods for linear systems. *BIT*, 35: 169–201, 1995.

[13] C. BREZINSKI, M. REDIVO-ZAGLIA, H. SADOK. A breakdown-free Lanczos type algorithm for solving linear systems. *Numer. Math.*, 63: 29–38, 1992.

[14] C. BREZINSKI, H. SADOK. Avoiding breakdown in the cgs algorithm. *Numerical Algorithms*, 1: 199–206, 1991.

[15] W. L. BRIGGS. *A Multigrid Tutorial*. SIAM, Philadelphia, 1987.

[16] W. BUNSE, A. BUNSE-GERSTNER. *Numerische lineare Algebra*. Teubner Verlag, Stuttgart, 1985.

[17] K. BURG, H. HAF, F. WILLE, A. MEISTER. *Höhere Mathematik für Ingenieure (Band II, Lineare Algebra)*. Springer Vieweg, Wiesbaden, 7. edition , 2011.

[18] T. F. CHAN, E. GALLOPOULOS, V. SIMONCINI, T. SZETO, C. H. TONG. A quasi-minimal residual variant of the Bi-CGSTAB algorithm for nonsymmetric systems. *SIAM J. Sci. Comput.*, 15 (2): 338–347, 1994.

[19] A. J. CHORIN. A Numerical Method for Solving Incompressible Viscous Flow Problems. *J. Comput. Phys.*, 2: 12–26, 1967.

[20] J. DEMMEL. *Applied Numerical Linear Algebra*. SIAM, Philadelphia, 1997.

[21] P. DEUFLHARD, A. HOHMANN. *Numerische Mathematik I*. de Gruyter, Berlin, New York, 4. überarb. edition , 2008.

[22] P. F. DUBOIS, A. GREENBAUM, G. H. RODRIGUE. Approximating the inverse of a matrix for use in iterative algorithms on vector processors. *Computing*, 22: 257–268, 1979.

[23] W. S. EDWARDS, L. S. TUCKERMAN, R. A. FRIESNER, D. C. SORENSEN. Krylov Methods for the Incompressible Navier-Stokes Equations. *J. Comput. Phys.*, 110: 82–102, 1994.

[24] J. H. FERZIGER, M. PERIĆ. *Computational Methods for Fluid Dynamics*. Springer, Berlin, Heidelberg, 2002.

[25] K. GRAF F. VON FINCKENSTEIN. *Einführung in die Numerische Mathematik*. Hanser, München, 1. edition , 1977.

[26] B. FISCHER. *Polynomial Based Iteration Methods for Symmetric Linear Systems*. Advances in Numerical Mathematics. Wiley-Teubner, Stuttgart, 1996.

[27] R. W. FLETCHER. Conjugate Gradients Methods for Indefinite Systems. In *Dundee Biennial Conference on Numerical Analysis*, (Ed.) G. A. WATSON, pp. 73–89, New York, 1975. Springer.

[28] R. W. FREUND. Conjugate Gradient-Type Methods for Linear Systems with Complex Symmetric Coefficient Matrices. *SIAM J. Sci. Stat. Comput.*, 13 (1): 425–448, 1992.

[29] R. W. FREUND. A Transpose-Free Quasi-Minimal Residual Algorithm for Non-Hermitian Linear Systems. *SIAM J. Sci. Comput.*, 14 (2): 470–482, 1993.

[30] R. W. FREUND, N. M. NACHTIGAL. QMR: A quasi-minimal residual method for non-Hermitian linear systems. *Numer. Math.*, 60: 315–339, 1991.

[31] A. L. GAITONDE. A dual-time method for the solution of the unsteady Euler equations. *The Aeronautical Journal of the Royal Aeronautical Society*, 98: 283–291, 1994.

[32] (Eds.) G. GOLUB, A. GREENBAUM, M. LUSKIN. Transpose-Free Quasi-Minimal Residual Methods for Non-Hermitian Linear Systems, volume 60 of *The IMA Volumes in Mathematics and its Applications, Recent Advances in Iterative Methods*, New York, 1994. Springer.

[33] G. H. GOLUB, C. VAN LOAN. *Matrix Computations*. The John Hopkins University Press, Baltimore, Maryland, 3. edition , 1996.

[34] A. GREENBAUM. *Iterative Methods for Solving Linear Systems*. SIAM, Philadelphia, 1997.

[35] M. J. GROTE, T. HUCKLE. Parallel Preconditioning with Sparse Approximate Inverses. *SIAM J. Sci. Comput.*, 18(3): 838–853, 1997.

[36] W. HACKBUSCH. *Multi-Grid Methods and Applications*, volume 4 of *Springer Series in Computational Mathematics*. Spinger, Berlin, Heidelberg, New York, Tokio, 1985.

[37] W. HACKBUSCH. *Integralgleichungen: Theorie und Numerik*. Teubner, Stuttgart, 1989.

[38] W. HACKBUSCH. *Iterative Lösung großer schwachbesetzter Gleichungssysteme*. Teubner, Stuttgart, 2., überarb. und erw. edition , 1993.

[39] M. R. HESTENES, E. STIEFEL. Methods of Conjugate Gradients for Solving Linear Systems. *NBS J. Res.*, 49: 409–436, 1952.

[40] M. HOCHBRUCK, C. LUBICH. Error analysis of Krylov methods in a nutshell. *SIAM J. Sci. Comp.*, 19: 695–701, 1998.

[41] E. ISSMAN, G. DEGREZ. Acceleration of compressible flow solvers by Krylov subspace methods. Von Karman Institute for Fluid Dynamics, Lecture Series 1995-14, Rhode-Saint-Genèse, March 1995.

[42] O. G. JOHNSON, C. A. MICCHELLI, G. PAUL. Polynomial preconditioners for conjugate gradient calculations. *SIAM J. Numer. Anal.*, 20: 362–376, 1983.

[43] W. JOUBERT. Lanczos Methods for the Solution of Nonsymmetric Systems of Linear Equations. *SIAM J. Matrix Anal. Appl.*, 13 (3): 926–943, 1992.

[44] C.T. KELLEY. *Iterative Methods for Linear and Nonlinear Equations*. SIAM, Philadelphia, PA, 1995.

[45] R. KRESS. *Linear Integral Equations*, volume 82 of *Applied Mathematical Sciences*. Springer, New York, Berlin, Heidelberg, 3. edition , 2014.

[46] C. LANCZOS. An iterative method for the solution of the eigenvalue problem of linear differential and integral operators. *J. Res. Nat. Bur. Standards*, 45: 255–282, 1950.

[47] C. LANCZOS. Solution of systems of linear equations by minimized iterations. *J. Res. Nat. Bur. Standards*, 49: 33–53, 1952.

[48] T. A. MANTEUFFEL. An incomplete factorization technique for positive definite linear systems. *Math. Comp.*, 34: 473–497, 1980.

[49] J. A. MEIJERNINK, H. A. VAN DER VORST. An Iterative Solution Method for Linear Systems of which the Coefficient Matrix is a Symmetric M-Matrix. *Mathematics of Computation*, 31 (137): 148–162, 1977.

[50] A. MEISTER. Comparison of Different Krylov Subspace Methods Embedded in an Implicit Finite Volume Scheme for the Computation of Viscous and Inviscid Flow Fields on Unstructured Grids. *J. Comput. Phys.*, 140: 311–345, 1998.

[51] A. MEISTER, C. VÖMEL. Efficient Preconditioning of Linear Systems arising from the Discretization of Hyperbolic Conservation Laws. *Advances in Computational Mathematics*, Vol.14 (1): 49–73, 2001.

[52] A. MEISTER, J. WITZEL. Krylov Subspace Methods in Computational Fluid Dynamics. *Surv. Math. Ind.*, 10 (3): 231–267, 2002.

[53] G. MEURANT. *Computer Solution of Large Linear Systems*. North-Holland, Amsterdam, 1999.

[54] N. M. NACHTIGAL, S. C. REDDY, L. N. TREFETHEN. How fast are nonsymmetric matrix iterations? *SIAM J. Matrix Anal. Appl.*, 13 (3): 778–795, 1992.

[55] G. OPFER. *Numerische Mathematik für Anfänger*. Vieweg+Teubner, Wiesbaden, 5. edition , 2008.

[56] C. C. PAIGE, M. A. SAUNDERS. Solution of Sparse Indefinite Systems of Linear Equations. *SIAM J. Num. Anal.*, 12 (4): 617–629, 1975.

[57] B. N. PARLETT, D. R. TAYLOR, Z. A. LIOU. A Look-Ahead Lanczos Algorithm for Unsymmetric Matrices. *Math Comp.*, 44: 105–124, 1985.

[58] G. RADICATI, Y. ROBERT, S. SUCCI. Iterative Algorithms for the Solution of Nonsymmetric Systems in the Modelling of Weak Plasma Turbulence. *J. Comput. Phys.*, 80: 489–497, 1989.

[59] J. K. REID. On the method of conjugate gradients for the solution of large sparse systems of linear equations. In *Large sparse sets of linear equations*, (Ed.) J. K. REID, London, New York, 1971. Academic Press.

[60] Y. SAAD. Practical use of some Krylov Subspace Methods for Solving Indefinite and Nonsymmetric Linear Systems. *SIAM J. Sci. Stat. Comput.*, 5 (1): 203–228, 1984.

[61] Y. SAAD. Practical use of polynomial preconditionings for the conjugate gradient method. *SIAM J. Sci. Stat. Comput.*, 6(4): 865–881, 1985.

[62] Y. SAAD. Krylov subspace techniques, conjugate gradients, preconditioning and sparse matrix solvers. *von Karman Institute of Fluid Dynamics, Lecture Series*, 1994–05, 1994.

[63] Y. SAAD. *Iterative Methods for Sparse Linear Systems*. SIAM, Philadelphia, 2. edition , 2003.

[64] Y. SAAD, M. H. SCHULTZ. GMRES: A generalized minimal residual algorithm for solving nonsymmetric linear systems. *SIAM J. Sci. Stat. Comput.*, 7: 856–869, 1986.

[65] H. R. SCHWARZ, N. KÖCKLER. *Numerische Mathematik*. Vieweg+Teubner, Wiesbaden, 8. edition , 2011.

[66] G. L. G. SLEIJPEN, D. R. FOKKEMA. BiCGSTAB(ℓ) for Linear Systems involving Unsymmetric Matrices with Complex Spectrum. *Electronic Transactions on Numerical Analysis*, 1: 11–32, 1993.

[67] A. VAN DER SLUIS. Condition Numbers and Equilibration of Matrices. *Numerische Mathematik*, 14: 14–23, 1969.

[68] A. VAN DER SLUIS. Condition, Equilibration and Pivoting in Linear Algebraic Systems. *Numerische Mathematik*, 15: 74–86, 1970.

[69] P. SONNEVELD. CGS: A fast Lanczos-Type Solver for Nonsymmetric Linear Systems. *SIAM J. Sci. Stat. Comput.*, 10 (1): 36–52, 1989.

[70] J. STOER, R. BULIRSCH. *Numerische Mathematik II*. Springer, Berlin, Heidelberg, New York, 6. edition , 2011.

[71] L. N. TREFETHEN, D. BAU. *Numerical Linear Algebra*. SIAM, Philadelphia, 1997.

[72] H. A. VAN DER VORST. The convergence behaviour of preconditioned CG and CG-S in the presence of rounding errors. *Lecture Notes in Mathematics*, 1457: 126–136, 1990.

[73] H. A. VAN DER VORST. BI-CGSTAB: A fast and smoothly converging variant of BI-CG for the solution of nonsymmetric linear systems. *SIAM J. Sci. Stat. Comput.*, 13: 631–644, 1992.

[74] H. A. VAN DER VORST. *Iterative Krylov Methods for Large Linear Systems*, volume 13 of *Cambridge Monographs on Applied and Computational Mathematics*. Cambridge University Press, Cambridge, 2009.

[75] H. F. WALKER. Implementation of the GMRES Method using Householder Transformations. *SIAM J. Sci. Comput.*, 9 (1): 152–163, 1988.

[76] R. WEISS. *Parameter-Free Iterative Linear Solvers*. Akademie Verlag, Berlin, 1. edition , 1996.

[77] R. WEISS, H. HÄFNER, W. SCHÖNAUER. LINSOL (LINear SOLver) Description and User's Guide for parallelized version. IB Nr. 61/95, Rechenzentrum der Universität Karlsruhe, 1995.

[78] J.H. WILKINSON, C. REINSCH. *Linear Algebra. Handbook for automatic computation, Vol 2*. Grundlehren der mathematischen Wissenschaften in Einzeldarstellungen, volume 186. Springer, Berlin, Heidelberg, New York, 1971.

Index

© The Editor(s) (if applicable) and The Author(s), under exclusive license
to Springer Fachmedien Wiesbaden GmbH, part of Springer Nature 2026
A. Meister, *Numerical Methods for Linear Systems of Equations*,
Mathematics Study Resources 26, https://doi.org/10.1007/978-3-658-50260-7